Farming Systems: Issues and Strategies

Farming Systems: Issues and Strategies

Dr. Sunil Kumar
Head, Crop Production Division

Dr. D. R. Palsaniya
Sr. Scientist, Agronomy

Dr. T. Kiran Kumar
Scientist, Agronomy

Dr. A. K. Dixit
Sr. Scientist, Agronomy

Dr. M. M. Das
Principal Scientist, Animal Nutrition

Dr. P. K. Ghosh
Director

ICAR - Indian Grassland & Fodder Research Institute
Jhansi-284 003 (U.P.), India

SATISH SERIAL PUBLISHING HOUSE
403, Express Tower, Commercial Complex
Azadpur, Delhi-110033 (India)
Phone : 011-27672852, Fax : 91-11-27672046
e-mail : info@satishserial.com, hkjain1975@yahoo.com
Website : www.satishserial.com

Published by :

SATISH SERIAL PUBLISHING HOUSE
403, Express Tower, Commercial Complex, Azadpur, Delhi-110033 (INDIA)
Phone : 011-27672852 Fax : 91-11-27672046
E-mail : info@satishserial.com, hkjain1975@yahoo.com

ISBN 978-93-85055-88-1

Composed, Designed & Printed in India

Preface

The challenge for Indian agriculture over the coming decades will be to meet the increasing demand for quality food in a sustainable way. Agriculture and livestock production are complementary to each other and both are crucial for overall food security. Although, great potential for growth of livestock sector exists, low livestock productivity is a matter of great concern in the country. Indian agriculture can largely be characterized as rainfed small holders' production system, with a special significance to livestock in attaining sustainability, hedging risk and creation of livelihood. The share of livestock for under-privileged marginal and landless livestock owner is as high as 70-80% during drought year.

The integrated farming systems approach holds the key for socio-economic upliftment of rural households as it judiciously combines crop, trees, livestock, fishery etc. in a holistic manner and has the potential to meet the diverse needs of the farming communities on sustainable basis. Farming systems approach relates to the whole farm rather than individual component and is driven as much by the overall welfare of farming households as by goals of yield and profitability. It minimizes the use of external inputs and provides opportunities for recycling of residues or by-products or wastes, enhances productivity of land or farmer as a whole, minimizes risks, optimizes the farm income and provides round the year employment.

In this book, focus has been given to accommodate all the key components of farming system approach; such as, concept, importance, status and scope, enterprise planning and management, input management in IFS for higher productivity and efficiency, mechanization, processing and value addition strategies for IFS, procedures and methodologies in IFS, social issues and policy aspects in IFS and other relevant issues and it will helpful to all stakeholders including farmers, researchers, teachers, planners and institutions.

Efforts have been made in presenting all the information available in a clear and concise manner. Editors thankfully acknowledge the support from scientistific fraternity in terms of the contribution of book chapters. The gratitude is also extended to all those who have contributed directly or indirectly in the success of this venture.

Editors

Contents

Chapter 1

Farming System Research in India: An Overview

U.K. BEHERA

ICAR - Indian Agricultural Research Institute, New Delhi-110012

Introduction

India is basically an agrarian country with more than 60% population relying on agriculture. Today, India has 1.25 billion people. Ensuing food, nutritional and income security for such a huge population under depleting and degrading natural resources is the tough challenge before the scientist, farmers and policy makers. To make the country self-sufficient in food and nutrition, systematic agricultural research started during 1950. India's agricultural production increased by 5-folds but at the cost of degradation of natural resources. The after effects of green revolution for example are not unknown to the scientists and the government. Also, the agricultural development occurred in well endowed regions and small and marginal farmers got bypassed or were less benefited. Thus to meet the multiple objectives of poverty reduction, food and nutritional security, competitiveness and sustainability, several researchers have recommended the farming systems approach to agricultural research and development. Farming systems research is considered as a powerful tool for natural and human resource management in developing countries including India. This is a multidisciplinary whole-farm approach and very effective in solving the problems of small and marginal farmers. The approach aims at increasing income and employment from small-holdings by integrating various farm enterprises and recycling crop residues and by-products within the farm itself (Behera and Mahapatra, 1999; Singh *et al.*, 2006; Mahapatra, 1994).

In India more than 80% farmers are small and marginal which are hardly economically viable under existing technological scenario.

Moreover, decline in per capita availability of land from 0.5 ha in 1950-51 to 0.15 ha by the turn of the century and a projected further decline to less than 0.1 ha by 2020, point to the urgent need of developing strategies and agricultural technologies that enable adequate employment and income generation, with small and marginal farmers at the fore front. The crop and cropping system based perspective of research needs to make way for farming systems based research particularly small farmers (Jha, 2003). In FSR, integration of land based enterprises such as fishery, poultry, duckery, apiary, field and horticultural crops, etc. within the bio-physical and socio-economic environment of the farmers is important to make farming more profitable and dependable (Behera *et al.*, 2004). Adoption of farm enterprise individually in isolation cannot sustain farm family, but the integrated farming systems approach holds the promise of addressing the issues of sustainable economic growth of farming communities in India. Integrated farming systems benefit from synergisms among different enterprises, diversity in produce, and environmental soundness (Lightfoot, 1990). For this reason, IFS models have been suggested by several workers for the development of small and marginal farms across the country (Rangaswamy *et al.*, 1996; Behera and Mahapatra, 1999; Singh *et al.*, 2006).

The basic aim of IFS is to derive a set of resource development and utilization practices, which lead to substantial and sustained increase in agricultural production (Kumar and Jain, 2005). However, there exists a web of interactions among the components within the farming systems and it becomes difficult to deal with such inter-linking complex systems. This is one of the reasons for a slow and limited progress in the field of farming systems research (FSR) in the country. This problem can be overcome by construction and application of suitable whole farm models (Dent, 1990). But, inadequacy of available data from the whole farm perspective currently constrains the development of whole farm models.

Conceptual definition

"Farming System is a complex inter-related matrix of soil, plants, animals, implements, power, labour, capital and other inputs controlled in parts by farming families and influenced to varying degrees by political, economic, institutional and social forces that operate at many levels" (Mahapatra, 1992). The term "Farming System" refers to a particular arrangement of farming enterprises that are managed in response to physical, biological and socio-economic environment and in accordance with farmer's goals, preferences and resources (Shaner *et al.*, 1982). "The household, its resources and the resource flows and interactions at the individual farm levels are together referred to as a farm system" (FAO, 2001).

"Systems" could be defined as an organised unitary whole composed of two or more inter dependant and interacting parts, components or subsystems delineated by identifiable boundary or its environmental super system (Behera, 2010). It is a set of interrelated elements each of which is associated directly or indirectly with other elements and no subset is under-related to any other sub-sets.

A farming system can be described and understood by its structure and functioning. The structure, in its wider sense, includes among others, the land use pattern, production relations, land tenures, size of holding and their distribution, irrigation, marketing including transport and storage, credit institutions and financial markets and research and education (Fig. 1), represents a typical farming system). Thus, the "farming system" is the result of a complex interaction among a number of interdependent components. In a farming system, the individual farmer allocates certain quantities and qualities of fundamental factors of production are land, labour, capital and management, to which he has access. The processes, like crop and livestock production, off- farm enterprises and activities etc. are taken place within the knowledge he possess to maximize the attainment of goal he is striving for.

Overall a farming system focuses on:

- The interdependencies between components under the control of household and,
- How these components interact with the physical, biological and socio-economic factors, which is not under the control of household.
- Farm household is the basic unit of farming system and interdependent farming enterprises carried out on the farm.
- Farmers are subjected to many socio-economic, bio-physical, institutional, administrative and technological constraints.
- The operator of the farming system is farmer or the farming family.

The primary inter-relationships at the farming system level are illustrated in Fig. 1. This highly simplified model puts the farmer the decision maker, at the center. Decisions are influenced by the priorities of the household, farmer's knowledge and experiences, and resource at his command. External factors - natural, economic and socio-cultural, also play significant roles.

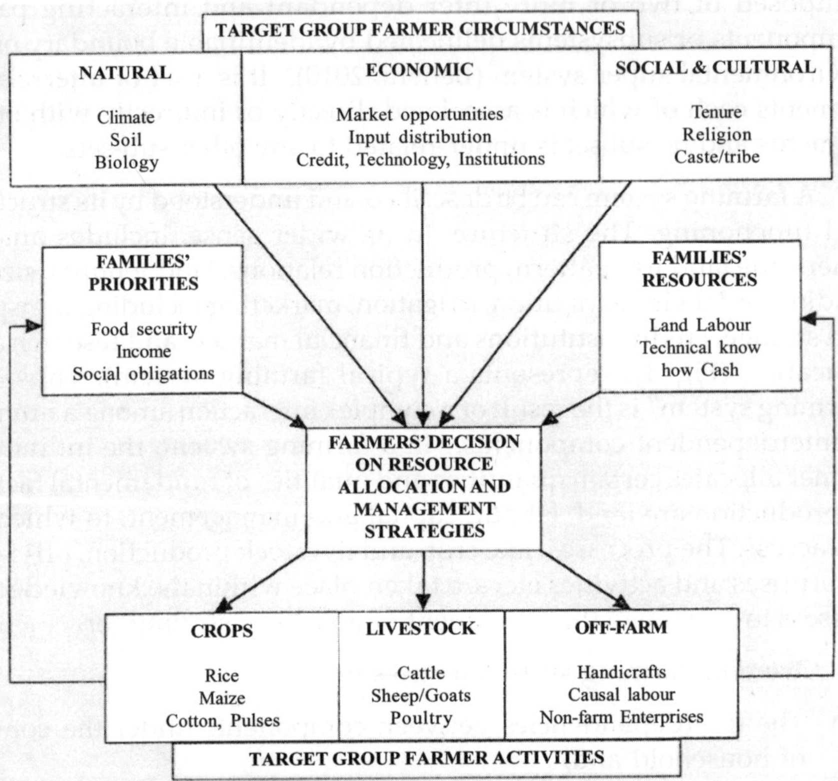

Figure 1. Farming system model showing interrelationships at the farming system level

Source : Collinson, M.P. (1987). Solid boxes are observable, broken ones are deduced from the observed facts or elicited from target farmers

Major focus in farming system

In the past few decades farming system research (FSR) has emerged as a popular and major theme in international agricultural research (Sands, 1986). Yet despite the widespread use of the term FSR, substantial ambiguity persists about its meaning and the types of research concepts, objectives, approaches, activities and methods to which it should be applied. FSR integrates the following key activities and concepts into a coherent research process designed to overcome the perceived weaknesses in main-stream agricultural research.

i) **It is problem solving:** FSR is an operational research which first identifies technical, biological and socioeconomic constraints to improved production in farming systems. It then endeavors to develop solutions which are appropriate for the management conditions of that system (Biggs and Stephen, 1995).

ii) **FSR is farmer-oriented:** FSR views small farmers as clients for research and technology development. Therefore, its fundamental objectives are to generate technologies relevant to their goals, needs and priorities. Several mechanisms are employed to attain these objectives: (i) farmers are integrated in to research process, the existing farming system is studied before proposing technological solutions, and (iii) technologies are adapted to local circumstances and needs of a specific group of farmers (Rhoads and Booth, 1982; Chambers and Ghildyal, 1985).

iii) **FSR is system oriented:** FSR views the farm in a holistic manner and focuses on interactions between components. In practice, the whole farming system serves as the frame-work for analysis, but specific components, sub-systems or interactions are targeted for interventions.

iv) **FSR is interdisciplinary:** FSR, by nature, cuts across conventional, commodity and disciplinary boundaries. Biological and social scientists must collaborate in order to understand the conditions under which small farmers operate, to diagnose constraints and to develop appropriate and improved technologies (Rhoades and Booth, 1982; Mahapatra and Behera, 2004).

v) **FSR complements mainstream commodity and disciplinary agricultural research; it does not replace it.** FSR draws on the "body of knowledge" of technologies and management strategies, generated by discipline and commodity research and adapts them to the specific environment and socio-economic circumstances of a target group of relatively homogeneous farmers (Sands, 1986).

vi) **On-farm research is central to FSR approach**: On-farm research provides the context for collaborations between farmers and researchers (Chambers and Ghildyal, 1985). Researchers get a deeper understanding of the farming system and the decision making context of the farm family. It revolves round the basic principle that successful agricultural research and development efforts should start and end with the farmers (Rhoades and Booth, 1982). Farmer participation is ensured at different stages of technology generation and transfer processes, such as system description, problem diagnosis, design and implementation of on-farm trials, and providing feed-back through monitoring and evaluation.

Farming system research

The FSR concept was developed in 1970s in response to the observation that groups of small-scale farm families operating in harsh environment were not benefiting from the conventional agricultural research and extension strategies. The farming system, as a concept, takes into account the components of soil, water, crops, livestock, labour, capital, energy and other resources with the farm family at the center managing agricultural and related activities. Bhatia and Venugopal (1991) reported that the farming system approach is needed in integrating trees, livestock and fisheries with the agriculture in formulating synergetic "mixed farming" models, appropriate to different agro-climatic regions and on-farm situations for a total upliftment and sustainability of output from the millions of small and marginal farms which account for a major share in Indian agriculture. The farm family functions within the limitations of its capability and resources, socio-cultural setting and interaction of these components with physical, biological and economic factors. The term FSR in its broadest sense is any research that views the farm in a holistic manner and considers interactions (between components and of components with environment) in the system. This type of research is most appropriately carried out by interdisciplinary teams of scientists, who, continuously interact with farmers in the identification of problems and in advising ways of solving them. It aims at generating and transferring technologies to increase the resource productivity for an identified group of farmers.

The FSR advocates that: (i) development of relevant and viable technology for small farmers should have the full knowledge of the existing farming system, and (ii) that technology should be evaluated not solely in terms of its technical performance, but in terms of its conformity to the goals, need and socio- economic circumstances of the targeted small farm system with special reference to profitability and employment generation.

Research and extension efforts with a farming systems perspective have varying objectives ranging from increasing the body of knowledge about farming systems to solving problems in different farming situations. Expectations are high in its problem solving role where the aim is to increase productivity of the farming system by generating and transferring agricultural technologies appropriate for farmers (Biggs, 1985). Primary objective of FSR is to improve the well being of individual farming families by increasing the productivity of their farming system, given the constraints imposed by resources and the environment (Norman and Collinson, 1985).

Methodologies of conducting farming system research

Collinson (1987) while considering procedures of farming system research has suggested that farming system adaptive research is a better approach for allowing better decision making on recommendation to farmers and on priorities for technical research. In less developed countries, the majority of the farmers (70%) are small farm holders with holding size of 0.5 ha. Since, in FSR, small farmers are considered to be clients for agricultural research and development of technology, the farm as a whole is viewed in a holistic manner and focus is on interactions between components and also considered to be problem solving operational research. Instead of a lengthy trial and error process to test innovation on the farmers' field, a wide range of experimental pattern could be pre-screened for compatibility with the existing farm resources and the community social and economic structure (Chambers and Ghildyal, 1985; Norman and Mediakgotla, 1990).

With the development of the modeling approach to farming systems research with due verification and validation using computer software, followed by on-farm experimentation, provides a sound base for agricultural research in the third world countries. The most important contribution of FSR is that it provides a focus for all the disciplines involved in agricultural research and development and that it attempts to classify farmers into relevant categories for agricultural research and policy, while involving them in an active way (Fresco, 1984). Moreover, FSR is an effective way of improving the focus of scientists on problems faced by farmers. Sands (1986) reported in detail that FSR by nature is a farmer oriented system research (takes the holistic view) which cuts across the disciplinary boundaries (interdisciplinary), centered to on-farm research, problem solving operational approach complements main-stream commodity research and also provides a feedback from farmers. Most farming system activities are associated with the use of teams comprised of bio-physical and socio-economic scientists, such as Agronomist, Soil scientists, Entomologists, Pathologists, Horticulturists, Sociologist, Economists, Animal production scientists etc. This approach involves interdisciplinary cooperation between the above mentioned scientists in close association with other production specialists and farmers in a collaborative and integrated ways (Collinson, 1987). Generally, farming system research is conducted by the following three possible ways:

a) FSR : On-farm Adaptive Research (OFAR)
b) FSR : On-station research
c) FSR: Study of farming system by modeling, using suitable computer software.

1. On-farm research

On-farm research refers to the research which is conducted at farmers' field in relatively larger plots compared to conventional on-station research with active participation of the farmers with the hope that technology generated through the combined efforts of researchers and farmers will be realistic to the socio-economic environment of the resource poor group and the problematic situations that the farmers practically face during the process of farming.

While conducting on-farm research in farming system perspective the following principles need to be considered.

i) The whole farm viewed as a system - the research is conducted with recognition and emphasis on choice of priorities that reflect the whole farm.
ii) Avoid complex procedures that require scarce and highly qualified individuals to collect and analyze data.
iii) Maximize the returns by making results more widely applicable. This means defining target groups of farmers (recommendation domains) in broad terms. The extent to which improved systems can be transferred or extrapolated to other areas directly affects their efficiency.
iv) Be open to using second best solutions or the best of those readily available. Therefore, the emphasis in FSR has been on developing improved technologies that are better than most but not necessarily best for each environment.

On-farm research processes

After identification of target area and research area, the following four important operational stages in on-farm research process need to be followed (Zandstra *et al.*, 1981):

i) Descriptive or diagnostic stage

In this stage, target area is picked, the frame of farming families are divided into target groups or recommendation domains. Then efforts are made to determine the constraints farmers face in increasing the farm productivity, the circumstances in which farmers work, the weakness, strength, the opportunities and the threats with the farmers. The main aim is to understand the farming system; to prepare an inventory of farm resources, production constraints and support services. Talking to knowledgeable people, examining relevant secondary sources of information, surveys and technical monitoring are the chief strategies of this stage. In general, however, the methods used should be based on criterion of the lowest possible cost commensurate with the degree of understanding that is necessary. Extra accuracy takes resources and time. This information through diagnostic survey help in improving experimental (trial) planning in: bounding treatment levels; verifying evaluation criteria; identifying special location characteristics to be observed in setting experiments and assessing current productivity levels. Other aspects important in this diagnostic or descriptive stage are participatory rural appraisal (PRA), agro-ecosystem analysis, establishing recommendation domains. Some of the information generated based on agro-ecosystem analysis and PRA.

ii) Design or planning stage

The priorities for research are identified at the descriptive/ diagnostic stage. Planning or design stage is recognized as crucial to the success of FSR in technology generation. On the whole, farmers' problems are readily identified. A range of strategies are identified that are thought to be relevant in dealing with constraints. The factors taken into considerations are: technical feasibility, economic viability and social acceptability. In this stage, suitable action plan for the selected farmers is formulated. The main variables included in designing and planning are: (i) potential for poverty reduction / income / employment generation; (ii) potential for agricultural growth, and (iii) demographic or ecological or economic importance etc.

iii) Testing stage

The objective of this stage is to evaluate the improved practices flowing from the design or planning stage to the farm. The evaluation criteria should be those found important to farmers in the descriptive/ diagnostic stage. Usually, the performance of the improved technology drops when it moves from the controlled conditions of the experimental station to the farm and drops even further when farmers manage and implement the final trials. In this stage, most promising strategies identified at the design stage are evaluated under local farmers' conditions.

At the testing stage, compromises have to be made in the experimental design; farm trials need to be less complex than those undertaken on experiment stations because of costs, worries about too much land being asked from farmers and the desirability of interaction between farmers and research workers. Researcher – farmer interaction is less likely when experiment become too complex. More emphasis is given on replication across farmers' fields rather than within farmers' fields at this stage of testing. The problem of experimental work is exposed to many additional sources of variation, including differences in management and non-treatment of variables by host farmers and often inability to explain differences between plots. During testing stage, generally, three types of trials are conducted with the participation of farmers, viz., researcher designed and researcher managed trial (RDRM), researcher designed and farmers managed trial (RDFM), farmers designed and farmers managed trials (FDFM).

iv) Recommendation and dissemination stage

The acceptable new technology is promoted in collaboration with the line department of the state Govt. and NGO. Once the technology or product is ready for extension, necessary supplies and support services must be ensured by the policy makers and planners and other involved, such as extension workers and researchers.

After the technology has been demonstrated and promoted to all the farmers in the target group, it is important that their experiences with it are monitored. The improvement suggested may not always suit the farmers' situation, especially as circumstances may change over time. There may need to be a number of options rather than single recommendation. Feedback of the farmers' reaction to the technology will determine if technology is suitable and also when changes are needed. This review phase is vital since it emphasizes the continuous nature of needed improvements.

2 On-station FSR

FSR is considered as highly farmers' participatory and conducted at the farmers" field by the interdisciplinary group of scientists. Farmer's participation is ensured at different stages of technology generation and transfer processes, such as system description, problem diagnosis, design and implementation of on-farm trials, and providing feedback through monitoring and evaluation (Rhoades and Booth, 1982). On-station experiments on farming system perspectives are also conducted at the research station by taking into consideration the farmers problems, resource availability with farmers, such as land, labour, capital etc. and farm constraints (physical and bio-physical) into consideration (Rangaswamy *et al.,* 1996; Behera and Mahapatra, 1999). Number of on-station studies on integration of different enterprises: lowland rice-cum- pisciculture farming system, rice-poultry-fish-mushroom integrated farming systems for low land, alternate system of land use through diversification of farming system etc. have been conducted in different parts of the country just by simulating the small and marginal farms situations (Rangasamy *et al.,* 1996; Rangaswamy *et al.,* 1992; Mahapatra, 1994, Rath, 1989, Rautaray, 2004).

3 Farming System Research through system modeling

A model is a simplified abstraction of the real world. It simulates the behaviour of a real system. Modeling begins with the analysis of the systems, its circumstances and purposes. Defining the model gives insight into the working of the system. So far, the farming systems research has been rather inadequate or slow, particularly in less developed countries. Perhaps the only way by which improvement can be achieved is by the construction and application of suitable whole farm models (Dent, 1990). Recent computer software development may provide the basis for a start in modeling of whole farm systems even with incomplete conceptual understanding and data sets.

3.1 Utility of farming system models

Farming system models are useful in the following ways:

i) **To improve the understanding of farming systems,** thereby helping in prioritization of enterprises, better planning and designing of FS experiments, and farm management and policy development.

ii) To analyze and explain behaviour of a complex system and to determine the relative importance of different components/ enterprises of the systems.

iii) To examine the different scenarios resulting due to integration or mixing of different components or modifying different components in the systems.

iv) To identify the areas where the knowledge of the system is fundamentally lacking.

v) Improvising the system for its wider application in varying situations i.e. under varying resource availability and resource constraints situations.

vi) **Models are cheaper than real life farming systems experiments**. Experimentation in real world is expensive, time consuming and there are problems in controlling variables exogenous to the experiment. Thus, model saves energy, time and resources. The FSR studies are very complicated and time consuming and involve huge expenditure. With the help of suitable software, the results can be simulated for important decision- making. With the due perfection of the technique, a series of options can be chalked out and a few most important ones can be tested under actual field conditions.

v) **Development of science.** The FSR models help in integration of knowledge of various disciplines, better understanding of the process and their linkages with others and in identifying gaps in knowledge. This will help in re-orienting the research priorities of commodity and discipline based research institutes.

3.2 Enterprise mix by using linear programming (LP) models

The purpose of constructing a LP farming system model is to identify which one of the new technologies are profitable at the farm level and on which type of farm they are likely to yield the best financial results. This model assesses the economic and production consequences of adoption of new technology at the farm levels (Yates, 2000). The linear programming models help in taking decisions for efficient allocation of the limited resources to optimize well-defined objective under a set of constraints. For example, a small farmer wishes to allocate his farm resources, such as labour, land, capital and other resources between various potential agricultural enterprises in order to maximize gross margins.

Conclusion

Farming system research is recognized as a potential tool for management of vast natural resources in developing countries. The farming system, as a concept, takes into account the components of climate, soil, water, crops, farm wastes livestock, land, labour, capital, energy and other resources with the farm family at the centre managing agriculture and related activities. Integrated farming system is extremely important for the efficient management of available resources at the farm level to generate adequate income and employment for the rural poor, and improvement of their livelihoods in a of sustainable manner. The synergistic interactions of the components of farming systems need to be exploited to enhance resource-use efficiency and recycling of farm by-products.

The farming system approach to agricultural research and development efforts will accelerate agricultural growth and thereby provide leverage for transforming poverty-prone rural India into a prosperous India. This approach might revolutionize agricultural development in the 21st century to make India a developed nation.

References

Behera, U.K. 2010. Manual on Farming Systems. Division of Agronomy, Indian Agricultural Research Institute, New Delhi, 110 p.

Behera, U.K. and Mahapatra, I.C. 1999. Income and employment generation of small and marginal farmers through integrated farming systems. *Indian Journal of Agronomy.* **44**(3): 431-439.

Behera, U.K. and Mahapatra, I.C. 1998. Income and employment generation for small and marginal farmers through integrated farming systems. *Indian Farming* **48**(3): 16-28.

Behera, U.K., Jha, K.P. and Mahapatra, I.C. .2004. Integrated management of available resources of the small and marginal farmers for generation of income and employment in eastern India. *Crop Research* **27**(1): 83-89.

Bhatia, P.C. and Venugopal, K. 1991. Farming systems researches in Indian context with special reference to humid tropics. Proc. Symp. Farming systems for sustained productivity in humid tropics. Proc. Symp. Farming systems for sustained productivity in humid tropics. Ed. Gangwar, B. and Rao, G.C. (1993): 14-21.

Biggs and Stephen, D. 1995. Farming system Research and Rural Poverty: Relationship between context and content. *Agricultural Systems* **47:** 161-174.

Chambers, R. and Ghildyal, B.P. 1985. Agricultural research for resource-poor farmers: the farmer-first and last model. *Agricultural Administration* **20**: 1-30.

Collinsion, M.P. 1987.Farming system research: Procedures for technology development. *Experimental Agriculture* **23**: 365-386

Dent, J.B.1990. *Systems theory applied to agriculture and the food chain*. Elsevier, Amsterdam.

Jha, D. 2003. An overview of farming systems research in India. *Annals of Agricultural Research* **24**(4): 695-706.

Kumar, S. and Jain, D.K. 2005. Are linkages between crops and livestock important for the sustainability of the farming system? *Asian Economic Review* **47**(1): 90-101.

Lightfoot, C. 1990. Integration of aquaculture and agriculture: a route to sustainable farming systems. Naga, The ICLARM Quarterly **13**(1): 9-12.

Mahapatra, I.C. 1994. Farming system research – A key to sustainable agriculture. *Fertilizer News*, **39**(11): 13-25.

Mahapatra, I.C. and Behera, U.K. 2004. Methodologies of farming systems research. Panda, D., Sasmal, S., Nayak, S.K., Singh, D.P. and Saha, S.(Eds), *Recent Advances in Rice-based Farming Systems*, 17-19 November 2004, Cuttack, Orissa, Central Rice Research Institute, pp. 79-113.

Rangaswamy, A, Venkataswamy, R., Premsekhar, M. and Palaniappan, S.P., 1992. Sustainable agriculture for rice (*Oryza sativa*) based ecosystem, *Indian Journal of Agronomy* **37**(2): 215-219.

Rangaswamy, A. Venkitaswamy R. Purushothaman and Palaniappan, Sp.1996. Rice-Poultry – Fish – Mushroom integrated farming systems for lowlands of Tamil Nadu – *Indian Journal of Agronomy,* **41**(3): 344-348.

Rath, R.K. 1989. Integrated fish farming system in China: An analytical Review. Fishing Chimes (Sept. issue): 20-27.

Rautaray, S.K., Das, P.C., Sinhababu, D.P. 2005. Increasing farm income through rice-fish based integrated farming system in rainfed lowlands of Assam. Indian J. of Agric. Sci. **75**(2): 79-8.

Rhoades, R.E. and Booth, R.H. 1982. Farmer-back to back farmers: a model for generating acceptable agricultural technology. *Agricultural Administration* **11:** 127-137.

Sands, M.D. 1986. Farming systems research: clarification of terms and concepts. Experimental Agriculture **22**: 84-104.

Shaner, W.W., Philipp, P.F. and Schmehl, W.R. 1982. *Farming Systems Research and Development: Guide for Developing Countries.* West View Press, Colorado, USA.

Singh, K, Bohra, J.S., Singh, Y. and Singh, J.P. 2006. Development of farming system models for the north-eastern plain zone of Uttar Pradesh. *Indian Farming* **56**(2): 5-11.

Yates, C.M. 2000. Modelling consequences of new reproductive technologies in the bovine sector of U.K. Agriculture. Ph.D. Thesis, The University of Reading, Reading, U.K.

❑❑❑

Chapter 2

Sustaining Indian Agriculture through IFS based Diversification: Issues and Strategies

R.V. KUMAR, H.V. SINGH AND D. DEB

ICAR - Indian Grassland and Fodder Research Institute, Jhansi-284003

Introduction

In recent years, food security, livelihood security, water security as well as natural resources conservation and environment protection have emerged as major issues worldwide. Developing countries like India, is struggling to deal with these issues. Besides the food security and sustainability of the agricultural systems, the burden of climate change and its globalization is another issue. It has been accepted by everyone across the globe that sustainable development is the only way to promote rational utilization of resources and environmental protection without hampering economic growth. Developing countries around the world are promoting sustainable development through sustainable agricultural practices which will help them in addressing socioeconomic as well as environmental issues simultaneously. Integrated Farming Systems (IFS) hold special position as in this system the byproduct of one system becomes the input for other and nothing is wasted. Integrated farming is an integrated approach to farming and it refers to agricultural systems that integrate livestock and crop production. Population growth, urbanization and income growth are imposing a substantial increase in the demand for food of animal source. An integrated farming system consists of a range of resource-saving practices that aim to achieve acceptable profits and high and sustained production levels which should be environmentally eco-friendly. Moreover, the system helps poor and small farmers, who have very small land holding for crop production and a few heads of livestock to diversify farm production. Integrated Farming System may be defined as a set of agricultural activities organized into a functional unit to profitably harness the solar energy while preserving land productivity, environmental quality and maintaining the desirable level of biological diversity and ecological stability. Integration of different agricultural allied enterprises with crop activity as base would provide ways to reuse and recycle produce/waste material of one component as input in the other linked component and to reduce the cost of production of the economic produce of the component two and finally to enhance the net-income of the farm as a whole.

The main component of IFS is given in Fig 1.

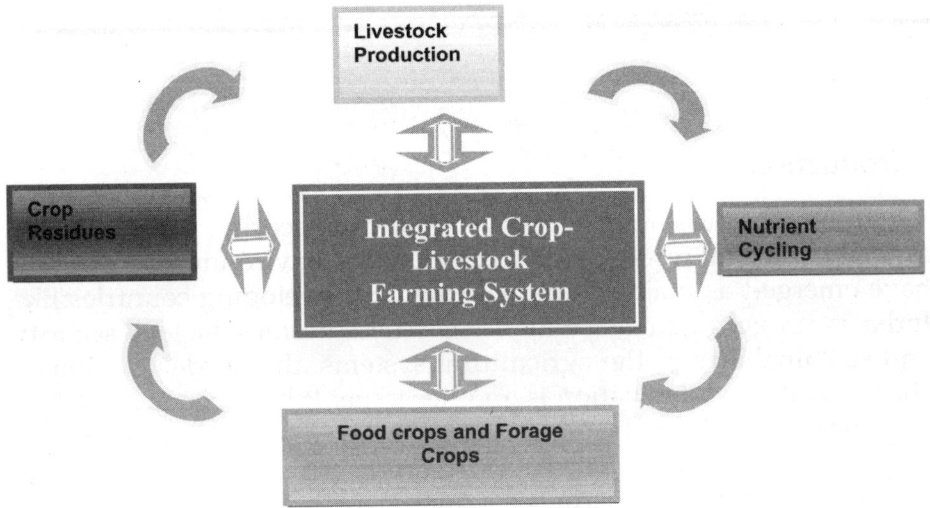

Fig. 1

For the last few decades, agricultural scientists in the developing and underdeveloped countries were looking for an alternate agriculture system to increase the net income of poor and small farmers so that the economic condition of these marginal farmers can be improved. Integrated farming system may helps and provides opportunity which would improve the economic condition of the marginal and small farmers up to certain level with the help of appropriate alternative innovation in this field. An integrated farming system represents multiple crops (cereals, legumes, tree crops, vegetables etc.) and multiple enterprises (animal farming, bee keeping, fish farming etc) in a single farm. It is widely accepted as a means of achieving sustainable agriculture to meet the multiple objectives of poverty reduction, food security and competitiveness and sustainability.

Objectives of the IFS

Integrated farming system has some of the objectives and these are:

1. To integrate different production systems (dairy, poultry, livestock, fishery, floriculture, horticulture, apiculture) etc. with agricultural crops production as the base.
2. To increase farm resource use efficiency (land, labour and production/by-products) so as to increase farm income and gainful employment opportunity.
3. To promote multi-cropping for multi-layered crops of economic value so as to sustain land productivity.
4. To uphold environmental quality and ecological stability.

Issues of Integrated farming system

Due to declining profitability as a result of high input demand in crop production, increasing cost of inputs, the farmers are considering farming as a secondary occupation which resulted large scale migration of rural mass to the nearby towns and cities. In India, more than 70 percent of the total population of the country still live in villages and mainly depend on Agriculture and/or related enterprises. Marginal and small farmers constitute more than 84 percent of the 115 million operational holdings in India which are cultivating only 29 percent of the arable land. Most of them are resource poor and work in diverse, risk prone environment. Further, small farm holders remained deprived of benefits of advancement in the field of agriculture, major beneficiary being resourceful well capable farmers. The livelihoods of the small and marginal house hold families are the major concern. One of the key challenges of the 21st century is developing ways of producing sufficient amounts of food while protecting both environmental quality and the economic well-being of rural communities. The potentialities of each technology of the agricultural production system are not being fully harvested / utilized and make the system unviable. Traditionally different farming systems have evolved independently and being practiced by the farmers especially by poor and marginal farming communities without any rationale for utilizing the wastes and residues arising out of cropping/animals of their farm and other associated enterprises at farm resulting wastage of resources. The income from average farmers from mono cropping is hardly sufficient to sustain his family. Dairy, irrespective of kind of animals and their breeds, has been an integral part of prevailing farming systems across the country. Due to undisruptive breed reared by resource poor farmers, the income from the livestock remained fluctuating and also uncertain. Infertility in cows and buffaloes and large scale mortality in poultry and goats are some of the major problems for such uncertainty. Because of poverty, the small land holders are reluctant to adopt new technologies and/ or enterprises. There is emergent need to :

- Build up strategies and promote crop livestock synergies and their interactions with aim to :
  a) Integrate crops and livestock effectively with careful land use.
  b) Increase the productivity of specific mixed crop-livestock systems.
  c) Facilitate expansion of food production .
  d) Keep the environment clean and healthy with careful and efficient use of natural resources.
- Systematic planning of efficient use of biomass, reducing pressures on natural resources; and develop a sustainable livestock manure management system to control environmental losses and contaminant spreading.

Strategies of Integrated farming system

Conventional agriculture as the result of green revoluation has much bad impacts not only on crop production but on environment also and is known to cause soil and pasture degradation because it involves intensive tillage and use of agrochemicals in particular if practiced in areas of marginal productivity. An integrated crop-livestock farming system represents a key solution for enhancing livestock production and safeguarding the environment through careful and efficient resource use. The increasing pressure on land to meet the ever-increasing demand of burgeoning population, different anthropogenic activities and the growing demand for livestock products makes it more and more important to ensure the effective use of feed resources, including crop residues. An integrated farming system consists of a range of resource-saving practices that aim to achieve acceptable profits and high and sustained production levels, while minimizing the negative effects of intensive farming and preserving the environment. Based on the principle of enhancing natural biological processes above and below the ground, the integrated system should be as a the combinations of many farming systems that helps in (i) reduce erosion (ii) increase crop yields, soil biological activity and nutrient recycling (iii) intensify land use, improving profits and (d) can therefore help reduce poverty and malnutrition and strengthen environmental sustainability. The waste products of one component serve as a resource for the other. For example, manure is used to enhance crop production; crop residues and by-products feed the animals, supplementing often inadequate feed supplies, thus contributing to improved animal nutrition and productivity. Integrating crops and livestock serves primarily to minimize risk and not to recycle resources. In an integrated system, crops and livestock interact to create a synergy, with recycling allowing the maximum use of available resources. Crop residues can be used for animal feed, while livestock and livestock by-product production and processing can enhance agricultural productivity by intensifying nutrients that improve soil fertility, reducing the use of chemical fertilizers. A high integration of different crops and livestock enterprises is often considered as a step forward, but small farmers need to have sufficient access to knowledge, assets and inputs to manage this system in a way that is economically and environmentally viable. Integrated farming systems should be planned as resource management strategy to achieve economic and sustained production to meet diverse requirements of farm households while preserving resource base and maintaining a high level environmental quality (Lal and Miller, 1990). Integrated farming system models developed in different parts of the country involving dairy, duckery, poultry , horticulture, apiary, pisciculture and plantation crops *viz;* mostly horticultural along with crops, have been found to increase net profit significantly as compared to cropping alone. The strategies of IFS should be with the objective that these IFS systems are more sustainable, environmental friendly and employment generative. Animals of the farms must be the integral part of the systems and should play key and multiple roles in the functioning of the farm, and not only because they provide livestock products or can be converted into prompt cash in times of need. Animals transform plant energy into useful work, animal power must be used for ploughing, transport and in activities such as milling, logging, road construction, marketing, and water lifting for irrigation. Animal's byproduct should also be used for manure and other types of animal waste.

Mechanism for maximization of integrated farming system

An integrated farming system represents multiple crops (cereals, legumes, tree crops, vegetables etc.) and multiple enterprises (animal farming, bee keeping, fish farming etc) in a single farm. It is widely accepted as a means of achieving sustainable agriculture. To meet the multiple objectives of poverty reduction, food security, competitiveness and sustainability, several researchers have suggested farming systems approach to research and development. A farming system is the result of complex interactions among number of interdependent components where an individual farmer allocates certain quantities and qualities of four factors of production, *viz*. land, labour, capital and management. Some of the farming systems and enterprises are listed below:

1. Agriculture
2. Livestock
3. Fishery
4. Sericulture
5. Silviculture
6. Horticulture
7. Mushroom culture
8. Bio-Gas
9. Apiculture

Advantage of IFS

Farming system research is considered as a powerful tool for the management of natural and human resource in developing countries like India. This is a multi-disciplinary whole-farm approach for solving the problems of small and marginal farmers. This approach aims at increasing income and employment from small-holdings by integrating various farm enterprises and recycling crop residues and by-products within the farm itself. Under the gradual shrinking of land holding, it is required to integrate the land-based enterprises like fishery, poultry, duckery, apiary, field and horticultural crops etc. within the bio-physical and socio-economic environment of the farmers to make farming more profitable and dependable. The research in integrated farming systems (IFS) for the last few decades in India has revealed that the enterprise planning and implementation are usually done in component approach in isolation and needs scientific and systematic approach. In this situation, optimization techniques are useful for resource allocation and designing IFS on scientific basis. Farming systems studies should involve a number of enterprises and take the physical, socio-economic and bio-physical environments into consideration.

Advantages

1. **Yield Enhancement:** Increase in economic yield per unit area - per unit time by virtue of intensification of crops canopy, agricultural crop rotation and allied enterprise.
2. **Sustaining the System:** In IFS, organic supplementation through effective utilization of by-products of linked components is possible and this will certainly provide opportunity to promote soil health and to sustain the potentiality of the soil.
3. **Balanced food:** In IFS, the components of different nature like crop, livestock and backyard poultry enabling to produce different sources of nutrition, namely, protein, carbohydrates, fats, minerals, vitamins, etc from the same unit area. It will provide opportunity to mitigate malnutrition problem of the farmers.
4. **Cost Effectiveness:** The system, as a whole provides opportunity to make use of produce/ waste material of one component as input on the other component at the least cost which ultimately increase farm profitability.
5. IFS provide opportunities as crop insurance cover as money round the year are obtained from different farm produces.
6. Avoid degradation of land resources.
7. Provide opportunities for Agri-oriented industries, tourism and related tourism based activities, etc.
8. Technology Infusion (R&D) integrated with indigenous/ Traditional knowledge.
9. Mitigating the wood, Fodder crisis, etc.
10. Mitigating energy crisis.
11. **Environmental Friendly:** In crop based activity, some of the organic inputs are left as waste materials which in turn pollute the environment on decomposition. Applications of huge quantity of fertilizers, pesticides, weedicides, insecticides, etc pollute soil,

water and air. Much of the wastes could be converted /recycled under the IFS.

12. Less vulnerable to Climate change from the perspective of adaptive & mitigation.

13. Checking rural- urban migration.

Integrated Crop-livestock farming systems

Population growth, urbanization and income growth in developing countries are imposing a substantial global increase in the demand for food of animal origin, while also aggravating the competition between crops and livestock. The livestock revolution is stretching the capacity of existing production, but it is also aggravating environmental problems. The increasing pressure on land and the growing demand for livestock products makes it more and more important to ensure the effective use of feed resources, including crop residues. An integrated farming system consists of a range of resource-saving practices that aim to achieve acceptable profits and high and sustained production levels, while minimizing the negative effects of intensive farming and preserving the environment. The advantages an integrated system, livestock and crops are produced within a coordinated framework. The waste products of one component serve as a resource for the other. The result of this cyclical combination is the mixed farming system, which exists in many forms and represents the largest category of livestock systems in the world.

Animals play key and multiple roles in the functioning of the farm, and not only because they provide livestock products or can be converted into prompt cash in times of need. Animals transform plant energy into useful work: animal power is used for ploughing, transport and in activities such as milling, logging, road construction, marketing, and water lifting for irrigation. Animals also provide manure and other types of animal wastes which play important role in nutrients cycling. Numerous interventions are available, or are being developed, that can be used and combined. Integration and capacity strengthening mechanisms are also needed for local successes to be scaled up. The importance of a horizontally and vertically integrated approach is recognized in many systems. Such approaches aim to identify, quantify and address the driving forces and interactions that shape and constrain farming systems and the management of natural resources (Lockeretz and Boehncke 2000; Roetter *et al.*, 2000).

Crop residues represent the other pillar on which the equilibrium of this system rests. They are fibrous by-products that result from the cultivation of cereals, pulses, oil plants, roots and tubers. They are a valuable, low-cost feed resource for animal production, and are consequently the major source of nutrients for livestock in developing countries. Livestock and crop production systems are an integral part of one another (Kallah and Adamu, 1988). Crop residues provide fodder for livestock (Van Raay and de Leeuw, 1971) while, occasionally, grain provides supplementary feed for productive animals. Animals improve soil fertility through manure and urine deposition and animal power for farm operations and transport. Sale of animals sometimes provides cash for farm labor and agricultural inputs. There are several examples of completely integrated crop-livestock production systems where sustainable increases in both crop and livestock production have been achieved after considerable periods (30–40 years) of continuous cropping without resulting in land degradation.

Crop diversification

Crop diversification is intended to give a wider choice in the production of a variety of crops in a given area so as to expand production related activities on various crops and also to lessen risk. Crop diversification in India is generally viewed as a shift from traditionally grown less remunerative crops to more remunerative crops. However crop diversification has its own limitation. It is area specific, need based and very much dependent on climatic factors of the region. Crop diversification and also the growing of large number of crops are practiced in rainfed lands to reduce the risk factor of crop failures due to drought or fewer rains. Crop substitution and shift are also taking place in the areas with distinct soil problems. For example, the growing of rice in high water table areas replacing oilseeds, pulses and cotton; promotion of soybean in place of sorghum in vertisols (medium and deep black soils) etc.

Sustainable agriculture through diversification of natural resource management

This is well known fact that there is little scope for further expansion of the net sown area (142 m/ha) and that land scarcity will become an acute feature of the rural economy. Water is a precious national asset and there are several concerns regarding water resources in the country. Therefore, a judicious use of land and water resources will have to be the central theme for sustainability of agricultural growth. There is a greater need to have an integrated approach in the management of plant nutrients, chemicals and taking effective measures to deal with the overall pollution problems. There are several possible technologies and alternatives to reduce the use of chemicals in agriculture. These alternatives are not perfect substitutes to chemicals but adoption of these can substantially reduce the adverse impact on environment. Proper land and water management policies would reduce environmental degradation. Community and village institutions should be encouraged to participate in protecting natural resources from degradation.

Poverty Alleviation through diversification

Elaborating sustainable livelihood approaches to reduce poverty is a continuing challenge for development planners throughout the developing world. In Southeast-Asian countries, small scale farming, rice farming in particular, represents the dominate source of income. The role of poor farmers in degrading natural resources while searching for accessible income sources is well known. Small-scale farmers are confronted with the problem of declining productivity in their farming and a limited farm output, which frequently is not enough to provide income or at least food for the farm household throughout the year. Farmers are forced to look for alternatives, which simultaneously favours farm diversification as a possible option. There is a great potential for small scale farmers to alleviate poverty through diversification in the form of integrated agriculture.

Silvipasture system for diversification

Silvipasture is the planned integration of fodder trees and perennial grasses (pasture grass and legumes) for grazing livestock operations on the same land. These systems are intensively managed for both forest products and forages as well. Well managed silvipastures employ agronomic principles for introduced or native pasture grasses, fertilization and nitrogen-fixing legumes, and rotational grazing systems that employ short grazing periods which maximize vegetative plant growth and harvest. Silvipasture is the combinations of pastures, trees, and animals which are purposefully managed to provide forage and timber products on a sustained basis. These are not individual practices that occur coincidentally together or are managed independently. Silvipastures systems could be implemented using different approaches:

(i) pastures where trees or shrubs can be added, (ii) forested areas where forages can be added or (iii) land on which neither the desired trees nor forages exist in sufficient quantity to meet the land use objectives. Appropriate establishment methods depend on the following: site conditions, tree species, age, and spacing, and existing pasture conditions (climate, terrain, type of livestock, labor requirements, fencing, water supplies and other vegetation). Tree spatial arrangement is an important factor to make a silvipasture system successful.

Rainfed agro-ecosystem has a distinct place in Indian agriculture, occupying 67% of the cultivated area, contributing 44% of the food grains and supporting 40% of the human and 65% of the livestock population (Venkateswarlu, 2005). Tree leaves and pods form a natural part of the diet of many ruminant species and have been used traditionally as sources of feed for domesticated livestock in Asia, Africa and the Pacific and serves as protein banks to supplement grass or crop residues in dry season (Bhatta *et al.*, 2005). It is suggested to develop silvopastoral systems/models by introducing trees/shrubs into natural pasturelands/waste lands to provide nutritious green foliage throughout the year (Singh, 1995). Similarly, horticulture and small ruminant (sheep and goat) production systems play a vital role in sustenance of livelihoods of rural poor of rainfed agro-ecosystem (Pasha, 2000) in arid and semi-arid regions, where crop production is a risk-prone enterprise due to uncertain rainfall and frequent draughts. In general, farmers develop plantation trees for timber and orchards for fresh fruit production and are not considered as potential grazing resources. Moreover, small ruminants are primarily maintained on natural pasturelands/waste lands with in situ grazing and the productivity is constrained by the low quality of native grasses as well as the shortage of good quality forage, especially during the dry season. Hence, it is suggested to develop silvipastoral/hortipastoral systems/models by introducing pasture and foliage component under trees so as to provide nutritious green forage and foliage (Pathak and Roy, 1994) to small ruminants for getting higher production from unit of land in rainfed areas.

Forages in reducing poverty and degradation of natural resources

The objective of agricultural research and development is to develop sustainable production systems. This goal can only be attained when environmental protection is balanced with social and economic sustainability. Land and water degradation decreases crop yields and increases food costs leading to poverty (Vosti and Reardon, 1997), which in turn can further increase the pressure on natural resources. Achieving food and income security is a prerequisite for poor farmers making decisions beyond immediate survival and investing in the sustainable use of natural resources. Intensification of production with a natural resource management focus is likely to be the only way to reverse degradation, alleviate poverty and improve food and income security of resource-poor farmers. Forages crops, mostly are perennial in nature are suited for rainfed areas and provides good quantity of forage only on monsoon water. These crops requires one time investment in term of its establishment and provide long term forage resources of the community by cut and carry method or by grazing.

Intensification of smallholder production systems

Opportunities for intensification in smallholder farming systems are limited because of the inability of farmers to pay for external inputs. Smallholder farmers are resource poor, socially backward and unable to bear the risk of newer agricultural technologies. Forages, in particular legumes, integrated with crops, trees and livestock can produce synergistic effects and minimise external inputs. The versatility of forages allows them to be used in different ways in the complex production systems (Schultze-Kraft and Peters, 1997). Nevertheless, the role of forages in reducing poverty and resource degradation is itself complex. Forages can have both direct and indirect effects in increasing resource and land use efficiency (Humphreys, 1994). Cultivated fodder and forages crops are multi-purpose plants and can be used as food and feed. Introduction of legumes as forage, direct effects on crop production include nitrogen fixation, weed suppression, pest and disease reduction (when used in rotation), while indirect effects include their use as green manures, improved fallows, cover crops and live barriers besides fodder production. Production costs are decreased due to the reduced need for external inputs such as fertilisers and pesticides and there are environmental benefits from less contamination of crops and water with pesticide residues, conservation of fossil energy as well as soil improvement through nitrogen fixation. Higher feed quality also results in improved manure and compost. Increased productivity can result in improved cash income and the ability to pay for household needs and education, or income derived from livestock as a market-oriented activity can improve cash flow and purchasing power for inputs and thus act as an 'engine for sustainable intensification'

(Delgado *et al.*, 1999). The increased land-use efficiency resulting from intensification can lead to protection of areas unsuitable for agricultural production where policies favor maximizing returns from labour and land rather than clearing new land for agriculture (Schultze-Kraft and Peters, 1997). Increased feed production on agricultural land marginal for cropping and reclamation of degraded land also contribute to increased land use efficiency (Schultze-Kraft and Peters, 1997 and Delgado *et al.*, 1999).

Diversity for productivity, profitability, environmental and weather events

One of the key challenges of the 21st century is developing ways of producing sufficient amounts of food while protecting both environmental quality and the economic well-being of rural communities [Foley *et al.* (2011), Robertson and Swinton (2005)]. Over the last half century, conventional approaches to crop production have relied heavily on manufactured fertilizers and pesticides to increase yields, but they have also degraded water quality and posed threats to human health and wildlife [Dubrovsky *et al.*, 2010, Rohr and McCoy (2010)]. Consequently, attention is now being directed toward the development of crop production systems with improved resource use efficiencies and more benign effects on the environment [Foley *et al.* (2011), Tilman (2002)]. Less attention has been paid to developing better methods of pest management, especially for weeds. Crop diversification is a less expensive option than other adaptation measures that may require a higher investment. Of course, it may not be effective if large- scale natural disasters occur. These facts remind us that farmers with less ability to employ adaptation measures may need more external assistance. In addition to households own ability to adapt to extreme weather events, external support from the government, such as releasing early warning information, will enhance farmers adaptive capacity to mitigate the negative risks from climate change in general and extreme weather events in particular (Chen *et al.*, 2013). Now a days to mitigate the effect of climate change, crop diversification is the best option to cop up the deleterious effect on crop production. Farmers are shifted from traditional cropping pattern to other climate resilient crop.

Summary

Sustainable development is the only way to promote rational utilization of resources and environmental protection without hampering economic growth and integrated farming systems hold special position as in this system nothing is wasted, the by-product of one system becomes the input for other. India has a considerable livestock, poultry population as well as crop wastes. All efforts have to be mobilized to reclaim the resources and to put them to use effectively. Suitable developed technologies help in sustainable and profitable crop production or the newer technology has to be developed for all round effective utilization of the wastes so that, it can help in to reducing the poverty and malnutrition and strengthen environmental sustainability. The increase in demand for livestock products presents opportunities for small farmers who can increase livestock production and benefit from related income. The growing number of livestock and the increase in livestock produce processing can have a positive impact on natural resource utilization. Farming practices that are economically and ecologically sustainable are puss to reach the poor resource farming communities for better utilization of natural resources. The highly improved integrated crop-livestock system can guarantee more sustainable production and therefore constitutes a valid new approach. This can be done with following ways in adopting sustainable management practices which can improve production while preserving the environment through residues, wastes and by-products of each component serve as resources for the others and resource poor farmers have the traditional knowledge needed to integrate livestock and crop production.

References:

Bhatta, R. , Vaithiyanathan, S. , Singh, N. P. , Shinde, A. K. and Verma, D. L. 2005. Effect of feeding tree leaves as supplements on the nutrient digestion and rumen fermentation pattern in sheep grazing on semi-arid range of India – I. Small Rumin. Res., **60**: 273-280

Chen H, Wang J, Huang J. 2013. Policy support, social capital and farmers' adaptation to drought in China. *Global Environmental Change*, **24**: 193-202.

Delgado, C., M. Rosegrant, H. Steinfeld, S. Ehui and C. Courbois. 1999. Livestock to 2020: The Next Food Revolution. Food, Agriculture and the Environment Discussion Paper 28, Washington, D.C.International Food Policy Research Institute.

Dubrovsky, N.M., Burow, K.R., Clark, G.M., Gronber, J.M., Hamilton, P.A. 2010. The quality of our nation's waters: Nutrients in the nation's streams and groundwater, 1992–2004. Circular 1350. Reston, VA: U.S. Geological Survey.http://pubs.usgs.gov/circ/1350/. Accessed 2012 Jul 10.

Foley, J.A., Ramankutty, N., Brauman, K.A., Cassidy, E.S. and Gerber, J.S. 2011. Solutions for a cultivated planet. Nature **478**: 337–342. doi: 10.1038/ nature10452

Humphreys, L.R. 1994. *Tropical forages: Their role in sustainable agriculture*. Essex, England: Longman Scientific and Technical.

Kallah, M.S. and A.M. Adamu. 1988. The importance of animal feces as fertilizer. National Animal Production Research Institute, Ahmadu Bello University, Zaria, Nigeria. 9 pp.

Lal, R. and Miller, F.P.1990. Sustainable farming for tropics. In: Singh, R.P. (Ed.) Sustainable agriculture: Issues and Prospective. Vol.1 , pp. 69-89, Indian Society of Agronomy, IARI, New Delhi.

Pasha, S. M., 2000. Economy and ecological dimensions of livestock economy. Commonwealth Publishers, New Delhi.

Pathak, P. S. and Roy, M. M. 1994. Silvipastoral system of production. Bulletin IGFRI Jhansi.

Robertson, G.P. and S.M. Swinton. 2005. Reconciling Agricultural Productivity and Environmental Integrity: A Grand Challenge for Agriculture. Frontiers in Ecology and the Environment **3**(1): 38-46.

Roetter, R. P., van Keulen, H., Laborte, A. G., Hoanh, C. T., & van Laar, H. H. 2000. Systems research for optimizing future land use in South and Southeast Asia. Proceedings of the SysNet'99 International Symposium. SysNet Research Paper Series 2. Los Baños: International Rice Research Institute.

Rohr, J.R. and McCoy, K.A.2010. A qualitative meta-analysis reveals consistent effects of atrazine on freshwater fish and amphibians. Environ Health Persp 118: 20–32. doi: 10.1289/ehp.0901164

Schultze-Kraft, R. and Peters, M. 1997. Tropical legumes in agricultural production and resource management: An overview.*Giessener Beiträge zur Entwicklungsforschung* No. 24, pp. 1–17.

Singh, G. 1995. An agroforestry practice for the development of salt lands using Prosopis juliflora and Leptochloa fusca. Agroforestry Systems. **29**: 61-75.

Tilman, D., Cassman, K.G., Matson, P.A,, Naylor, R. and Polasky, S. 2002. Agricultural sustainability and intensive production practices. Nature 418: 671–677. doi: 10.1038/nature01014.

Van Raay, J.G.T. and P.N. de Leeuw. 1971. The importance of crop residues as fodder: a resource analysis in Katsina Province, Nigeria. ABU, Zaria Samaru Research Bulletin 139.

Venkateswarulu, B., 2005. Rainfed agro-ecosystem production system research completion report, National Agricultural Technology Project, Agro-eco system directorate (Rainfed), Central Research Institute for Dryland Agriculture, Hyderabad, 211p.

Vosti, S.A. and Reardon, T. 1997. 'Introduction: The critical triangle of links among sustainabilty, growth, and poverty alleviation' in S.A. Vosti, and T. Reardon (eds) *Sustainability, growth, and poverty alleviation. A Policy and Agroecological Perspective.* Published for the International Food Policy Research Institute. Baltimore and London: Johns Hopkins University Press.

□□□

Chapter 3

SWOT Analysis of Farming System Research in India

SUNIL KUMAR, G. PRABHU AND P.K. GHOSH

ICAR - Indian Grassland and Fodder Research Institute, Jhansi-284003

Introduction

In recent years, food security, livestock security, water security, economic sustainability, balanced nutrient as well as enhancing resource use efficiency and environmental safety have emerged as major issues worldwide. Developing countries are struggling to deal with these issues and also content with dual burden of climate change and globalization. Throughout the world sustainable development through sustainable agricultural practices which will help in achieving livelihood security and environmental sustainability. Within the broad concept of sustainable agriculture "Integrated Farming System (IFS)" hold special status as in this system nothing is waste, everything is recycled, by product of one system become input of another system (Sharma *et al.,* 2012).

Farming system research

Farming systems research (FSR) is an approach introduced in the 1970's. It has evolved over time as the understandings of systems approaches and extension practice have changed, and as deficiencies in the FSR approach being used have become apparent. Farming system is an integrated set of activities that farmers perform in their farms under their resources and circumstances to maximize the productivity and net farm income on a sustainable basis. The farming system takes into account the components of soil, water, crops, livestock, labour, capital, energy and other resources, with the farm family at the centre managing agriculture and related activities. The farming system conceptually is a set of elements or components that are interrelated which interact among themselves. At the center of the interaction is the farmer exercising control and choice regarding the types of results of interaction. The income from cropping alone from small and marginal farm is insufficient now to sustain the farmers' family. A judicious mix of any one or more of these enterprises with agronomic crops. should complement the farm income and help in recycling the farm residues/wastes (Gupta *et al.*, 2012).

The selection of enterprises must be based on the cardinal principles of minimizing the competition and maximizing the complementary between the enterprises. Of late, the researchers on multi disciplinary approach greatly realized and started developing the various farming systems models in accordance with the agro-eco systems zones. The FSR has been clarified as an academic activity comprising of theory, concepts, principles, approaches etc. It creates an opportunity for developing diversified models for different type of farmers and different category of farmers. New farming system approach models could be developed by means of on farm research and extension. It causes consequential a complex change which demands for Government interventions for farming systems development. Infact it is a problem solving approach for the farmer. Farming system approach requires commonly homogenous type of farmers. It is an inter-disciplinary approach. It is a participatory and bottom up planning. It requires on farm trials. It depends on the concept learning by doing and farming system approach needs socially desirable technologies. Thus the concept of Farming System Approach can be summarized as it is a holistic approach, complex in nature, interrelated of components, matrix of soils, plants, animals, power, implements, labour, capital and other inputs, influenced by political, economic, institutional and social forces.

Definitions

Farming systems approach relates to the whole farm rather than individual elements; it is driven as much by the overall welfare of farming households as by goals of yield and profitability. Farming systems are closely linked to livelihoods because agriculture remains the single most important component of most rural people's living and also plays an important role in the lives of many people in semi-urban areas. Farming systems involve a complex combination of inputs, managed by farming families but influenced by environmental, political, economic, institutional and social factors. Research and extension institutions are increasingly aware that a holistic approach, drawing on both local and external knowledge, is necessary if they are to be effective in addressing poverty and sustainability. In nutshell " Farming System is defined as a complex inter related matrix of soil, plants, animals, implements, power, labour capital and other inputs controlled in part by farming families and influenced to varying degrees by political, economic, institutional and social forces that operate at many levels. The farming system therefore, refers to the farm as an entity of inter dependent farming enterprises carried out on the farm". The farm is viewed in a holistic manner (Evenson, 2000). The farmers are subjected to many socio-economic; biophysical, institutional, administrative and technological constraints.

Integrated farming system

The word integrated farming system is derived from Latin verb "integrare" which means to make whole, to complete by addition of parts, or to combine parts into a whole. The crop, livestock, fish subsystems may function independently in certain farming system and their products be only additive, however, an output from one system in integrated farming system, which otherwise may have been wasted becomes an input to another subsystem resulting in a greater efficiency of output of desired products from the land / water area under farmer control (Edwards *et al.*, 1988).

Key principles

Cyclic: The farming system is essentially cyclic (organic resources - livestock-land-crops). Therefore, management decisions related to one component may affect the others.

Rational: Using crop residues more rationally is an important route out of poverty. For resource-poor farmers, the correct management of crop residues, together with an optimal allocation of scarce resources, leads to sustainable production.

Ecologically sustainable: Combining ecological sustainability and economic viability, the integrated livestock -farming system maintains and improves agricultural productivity while also reducing negative environmental impacts.

Farming systems strategy

In view of serious limitations on horizontal expansion of land and agriculture, only alternative left is for vertical expansion through various farm enterprises required less space and time but giving high productivity and ensuring periodic income specially for the small and marginal farmers located in rainfed areas, dry lands, arid zone, hilly areas, tribal belts and problem soils (Carloni, 2001).

The following farm enterprises could be combined:

i) Agriculture alone with different crop combinations

ii) Agriculture + Livestock

iii) Agriculture + Livestock + Poultry

iv) Agriculture +Horticulture + Sericulture

v) Agro-forestry + Silvipasture

vi) Agriculture (Rice) + Fish culture

vii) Agriculture (Rice) + Fish + Mushroom cultivation

viii) Floriculture + Apiary (beekeeping)

ix) Fishery + Duckery + poultry

For meaningful execution of integrated farm-enterprises, the following activities should be undertaken by a multi-disciplinary team of extension professionals with farmer's participation and involvement at all stages.

- Thorough understanding of existing farming systems and their components.
- Assessment of resource availability in the farm environment and identification of bio-physical, socio-economic, institutional, administrative and technological constraints.
- Developments of economic viable and efficient integrated farming systems suitable for various domains.
- Diffusion of improved technology and receiving 'feed back' for further improvement of the system as a whole.
- Continuous improvement in components technology to fit into a given farming system.

- Improvement in quality of farming system.
- Research Extension linkage through "On farm Adaptive Research".
- Development of National and International linkages.

SWOT analysis of integrated farming system in India

Integrated farming may occur directly on-site, or indirectly through off-site needs and opportunities, or both. Benefits of integration are synergistic rather than additive; and the fish and livestock components may benefit to varying degrees. The term "waste" should not been omitted because of common usage but philosophically and practically it is better to consider wastes as "resources out of place. Small farmers in developing countries are poorer than the rest of the population, often not getting enough food to lead normal, healthy and active lives. Dealing with poverty and hunger in much of the world therefore means confronting the problems that small farmers and their families face in their daily struggle for survival. One option for economically and ecologically sustainable development of farming systems is the integration of agriculture and aquaculture.

The analysis of strengths, weaknesses, opportunities and threats (SWOT) is considered to be a pre-requisite for the formulation of a full proof strategy for development of each sector. An attempt has been made to analyse the farming system research in India and its strengths, weaknesses, opportunities and threats towards maximum achievement and adaptation of integrated farming system. The SWOT of IFS in Indian perspective has to be understood in the historical, social and economical boundaries. In general following philosophy revolves around the fast emerging phenomenon of IFS in Indian agriculture (Sunil kumar *et al.*, 2015).

Strengths

IFS have strong components of sustainability, profitability and employment generation in an individual farm household. This offers many benefits for the future agriculture as well.

- Increase in food production per unit of land to meet the demand of the exploding population of India.

- Improvement in farm income through effective recycling and use of residue and allied components.
- Improvement in crop productivity and sustainable soil fertility through organic waste recycling.
- Facilitating the availability of protein, carbohydrate, fat, minerals and vitamins through Integration of allied activities.
- Prevention of environmental pollution through effective recycling of farm wastes from animal components like piggery, poultry and pigeon rearing and crop residues.
- Saving of cost of production through Input recycling of allied components.
- Round the year income through enterprises like egg, milk, mushroom, vegetables, honey and silkworm cocoons from the linked activities in integrated farming.
- Solving the energy crisis by inclusion of biogas & agro forestry in integrated farming system.
- Adequate nutritious fodder for animal components like milch cow, goat / sheep, camel, pig and rabbit by cultivating the fodder crops as intercropping and as border cropping with food crops.
- Avoidance of soil loss through erosion by agro-forestry and proper cultivation of each part of land by integrated farming particularly in hill ecosystem.
- Round the year employment generation for the farm family members of small and marginal farmers.
- Adoptable to different categories of farmers (small/medium/large) across the regions.

Weakness

Despite of all advantages there are still certain limitations coming all the way in fullest utilisation of IFS by the practising farmers and communities where being adopted.

- Animals are in integrated farming system in general fed by crop residues of component crops, which are low in digestibility, nutrition value and protein content.

- To improve the digestibility and intake of crop residues require physical and chemical treatments is technically possible but not feasible for poor small farmers because they require machinery and chemicals that are expensive or not readily available.
- Crop residues are primarily soil regenerators, but too often they are either disregarded or misapplied.
- Intensive recycling of crop residues can result in nutrient losses.
- Application of manures could be resulted in poor in nutrient use efficiency and loss of nutrients, due to improper transportation, application and storage.
- The preference of farmers would be to use chemical fertilizer instead of manures because it availability, faster acts and is easier to use.
- Manure transportation is an important factor affecting manure use.

Opportunities

IFS offers many opportunity to address issues related to food and nutritional security. The challenges of yield stagnations, resource use fatigue, diminishing agricultural returns and climate related vulnerability are well taken by holistic approach of land use and its management.

- Agricultural intensification is being a current burning topic in most farming systems favours crop- livestock integration.
- Increases in prices of fertilizers, poor soil fertility, soil erosion, nutritional requirements and labour shortages, or unavailability have forced farmers to rely on alternatives such as manures and crop residues.
- Farmers can grow crop in the wet season and engage in livestock enterprises in the dry season.
- Livestock enterprises are more lucrative than crop farming so it is advantageous to integrate livestock into farm activities.
- Many indigenous, emerging, and developed technologies are available to support sustainable crop-livestock integration.
- These include improved cereal and grain legume varieties, cropping systems, weed and nutrient management strategies,

the eradication of most livestock diseases, and the development of modelling and all-year-round feed packages for animals.

Threats

The strengths and advantages are many times governed by some inbuilt constraints in any production system. These are further aggravated by social, economical and regional boundaries as well as government policy's. In integrated farming systems, some of the threats are encountered frequently by the farmers.

- Conflict on use of crop residues for recycling as there is high demand for livestock feeding.
- Social issues/ social acceptance related to the selection of components.
- Lack of credit facilities at easy and reasonable interest rate.
- Lack of interest among young farmers in the agricultural and allied activities due to increasing avenues in other occupations.
- Diversion of agricultural land to other economic activities.
- Decreasing trend of agricultural labour force due to migration.
- Reluctance of poor income group farmers to adopt modern agricultural technologies.

Challenges

At the farm level, for integrating components, it becomes necessary to link to a number of schemes, if available, to avail the resources. In the absence of a single source/scheme providing support on various aspects, it is a big challenge for small farmers to integrate various components on their farms. There is need to develop strategies and promote crop livestock synergies and interactions that aim to:

- Careful planning in land use while integrating crops and livestock effectively.
- Approach to enhance the productivity of specific mixed crop livestock systems;
- Focus to facilitate the expansion of food production; and simultaneously safeguard the environment with prudent and efficient use of natural resources.

- Facilitating large scale dissemination of bio-digesters to implement a more efficient use of biomass, reducing pressures on natural resources; and
- Develop a sustainable livestock manure management system to control environmental losses and contaminant spreading.

Limitations

The IFS is prevalent since centuries in India but scientific understanding and execution is still lacking in most of the regions. There are certain aspects which limit the potential of IFS as an approach to be promoted on large scale for desired benefits.

- Unavailability of suitable farming system models for different eco-regions
- Lack of awareness about the sustainable farming systems
- Lack of ensured market facilities & intelligence among farmers
- Lack of deep freezing and storage facilities to avoid spoilage.
- Lack of banking facility, access and aptitude to take full advantage of credit facilities.

Future thrusts

Integrated Farming System (IFS) is a labour intensive system, thereby engaging the farmer family enhances the productivity on their own farms, throughout the year. IFS will lead to collective efforts among the farmers like collective purchase of inputs and collective marketing of produce, thus reducing their costs of production (Agarwal & Narain, 1998). It takes 3-4 years to establish a good integrated farm with market linkages to ensure nutrition and livelihood of a family. When we have many such farms in the village, there will be enormous scope for employment and business opportunities, especially for the youth for selling seeds, seedlings, manures etc. Following are the core areas where future of IFS will depend.

a) Quantification of biomass production under IFS and its overall efficiency in attaining sustainability on individual farm.

b) Impact of IFS on the carbon sequestration in different land use systems.

c) Identification of efficient cellulolytic microorganisms for recycling of crop residues for bringing desired benefits within the system.

d) Development of indigenous technology know how (ITK) of IFS existing in the farming community and its scientific validation and popularization.

e) Development of On Farm Research (OFR) to identify and adopt technologies for solving site specific problems in different agro-climatic regions of the country.

References

Agarwal, A. and Narain, S. 1989. Towards Green Villages: A strategy for Environmentally, Sound and Participatory Rural Development, Center for Science and Environment, New Delhi, India.

Carloni, A 2001. Global Farming Systems Study: Challenges and Priorities to 2030 – Regional Analysis: Sub-Saharan Africa, Consultation Document, FAO, Rome, Italy.

Edwards, P., Pullin, R. S. V. and Gardner, J. A. 1988. Research and development for the development of ICLF farming system in the tropics, ICLARM studies and reviews 1653 p. International Centre for Living Aquatic Resource Management, Manila, Philippines.

Evenson, R.E. 2000. Agricultural Producitivity and Production in Developing Countries'. In FAO, The State of Food and Agriculture, FAO, Rome, Italy,

Gupta Vinod, Pradeep Kumar Rai and Risam, K.S. 2012. Integrated Crop-Livestock Farming Systems: A Strategy for Resource Conservation and Environmental Sustainability. *Indian Res. J. Ext. Edu.*, Special Issue (Volume II); 49-54.

Sharma P., Sunil Kumar, Maharaj Singh and Satyapriya, 2012. Change in livelihood option in FSRP village clusters of District Tikamgarh, Madhya Pradesh, India, IGFRI News Letter, 18(1): 5

Sunil Kumar, D. R. Palsaniya, A. K. Dixit, T. Kiran Kumar, M. M. Das and G. Prabhu, 2015 : *E-Compendium* on Livestock based Integrated Farming Systems for Enhancing Resource use Efficiency and Improving Livelihood of Small and Marginal Farmers, ICAR- Indian Grassland and Fodder Research Institute, Jhansi (UP), India.

❑❑❑

Chapter 4

Enterprise Selection in Integrated Farming Systems: Issues and Strategies

D. R. PALSANIYA
ICAR - Indian Grassland and Fodder Research Institute, Jhansi-284003

Introduction

The integrated farming systems approach holds the key for upliftment of rural households as it combines crop, trees, livestock, fishery etc. in a holistic manner and has the potential to meet the diverse needs of the farm communities (such as food, fodder, fruits, fibre, fuel, small timber etc.) on sustainable basis. Farming systems research approach relates to the whole farm rather than individual; it is driven as much by the overall welfare of farming households as by goals of yield and profitability. It minimizes the use of external inputs and provides for recycling of residues or byproducts or wastes produced in one enterprise as inputs in the other enterprise. It enhances productivity of land or farmer as a whole, minimizes risks, optimizes the farm income and provides employment round the year. Therefore, integrated farming system has special significance and is capable of imparting drought proofing, sustainability and livelihood security to farmers in vulnerable areas. Integrated farming system comprises a mix of enterprises and farming activities. The success of the farming system depends on the careful selection of the appropriate enterprises for it. This chapter provides a step by step guideline to the enterprise selection process.

Steps for selection of enterprises for farming systems

The selection of enterprises is critical in determining whether or not the goals of our family will be met through farming system. Therefore, the first step should be the determination of the goals or objectives of our farm family and business. The next step is to find out whether or not they are feasible by making an inventory of the resources available and the resource requirements of each enterprise we are considering. Then, realistically compare the resources required to the resources available and finally, select the enterprises for our farming system we wish to pursue and design a plan for implementation.

Guiding principles

Further, the selection of enterprises for farming system must be made by taking in to consideration the overall farm business strategy and with a long run vision. In addition, the proposed crop/enterprise should be a reasonable choice for farm with respect to local agro climatic conditions, edaphic factors, resources at hand, socio-economic situations, market facilities, demand of produce, and raw material availability, etc. Moreover, it is critical to have a clear business strategy for your farm to remain profitable and competitive besides meeting family needs. Our farming system should be a low cost producer and competing on price.

Farming is considered as a family tradition in India and majority of the farmers continue to practice what their fore fathers practiced. Therefore, it is the need of the time to make the agriculture profitable and economically viable through appropriate change in farming system, farming practice and diversify from traditional agriculture to modern agriculture, more importantly it should be in business mode and should have a sound management backup. Like any other enterprises, it should also have proper planning about demand forecast, choice of technology, inventory of resources, need and requirement of external inputs, skill level of available human resources and their training needs, infrastructure and services needed for carrying out various operations and marketing, etc. The mindsets of the farmers need to be changed.

Importance

The selected enterprises and their interactions should lead

- to utilize leisure periods of labour (family/hired) and other resources to fullest capacity.
- to utilize/recycle waste/by products adequately.
- to coordinate and utilize the resources available with farmer to best advantages.

- to increase income/making farming business most profitable.
- to increase employment opportunities.
- to improve livelihood security.
- to improve nutritional security of the farm family.
- to improve and maintain natural resource base of the farm.

Key factors for enterprise selection in farming systems

Physical factors

Land: The amount of land available, its physical profile and topography of the land, the soil type, fertility and drainage capability, soil weed flora and suitability of soil for crops etc. largely determine enterprise selection.

Climate: The average rainfall and its distribution, temperature (average, maximum and minimum), frost occurrence, aridity, photo period, wind characteristics of the region, etc.

Water availability: The source, amount, cost and quality of irrigation water, time of water availability, irrigation system, etc. There will be different set of farm enterprises for irrigated and rainfed farming systems.

Farm Structures: Type of farm buildings, animal sheds, stores, threshing floors, irrigation channels, fences, etc. and their cost.

Farm machinery: Type of farm power machinery (tractor, bullock, engines, and pumps), farm implements (cultivators, seed drills, ploughs, harrows, weeders, harvesters, processing machinery), transportation equipment, leasing/renting/custom hiring possibilities, etc.

Financial factors

- Amount of capital
- Borrowing capacity
- Cash flow situation
- Interest rates
- Risk bearing capacity
- Insurance needs.

Management factors

- Record keeping skills
- Labor management
- Budgeting capability
- Familiarity with tax and other relevant laws
- Mechanical skills
- Knowledge of agrotechniques, animal health, pest management and greenhouse production
- Familiarity with the agricultural information delivery systems.

Labor factors

- Labor needs on a monthly basis
- Family or hired labor
- Labor cost
- Indian labor laws
- Opportunity cost of using your own labor.

Marketing factors

- Preferred marketing method (broker, retailer, direct, roadside stand, farmers market, cooperative, contract with processor, contract farming)
- Proximity to various potential markets
- Advice of contractor on crop selection (contract farming)
- Cooling and storage facilities for perishable products
- Processing facilities
- Marketing regulations and laws
- B:C ratio and potential returns.

Other factors

- Family needs and requirement
- Socio-religious factors
- Compatibility with other enterprises
- Policy support.

List of some of the enterprises for farming systems

Farming systems may consist of few main enterprises, some subsidiary enterprises and some other related activities as per the above factors. Some of such enterprises are listed in Table 1.

Table 1. Main and subsidiary enterprises for farming systems

Category	Items/particulars
A. Main enterprises	
1. Food crops	
Cereals	Wheat, rice, sorghum, maize
Pulses	Chick pea, pigeonpea, greengram, blackgram, rajma
Oilseeds	Mustard, groundnut, soybean
2. Cash crops	Sugarcane, cotton
3. Vegetables	Vegetable pea, brinjal, tomato, chilli, cauliflower, cabbage, leafy vegetables, radish, cucurbits, okra etc.
4. Fodder crops	Berseem, lucerne, oat, sorghum, pearlmillet, cowpea, BN Hybrid, guinea, range grasses and legumes, stylo
5. Trees	As agroforestry, block plantation or boundary plantation
Agrihorticulture	Fruit trees like guava, ber, Aonla, citrus, banana, papaya etc.
Agrisilviculture	Timber trees like teak, shishame, acacia, etc
Silvipasture	Fodder trees like Prosopis, Albizia, Leucaena, Hardwickia, Ailenthus
6. Animals	
Large ruminants	Cattle and buffalo
Small ruminants	Sheep and goat
B. Subsidiary enterprises	Fishery, poultry, piggery, duckery, mushroom, bee keeping, sericulture, rabbit farming, hydroponics, green house farming, nursery
C. Processing and value addition	Achar, Murrabba, ketchup, squash, jam, jelly, drying, grading, packaging
D. Other activities	Resource recycling, composting, vermicomposting, biogas, resource conservation strategies (RCTs), rain water harvesting, latest technologies and ICMP, etc.

Principle of enterprise integration

A farm manager is often confronted with the problems as to what enterprises to select and the level at which each enterprises should be taken up. How far he can go far or should go in combine enterprises with another depends partly on the inter- enterprises with another depends partly on the interrelationships, between different enterprises and the prices of products and inputs. In agriculture, crop husbandry is the main activity. The income obtained from cropping is hardly sufficient to sustain the farm family throughout the year. Assured regular cash flow is possible when crop is combined with other enterprises. Judicious combination of enterprises, keeping in view of the environmental condition of locality will pay greater dividends. At the same time, it will also promote effective recycling of residues/ wastes.

Livestock is the best complementary enterprise with cropping, especially during the adverse years. Installation of bio-gas plant in crop-livestock system will make use of the wastes, at the same time provides the valuable manure and gas for cooking and lightning. In a wetland farm there is a greater avenue for fishery, duck farming and buffalo rearing. Utilizing the rice straw, mushroom production can be started. Under irrigated conditions (garden lands), inclusion of sericulture, poultry and piggery along with arable crop production is an accepted practice. The poultry component in this system can make use of the grains produced in the farm component in this system can make use of the grains produced in the farm as feed. Pigs are the unique components that can be reared with the wastes which are unfit for human consumption. In rainfed farming, sheep and goat rearing form an integral part of the landscape. Sericulture can be introduced in rainfed farming, provided the climatic conditions permit it. Agro-forestry (Silviculture and Silvi-horticulture) are the other activities which can be included under dryland conditions. In the integrated system, selection of enterprise should be on the cardinal principle that there should be minimal competition and maximum complementary effect among the enterprises. Some enterprises suitable for integration in farming systems in different regions of India are given in Table 2.

Prominent farming systems of India

Following are the main farming system practicing in India

- Crop – Livestock/Dairy farming system
- Agri – Horticulture farming system
- Agro – Forestry system
- Crop – Livestock- Pastural system

- Mixed farming system
- Sole crop farming system
- Rice based farming system
- Horticulture based farming system (fruit crop/coconut/cashew/areca nut/vegetable base farming system).

Table 2. Enterprises for farming systems in different regions of India

S.No.	Region	Suitable enterprises for farming systems
1	High altitude cold deserts	Pastures with forestry, goats, angora rabbits and limited settled agricultural crops like millets, wheat, barley and fodder.
2	Arid and desert region	Centering mainly in animal husbandary with the camels, sheep and goat and with moderate cropping components involving bajra, wheat, pulses, gram and fodder crops.
3	Western and Central Himalayas	Horticultural crops as a major component and have a less intensive agriculture mainly on the hill terraces an slopes with maize, rice, wheat, pulses and fodder crops.
4	Eastern Himalayas	Primitive crop husbandary with rice, millets, pulses etc. Agro-forestry systems are also common. Piggery and poultry are the chief livestock activity.
5	Indo-Gangetic Plains	Intensive crop husbandary Indo-Gangetic plains: Intensive crop husbandary involving rice, wheat, maize, mustard, pulses and livestock inclusive of dairy cattle and buffaloes.
6	Central and Southern Highlands	Cotton, sorghum, millet, pulses with dairy cattle, sheep and goats and poultry are the secondary livestock and animal husbandry enterprises.
7	Western Ghats	Major activity on plantation crops, cultivation of rice and pulses are the secondary agricultural activity. Cattle, sheep and goats are the livestock components which in most parts are maintained as large herds and allowed to range.
8	Delta and Coastal Plains	Rice cultivation along with fish culture, poultry and piggery enterprises, capture fisheries of the marine ecosystem is a specialized enterprise and does not mix with cropping activity.

Enterprises for diversification of existing farming systems

Possible options for diversification of existing farming systems are given in Table 3.

Table 3. Some possible enterprises for diversification of existing prominent farming systems

Farming System	Main Enterprises	Possible addition for diversification
Mixed Farming System	Field crops, Vegetables, Fruit crops, Livestock, Fish, fodder	Forest tress, Composting, Processing, Sericulture, Mushroom, Medicinal and Aromatic crops, Nursery, Apiculture.
Rice Based Farming	Rice, Wheat, pulses, oilseed, Vegetables	Dairy, Poultry, Mushroom, Vermicomposting, Processing, Packaging, Pisciculture, Livestock, Fodder.
Horticulture Based Farming	Fruit, Vegetables, Medicinal and Aromatic Plant	Fruit, and Vegetable processing Unit, Vermicomposting, Mushroom, Apiculture, Sericulture, Pisciculture, Livestock, Fodder, Nursery.
Livestock Based Farming	Dairy, Poultry, Piggery, Rabbitery, Goatery, Fodder crop	Milk and meat processing unit, Nursery, Vermicomposting and bio-composting.
Crop- Livestock	Field/fodder crops, Dairy animals	Vermicomposting, Milk processing.
Agri- Horti	Field and Fruit Crops	Vegetable/Ornamental crops, Apiculture, Preservation unit.
Agro- Forestry	Crops and Forest trees	Multi storeyed cropping, Apiculture, Bio-composting.
Crop Livestock Pastural	Field and Pasture crops, along with dairy animals	Poultry, Goatery, Rabbit farming.
Agro Aquaculture	Paddy and Pisciculture	Duckery , Bud planting, Processing unit (Rice milling, fish packing, etc).

Examples of IFS models

Table 4. Rainfed IFS model (1 ha)

Enterprise/Activity	Area (ha)	Crop (ha)	
A. Crops	0.55	*Kharif* Groundnut, 0.3 Sorghum or Maize (cob), 0.25	*Rabi* Barley, 0.2 Gram, 0.25 Oat, 0.1 (F)
B. Agri-horticulture	0.2	Ber + Sesame-gram	
C. Silvipasture	0.2	Subabool+TSH + S. Seabrana	
D. Animals	-	5 goats with kids up to 6 months	
E. Farm pond	0.05 (20x25m)	For rain water harvesting and recycling	
Total	1.0	-	

Boundary plantation of ber + subabul + caronda and bund plantation of napier/guinea/TSH

* If rainfall delayed, groundnut will be replaced by blackgram/sesame

Table 5. Irrigated intensive IFS model with balanced enterprises (1 ha)

Enterprise/Activity	Area(ha)	Crop (ha)		
		Kharif	**Rabi**	**Zaid**
A. Food Crops	0.55	Maize/sorghum (0.25), cob+grain Groundnut (0.3)	Veg. pea (0.25) Wheat (0.3)	Okra Greengram
B. Round the year forage production (Crop + napier/guinea)	0.2	Sorghum(0.1) Cowpea (0.1)	Berseem(0.1) Oat (0.1)	Sorghum(0.1) Cowpea (0.1)
C. Agri-horticulture	0.2	Guava+vegetables (okra/cowpea/cucurbits-brinjal/tomato/cauliflower/cabbage/veg pea)		
D. Animals	-	2 large ruminants (B)		
E. Fish	0.05 (20x25m)	Seasonal (July – Feb)		
F. Other activities	-	Hay/Silage, Composting		
Total	1.0			

Boundary plantation of Karonda and Subabul and pond dykes and field bunds will be utilized for perennial forages like Napier hybrids+guava

Constraints and limitations

There may be limitations beyond our control that may prevent us from adoption of farming system enterprises. Some of such limitations may be

- Legal, regulatory, moral or ethical limits that may limit the farming activity.
- Social or religious or cultural limitation.
- Timber, fruit trees, and some livestock enterprises take years in development before they return positive revenues. How long one can wait?
- There may be some physical, environmental, regulatory or financial constraints.
- Infrastructural limitation.

Conclusion

The selection of appropriate enterprises for farming system is most crucial and needs a lot of energy and enthusiasm. Therefore, the single most important requirement is careful advance planning for it. The successful IFS entrepreneur will take into consideration the key points outlined in this chapter to make strong, informed decisions, avoid pitfalls and ensure the success of the farming system. Protect yourself, your family, and your new farm by taking the time to consider all the factors completely before you make a decision as planning is essential to the success of your IFS venture.

References

http://www.sare.org/publications/business/business.pdf

http://www.ardictionary.com/farming

http://www.nslocal.ca/agricultural-enterprisees

http://smallfarms.ifas.ufl.edu

http://www.fao.org

http://www/agmr.org

http://www.cari.res.in

http://edis.ifas.ufl.edu/hs338

□□□

Chapter 5

Vegetable Production Strategies in IFS for Profitability and Nutritional Security

SUNIL KUMAR AND VIKAS KUMAR

ICAR - Indian Grassland and Fodder Research Institute, Jhansi-284003 (UP)

Introduction

The concept of sustainability is an important element in the development of integrated systems. The MEA (2005) defined it as a characteristic or state whereby the needs of the present population can be met without compromising the ability of future generation or population in other locations to meet their needs. Developing countries around the world are promoting sustainable development through sustainable agricultural practices which will help them in addressing socioeconomic as well as environmental issues simultaneously. Within the broad concept of sustainable agriculture "Integrated Farming Systems (IFS)" hold special position as in this system nothing is wasted, the byproduct of one system becomes the input for other. Sustainable development in agriculture must include integrated farming system with efficient soil, water crop and pest management practices, which are environmentally friendly and cost effective. Integrated farming system are often less risky, if managed efficiently, they benefit from synergisms among enterprises, diversity produce environmental soundness (Lightfoot, 1990). Moreover, based on the principle of enhancing natural biological processes above and below the ground, the integrated system is the combination that reduces erosion, increases crop yields, soil biological activity and nutrient recycling, helps in efficient use of water, reduces pest and diseases, intensifies land use, improving profits and can therefore help reduce poverty and malnutrition and strengthen environmental sustainability.

Integrated crop livestock farming system

An integrated farming system consists of a range of resource-saving practices that aim to achieve acceptable profits and high and sustained production levels, while minimizing the negative effects of intensive farming and preserving the environment (Lal and Miller, 1990; Gupta *et al.*, 2012). Within this framework, an integrated crop-livestock farming system represents a key solution for enhancing livestock production and safeguarding the environment through prudent and efficient resource use. In integrated crop livestock farming system the waste of one enterprise becomes the input of another for making better use of resources (Tiwari, 1993). For example, crop residues can be used for animal feed, while manure from livestock can enhance agricultural productivity by intensifying nutrients that improve soil fertility as well as reducing the use of chemical fertilizers (Gupta *et al.*, 2012). For agricultural use animal excreta can be used for fertilizer, feed and fuel. Excreta have two crucial roles in the overall sustainability of the system. In the farm level, the farmers/gardeners are motivated to change the shape/style/design of the land so that it could be developed in to integrated farming system that utilizes the waste of one component as are source for the other and sets up a network of nutrient flow. The diversification that comes from integrating crops, vegetables, livestock, trees and fish imparts stability in production, efficiency in resource use and conservation of the environment. In integrated farming, wastes of one enterprise become inputs to another and thus optimize the use of resources and minimize pollution.

Integrated vegetable crops for profitability and nutrition security

Under IFS if food and nutritional security are arguing how can ignore the vegetables. Vegetables are important constituents of Indian agriculture and nutritional security due to their short duration, high yield, nutritional richness, economic viability and ability to generate on-farm and off-farm employment. Our country is blessed with diverse agro-climates with distinct seasons, making it possible to grow wide array of vegetables. India is the second largest producer of fruits and vegetables in the world. Total area under horticultural crops is 21.83 million ha and production is 240.53 million t. Fruits and vegetables together contribute about 92% of the total horticultural production in the country Vegetables are vital sources of proteins, vitamins, minerals, dietary fibers, micronutrients, antioxidants and phytochemicals in our daily diet. Apart from nutrition, they also contain a wide array of potential phyto-chemicals like anti-carcinogenic principles and anti-oxidants (e.g. flavonoids, glucosinolates and isothyocynates). In India, vegetables are valuable biological assets especially genetic resources. India is rich in biodiversity of vegetables and is the primary/secondary center of origin of many vegetables.

India produces 14% (146.55 million tonnes) of world's vegetables on 15% (8.5 million hectares) of world area under vegetables. Productivity of vegetables in India (17.3t/ha) is less than the world average productivity (18.8t/ha). The country has witnessed tremendous progress in vegetable production, especially during the post green revolution period. Development of improved vegetable varieties/ hybrids/technologies through systematic research coupled with their adoption by the farmers and developmental policies of the government culminated in tremendous increase in area under vegetables (8.5 million ha), production (146.5 million t) and productivity (17.3 t/ha) in the country. Compared to area (2.84 million ha), production (16.5 million t) and productivity (5.8 t/ha) in 1950-1951, there had been phenomenal increase in area (2.99 folds), production (8.88 folds) and productivity (2.98 folds) of vegetables in our country during the last 6 decades. The per capita availability of vegetables is also on the increase (230.4 g/ day/head in 2013) .Increasing per capita income, health consciousness, urbanization, increasing working women, shifting of farmers to high value vegetables due to higher income favourable income elasticity of demand and annual growth rate for domestic demand for fruits & vegetables are also important ingredients for fuelling vegetable growth in the country.

These production and productivity may have been achieved from the unit piece of land by applying tons of chemicals in to the soil is not going to put a milestone in Indian agriculture system. According to The International Food Policy Research Institute (IFPRI), Washington there is an urgent need to produce nutritious food in a sustainable manner and to improve farm family income in order to ensure household food security while at the same time protecting the natural ecosystem and conserving the natural resource base. That is possible only through vegetable production in IFS. Production of vegetable through organic farming is generally recognized by practitioners as a method of production that uses, practices or substances which are biologically enhancing to the soil, plant life, animal and human consumers, and growers. The principles of organic agriculture include: replenishing and maintaining long-term fertility by providing optimal conditions for biological activity; producing viable quantities of high quality, nutritious food and feed; reducing the use of fossil fuels in agriculture and pollution that may result from farming; encouraging "closed cycle" farming systems that use local resources and recycled nutrients; enhancing ecological cycles within the food production system; maintaining genetic diversity of the agricultural system and its surroundings, including protection of plant and wildlife habitats; ensuring decent and non exploitative treatment of farm workers; creating conditions for farm livestock that ensures them a life free of undue stress, pain, and suffering; maximizing the farmers' return and satisfaction for their work; sustaining the land in a healthy condition for future generations; and optimizing multiple use capacity of the land.

Diversified nature of IFS and varied type of farm produces viz; milk, fruits, vegetables , fish and green fodders etc. in the system made the human and animal diet more nutritious compared to existing crop based farming system involving crops + dairy. Recycling of all the crop residues, animal and farm wastes and use of leguminous crops as green manure or dual purpose crops and bio-fertilizers could save more than 36% of plant nutrients. In addition, IFS approach generated more than the double man days as compared to crops alone which in turn can solve unemployment problem in rural youths. India has to go a long way to accelerate the vegetable production, considering the national and international demand. There is a need to achieve the target of 225 million tonnes by the end of 2020 and 350 million tonnes by 2030. Existence of adequate eco-diversity in the country has enabled to grow more than 60 vegetables belonging to thirteen plant families in the country like solanaceous, cucurbitaceous, leguminous, cruciferous (cole crops), root crops and leafy vegetables are grown in Indian tropical, subtropical and temperate regions. Important vegetable crops are onion, tomato, brinjal, peas, beans, okra, chilli, cabbage, cauliflower, pumpkin, bottlegourd, cucumber, muskmelon, water melon, radish, carrot, palak and methi.

In addition, about 30 lesser-known and underutilized vegetable crops needs to promote their cultivation under suitable climatic conditions. In India the production of cereals are likely to constant but area and production under vegetable crops are increasing year wise as well as state wise (Table 1 and 2). It is due to so many reasons. To get the maximum output in the minimum time, there is scope of intensive vegetable farming that provide quickly and high yield per unit area. Most of the vegetables, if properly grown, can give a yield which is five to ten times higher than cereal crop. India is next only to China in area and production, contributing 13.5 % to the total world production. India occupies first position in Cauliflower second in onion and third in cabbage in world. The productivity of vegetable crops is 14.9 t/ha. West Bengal, orrisa , Uttar Pradesh, Bihar and Maharashtra are major vegetable growing states in India.

Vegetable production strategies in IFS

1. Make timely availability of suitable cultivar of vegetables as per agro climatic zone in sufficient quantity and quality of latest varieties/hybrids.
2. Vegetable nursery management.
3. Replacing the old varieties/hybrids in particular regions with the latest ones.
4. Protected cultivation and nursery raising.
5. Use of insect, pest and disease resistant varieties.
6. Botanical, microbial and pheromones should be incorporated in IPM schedule to reduce dependency on chemical insecticides.
7. Quality control and development of bio-agent release techniques need to be emphasized to improve the performance of bio-agents and microbial control agents.
8. Effective promotional use of predators, parasitoids and entomo-pathogen. Emphasis must be given to develop behavioral management practices with special reference to sex pheromone and trap crop based management practice.
9. Adoption of improved technologies from the research institutes.
10. Effective market intervention for remunerative prices.
11. Increased awareness about the nutritive value vegetables in food habit.

Livestock, crop and vegetable crop relations in IFS

There has been increased diversification and re-integration of crop and livestock operations for (1) within farm integration where a crop or dairy operation diversifies and adds another enterprise and (2) among farm integration or "coupled" farms where two or more operations share land, manure nutrients, forage and grain crops grown in rotations. These coupled operations allow individual specialization, but allow sharing of nutrient and land resources, hopefully benefiting both operations. When evaluating the benefits of integrated systems, it is obvious that organic vegetable producers would benefit from the improved flow of nutrients from manure produced from a livestock enterprise. This is especially true when one considers areas that are deficient in soil phosphorus as feeding forages and grains through ruminants will improve phosphorus availability. Nitrogen from manure is also fairly rapid in plant availability during the growing season. This is of course dependent on the type of manure and bedding content, with poultry manure (cage layer) containing very little carbon and high nitrogen availability, to beef, sheep or horse manure which contains high amounts of carbonaceous material such as sawdust or shavings that may "tie up" nitrogen when applied to the soil.

Table 1. Nutritional requirement of vitamins a person/day for good health metabolic role and source as vegetable

Composition	Requirement	Role	Source	Deficiency
Vitamin B_1 (Thiamin)	0.5-20 mg	Utilization of sugar and starch	Green leafy vegetable	beri beri
Vitamin B_2 (Riboflavin)	07-202mg	Protein metabolism	Green leafy vegetable	Sore tongue, cracking in mouth angles, redness and burning of eyes, scaliness of skin
Vitamin B_6	6-2.5 mg	Protein & fat metabolism	Vegetable	Convulsion in infants
Folic acid	50-100 mg	Multiplication and maturation of RBC	Green vegetable	Anemia in children & pregnant women
Vitamin B_{12}	1 mg	Importance for DNA Synthesis	Animal food	Anemia in children & pregnant women

Vitamin D	3000 μg)	Promoting healthy bones, absorption of calcium.	Animal Food, veg. & Fruit	Osteoporosis or osteomalacia
Niacin	21 mg	Component of coenzyme	Rice, Wheat bran, nuts	Pellagra
Ascorbic acid	30 to 50 mg),	Resistance	Fruit & Veg finished 90% of the world, spinach, cabbage, tomato, chili	Immunity loss, teeth problem
Carotene	(3000 I.U.)	Eye sight & immunity	Fruit & Veg finished 60% of the world	Vision & Bone problem

Table 2. Nutritional requirement of minerals of a person/day for good health metabolic role and source as vegetable

Mineral	Metabolic function	Deficiency leads to	Daily require-ment	Source
Calcium	Maintenance of skeleton and teeth	Week bone and teeth	0.4-0.6g	Green leafy vegetable, milk and animal product
Iron	Formation of Hemoglobin	Anemia	1-3 mg	Drumstick leaves, leafy vegetable
Electrolytes (Na, k and Mg)	Maintenance of cellular osmotic balance	Cellular turgidity and muscular cramps	-	Drumstick, leaves, leafy vegetable
Phosphorus	Component of nucleic acid, ATP	Week bone and teeth		Cucurbits, beans, ginger & onion

Balanced diet of an Indian adult

Cereals (500 g), pulses (75 g),fruits (120 g), milk (300 ml), leafy vegetables (125 g) other vegetables (80 g), roots and tubers (90 g), = 300 g, fats and oils (40 g) and sugar (40 g).

Balanced diet, should constitute of 25 to 30% of fruits and vegetables.

Vegetable as a source of mineral and Vitamins

Table 3. Nutritive Value of vegetables per 100 g of edible portion

Vegetable	Moisture (%)	Carbo-Hydrates (g)	Protein (g)	Fat (g)	Vit. A (I.U)	Vit. B_1 (mg)	Vit. B_2 (mg)	Vit. C (mg)	Minerals Ca (mg)	P (mg)	Fe (mg)	K (mg)
Cabbage	91.9	4.6	1.8	0.1	2000	0.06	0.03	1241	39	44	0.8	114
Cauliflower	90.8	4.0	2.6	0.4	51	0.04	0.10	56	33	57	1.5	113
Chilli	85.7	3.0	2.9	0.6	292	0.19	0.39	111	30	80	1.2	217
Tomato	93.1	3.6	1.9	0.1	320	0.07	0.01	31	20	36	1.8	114
Brinjal	92.7	4.0	1.4	0.3	124	0.04	0.11	12	18	47	0.9	2
Cucumber	96.3	2.7	0.4	0.1	00	0.03	0.01	7	10	25	1.5	50
Bottle gourd	96.1	2.5	0.2	0.1	0	0.03	0.01	6	20	10	0.7	87
Biitergourd	92.4	4.2	1.6	0.2	210	0.07	0.09	88	20	70	1.81	152
Okra	89.6	6.4	1.9	0.2	88	0.07	0.10	13	66	56	1.5	103
Dolichos bean	86.1	6.7	3.8	0.7	312	0.1	0.06	9	210	68	1.7	74
Muskmelon	94	5.0	1.0	0	3420	0.04	0.05	33	0.017	0.016	-	-
Watermelon	95.8	3.3	0.2	0.2	590	0.05	0.05	6	0.07	0.07	-	-
Pea	72	15.8	7.2	0.1	139	0.25	0.01	9	20	139	1.5	79
Onion	86.8	11	1.2	-	0	0.08	0.01	11	180	50	0.7	-
Garlic (dry)	62	29.8	6.3	0.1	0	0.06	0.03	13	30	310	1.3	-
Radish White)	94.4	3.4	0.7	0.1	5	0.06	0.02	15	50	22	0.4	138
Carrot	86	10.6	0.9	0.2	3150	0.04	0.02	3	80	30	2.2	108
Pumpkin	92.6	4.6	1.4	0.1	84	0.06	0.04	2	10	30	0.7	139
Tomato green	93.1	3.6	1.9	0.1	320	0.07	0.01	31	20	36	1.8	114
Pointedgourd	92	2.2	2.0	0.3	22	0.05	0.06	29	30	40	1.7	83
Cowpea	84.6	8	4.3	0.2	941	0.07	0.09	13	80	74	2.5	-

Source: Aykroyd (1963)

A. Role of livestock production in IFS

The various benefits that induces farm producers to incorporate livestock production along with vegetables. Some important of them are given below.

1. A source of nutrients that can be used for direct application and/or composting. By combining livestock and vegetable production, whole farm nutrient balance of imports and exports becomes more even.

2. Along with nutrients, manure and compost applications tend to improve soil organic matter, biological activity and potential disease suppression. This improved "soil health" will manifest itself quickly and include improved soil nutrient cycling, improved soil structure, better water holding capacity in droughty soils and improved drainage in heavy soils.
3. Livestock operations improve the potential for profit in lands that are in a "sod" rotation. Sod crops help to build soil structure (grass roots) and soil drainage (legumes/alfalfa). Sod crops high in legume content will also provide a source of nitrogen when those fields are returned into the row crop rotation.
4. Livestock provide a use for crop residue and waste or cull vegetable crops. This can help to reduce disease prevalence while providing a "cheap" source of feed for livestock. Cows turned into a field of pumpkins in November provide great feed for the cows and help vegetable producers clean up a field.
5. Grain crops used by vegetable operators as cover crops can fit well into livestock rations. Winter grain crops provide fall nutrient catch, weed control in both fall and spring, and can be under sown with clover or other legumes to provide nitrogen in subsequent rotations and a sod crop establishment with minimum tillage.
6. Adding livestock products to your marketing can help improve cash flow in the winter and add a new aspect to CSA operations.
7. While not always discussed, successful "coupled" animal/vegetable operations can also help to build community with a farm region. Coupled operations also tend to share machinery resources and labor resources during busy periods. While there are many positives to integrating livestock and vegetable operations, there are also some risks and issues. Most involve the use and handling of livestock wastes and effluent from feeding operations. Organic operations have specific guidelines for the use of manure and manure composts.

Horticultural crops play a unique role in India's economy by improving the income of the rural people. Cultivation of these crops is labour intensive and as such they generate lot of employment opportunities for the rural population.

B. Role of vegetable production in Integrated farming system

Vegetables are always more profitable than paddy," as a result echoing farmers across the country who are rapidly shifting from the safer but less profitable option of food grain to vegetables. The central agriculture ministry estimates that fruits and vegetable crops are four to eight times more profitable than other crops.

- Improved irrigation, availability of disease-resistant hybrid seeds and new marketing channels are making it easier for farmers to shift to vegetables, which have been traditionally considered a riskier proposition because of shorter shelf life.
- Rising incomes have led to a spurt in demand for fresh vegetables and the farmers have responded by increasing area under production more than 40% over the past decade, contributing to output that has more than doubled during this period. As a result, the total production of vegetables and fruits, increasing at 6% annually, now exceeds that of wheat, rice and corn.
- Vegetables are becoming the top choice of Indian farmers, especially those who live closer to metros and big towns.
- Vegetables and fruits already make up a third of the output in Himachal Pradesh, Orissa, West Bengal, Jammu & Kashmir, Bihar and the Northeast. Eastern India has become the largest supplier of green vegetables, contributing 45% of the total output.
- With such rapid growth, new competitors are fast emerging to challenge the traditional centers such as Maharashtra for onions and Uttar Pradesh for potatoes. Madhya Pradesh, Rajasthan and Bihar are all challenging Maharashtra, for instance. So much so that, as a big exporter in Maharashtra lamented, this year all the export of onions to Bangladesh was from Bihar, which enjoys a critical locational advantage.
- Farmers in Himachal Pradesh have started growing exotic vegetables that cater to more sophisticated palates. "A sizeable area in the state has started cultivation of vegetables that are of European and Oriental origin.
- Jammu has likewise emerged as a prominent producer of tomatoes during the summer months. Area under vegetable crops has almost doubled in Jammu region during last five years.

- State-run institutions have contributed in a major way in a few other states as well. In Kerala, for instance, the Vegetable and Fruit Promotion Council Keralam, with farmers as majority shareholders, subsidises seeds and takes the produce to its own auction centres where wholesalers buy directly. "Every year we see an addition of 10,000 farmers.
- Last month, Punjab government announced a 50% subsidy on vegetable farming to boost production in the state. The centre's Rs 300-crore Vegetable Initiative proposes to encourage vegetable cultivation in the peri-urban areas around 29 state capitals and other cities with population of more than one million.
- Vegetables are especially attractive to those with small farmland, cultivates potatoes and tomatoes on two hectares in West Bengal's Hooghly district. Unlike row crops, egetables don't need a minimum sized plot for optimum yields and profits. The most popular choices are potatoes, tomatoes, brinjal, lady's finger, cauliflower, peas and cabbage.
- Increase in area under irrigation has allowed even tribal farmers to grow vegetables and availability of hybrid seeds allow farmers to shorten the crop cycle, grow a vegetable even in its off-season, get higher tonnage from the same field and better utilize crop inputs.
- Some seeds even include labour-saving technology such as tomatoes that ripen at the same time to reduce the number of pickings.
- The last decade has been the one of agrarian crisis, as food grain production has steadily decreased. The crisis made marginal farmers shift from cereal to vegetable cultivation, a move that has brought big gains to farmers and manifold increases in production.

References

MEA (Millennium Ecosystem Assessment) (2005). Current State and trends (Eds. Hassan, R., Scholes, R. and Ash,N.) **1**: 133-134.

Lightfoot, C. (1990). Integration of aquaculture and agriculture: a route to sustainable farming systems. NAGA:TheICLARM Quarterly, **13**(1): 9-12.

Lal, R. and Miller, F.P. (1990). Sustainable farming for tropics. In: Singh, R.P. (Ed.) Sustainable agriculture: Issuesand Prospective. Vol. **1**, pp. 69-89, Indian Society of Agronomy, IARI, New Delhi.

Tiwari, P.N. (1993). Integrated Farming Research for sustaining food production. Journal of Nuclear Agriculture Biology, **20**: 1-13.

Gupta, V., Rai, P.K. and Risam, K.S. (2012). Integrated Crop-Livestock Farming Systems: A Strategy for ResourceConservation and Environmental Sustainability. Indian Research Journal of Extension Education, Special Issue, **2**: 49-54.

Aykroyd, W.R. (1963). The nutritive value of Indian Foods and planning of satisfactory diets. ICMR Special Rpt. 42, New Delhi.

□□□

Chapter 6

Fodder Production Strategies for Integrated Farming System under Irrigated Conditions

T. KIRAN KUMAR, D.R. PALSANIYA, A.K. DIXIT, G. PRABHU and MANOJ CHAUDHARY

ICAR - Indian Grassland and Fodder Research Institute, Jhansi-284003 (UP)

Introduction

The Indian economy is predominantly agriculture oriented and livestock rearing forms an integral part of rural living. The contribution of agricultural sector in the Indian economy is steadily declining (from 36.4% in 1982-83 to 14% in 2012-13), the agriculture and livestock sector still provides employment to 52% of the work force. Crop and livestock are the two main components of a mixed farming system in India, which influence our agricultural economy and provide sustenance. The importance of livestock in India's economy can be gauged from the fact that 90 million farming families, cultivating 140 million hectare area, rear 90 million milch animals. Livestock is also important as a part of agriculture diversification and income enhancement, and crucial for nutrition enhancement. Livestock plays a vital role in the economic development, 25 % of the agricultural GDP is contributed by this sector. Despite the enormous contribution to the national economy and remarkable growth in milk production in last few decades, the productivity of our animals is low due to improper nutrition, health care and management. The availability of quality feed and fodder is one of the main reason for poor performance of livestock. Since last two decades the total area under cultivated fodder is only 8.4 m ha (5.23%), which is continued to be static. Due to ever-increasing population pressure of human beings, arable land is mainly used for food and cash crops, thus there is little chance of having good-quality arable land available for fodder production, unless milk production becomes remunerative to the farmer as compared to other crops. The fodder production in the country is not sufficient to meet the requirements of the growing livestock population and also the forages offered to animal are mostly of poor quality. The requirement of fodder is mainly governed by the population and species of livestock, productivity levels and climatic conditions. In the country, livestock in intensive cropped areas obtain only about 25 to 30 percent of their feed from grazing in nearby uncultivated lands as well as some fodders species grown in the patches; the balance comes from crop residues and byproducts. The major fodder and feed resources for livestock are the variety of herbage from cultivated forages, i.e. cultivated forage species, crop residues from crop lands, green/dry fodder from self owned or community land and grazing on wasteland, degraded land and forest lands. The increasing pressure on land, aggravating competition between the crop and livestock and the growing demand for livestock products makes it more and more important to ensure the effective use of natural resource base. To meet the requirement of food and fodder for increasing human and livestock population, the only option open is through time and effective space utilization in agriculture. The time concept relates to increasing the intensity of cropping under assured irrigated conditions, whereas space utilization pertains to building up of vertical dimension through multi-tier cropping and farming system approach. Thus by making use of these time and space concept either in irrigated or in rained areas, the productivity per unit area per unit time can be substantially enhanced.

Farming system especially refers to a combination of enterprises in which the products and the byproducts of one enterprise serve as the inputs for production of other enterprise. This is a multidisciplinary whole farm approach and very effective in solving the problems of small and marginal farmers. The approach aims at achieving food and nutritional security at house hold level besides increasing income and employment from small holdings by integrating various farm enterprises and recycling crop residues and byproducts within the farm (Behera and Mahapatra, 1999). An integrated crop-livestock farming systems represents a key solution for enhancing livestock production and safeguarding the environment through prudent and efficient resource use. Integrated farming system is based on the principles of enhancing natural biological process above and below ground. In integrated crop livestock farming system the waste of one enterprise becomes the input of another for making better use of resources (Tiwari, 1993). The main purpose of integrated farming system is to reduce erosion, optimum use of land, increase crop yields, soil biological activity and nutrient recycling, efficient use of water, reduce pest and diseases, maximum returns per unit area per unit time in terms of produce and money and strengthen environmental sustainability.

The need for Integrated farming systems

The last four decades has witnessed s sharp decline in the average size of operational land holdings in India. This is reflective of the immense population pressure on the limited land resource available for cultivation. Small and marginal holdings together, constitute 85 percent in terms of number of operational holdings and 44 percent of the operated area in the country. Thus over the period, the marginal category has emerged as a distinct and dominant class by itself with its average size dwindling to mere 0.38 ha. The majority of farmers hold less than two hectares of land. These farmers generally practice subsistence farming where they need to produce a continuous, reliable and balanced supply of foods, as well as cash for basic needs and recurrent farm expenditure. Crop intensity remained higher with marginal holders and decline with increase in size of holding (Ramesh Chand *et al.*, 2011). Besides improving the diversification, there is need to develop suitable integrated farming systems for such farmers since single crop production enterprises are subject to a high degree of risk and uncertainty because of seasonal, irregular and uncertain income and employment to the farmers. Sporadic but location specific research efforts in this regard have been encouraging (Jayanthi *et.al.*, 1994). The IFS introduces a change in the farming techniques for maximum production in the cropping pattern and takes care of optimal utilization of resources. The farm wastes are better recycled for productive purpose in which waste of one enterprise becomes the input of another thus making efficient use of resources. Optimum mixture of agricultural enterprises like dairy, poultry, mushroom utilization, piggery, fishery etc. suited to local agro-climatic conditions and socio-economic status of farmers would bring in prosperity to farmers. This double up the challenge of making farming on these fragmented pieces of land gainful, if not viable. In India, crop+livestock is the pre-dominant farming system and around 85% of farm house holds practice it (Gangwar and Ravisankar, 2014). Fodder needs of the livestock in this system can be addressed through effective and remunerative forage production strategies.

Components of IFS

Integrated farming is a system which tries to imitate the nature's principle, where not only crops but, varied types of plants, animals, birds, fish and other aquatic flora and fauna are utilized for production. These are combined in such a way and proportion that each element helps the other that the waste of one is recycled as resource for the other through integration. The basic principle is to enhance the ecological diversity by choosing the appropriate cropping methodology with mixed cropping, crop rotation, crop combination and inter cropping so that there is less competition for water, nutrition and space. Adopting eco-friendly practices by following multistory arrangement so that the total available area is effectively used and there is a high level of positive interaction among biotic and abiotic components and the whole farm productivity is increased (Fig. 1). IFS will normally have combination of three types of enterprise.

Table 1. Size Group wise distribution of Average Holdings in the country

S. No	Size groups	1970-71	1976-77	1980-81	1985-86	1990-91	1995-96	2000-01*	2005-06*	2010-11P
1.	Marginal (<1.0 ha)	0.40	0.39	0.39	0.39	0.39	0.40	0.40	0.38	0.38
2.	Small (1.0-2.0 ha)	1.44	1.42	1.44	1.43	1.43	1.42	1.42	1.38	1.42
3.	Semi-medium (2.0-4.0 ha)	2.81	2.78	2.78	2.77	2.76	2.73	2.72	2.68	2.71
4.	Medium (4.0-10.0 ha)	6.08	6.04	6.02	5.96	5.90	5.84	5.81	5.74	5.76
5.	Large (>10.0 ha)	18.1	17.57	17.41	17.21	17.33	17.21	17.12	17.08	17.38
All size classes		2.28	2.00	1.84	1.69	1.55	1.41	1.33	1.23	1.16

Note: Total may not tally due to rounding off. * excludes Jharkhand

P: Provisional.

Source: Agriculture Census 2010-11.

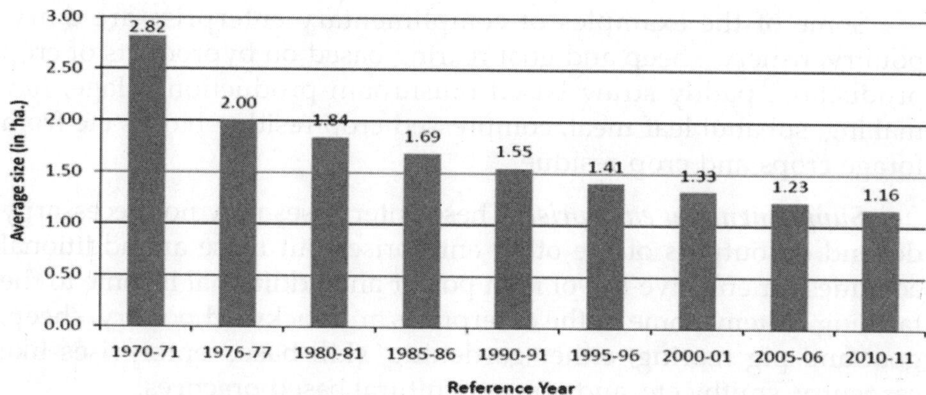

Fig. 1. Average size of operational holdings as per different agriculture Censuses in India

Primary enterprise: This is main component of the farm and they produce major saleable/utilization product of the farm and provide the main source of farm income. To undertake this component farmer should have necessary knowledge, skills, enough input and adequate labour availability.

Food crops: Rice, wheat, maize, sorghum, pearlmillet, pigeonpea, chickpea, blackgram, greengram

Forage crops: Napier bajra hybrid, Guinea grass, Tri specific hybrid, sorghum, maize, cowpea, oat, berseem

Agroforestry: Crop and tree crops eg. *Sesbania, Subabul*

Seed production: Field crops, vegetables, forage and flowers crops etc;

Vegetable crops: Bhendi, brinjal, tomato, field pea, cabbage, cauliflower;

Fruit crops: Guava, ber, mango, pomegranate, banana;

Plantation crops: Coffee, rubber

Other horticulture crops: Coconut, arecanut; Animal husbandry: Dairy, poultry, piggery, sheep and goat rearing etc.

Complementary enterprise: These enterprises generally based on the effective utilization/value addition to the primary product and other enterprise. The output of these enterprises is valuable input for other enterprise.

Some of the examples of complimentary enterprise are dairy, poultry, fishery, sheep and goat rearing based on byproducts of crop production, paddy straw based mushroom production, silage, hay making, subabul leaf meal, compressed crop residue blocks etc from forage crops and crop residue.

***Supplementary enterprise*:** These enterprises may not necessarily depend on outputs of the other enterprises but these are additional activities for effective use of man power and additional income to the farming system. Some of the enterprises are: backyard poultry, sheep, goat and pig rearing, either gardening, skill based enterprises like carpentry, smithy etc. and non agricultural based practices.

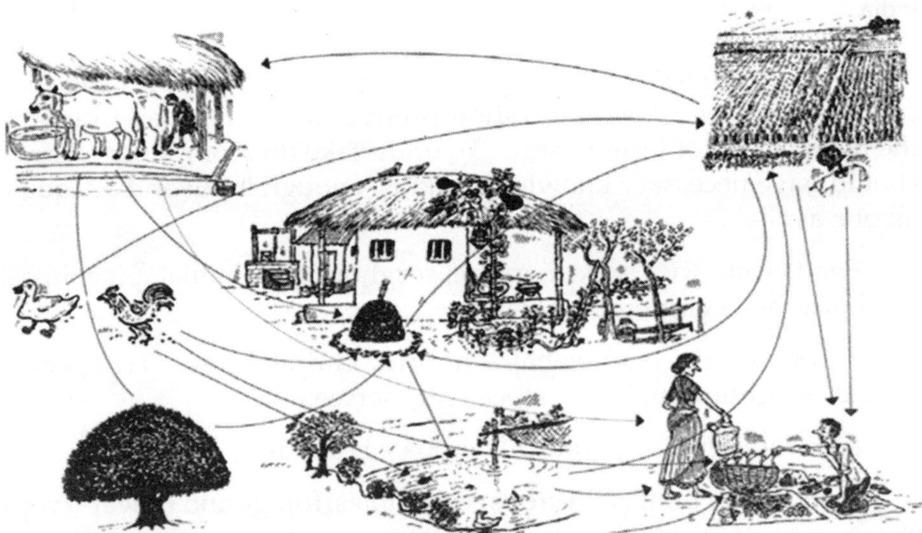

Fig. 2. Schematic representation of integration of the several components of the farm

Role of Forages in Integrated Farming Systems

Though forages faces tough competition from food and commercial crops for land and other inputs, but they are unique from other crops in terms of their wider adaptability and variability in terms of growth, regeneration potential, yield and quality of green biomass. The important characteristics forage crops suitable for livestock based farming systems are: Short growth period and grown in closer spacing with high seed rate which results in higher green biomass for feeding dairy animals throughout the year i) crop duration can be adjusted and risk due to aberrant weather conditions minimized, ii) due to dense stand, forages smother weeds and prevent soil erosion, iii) improve soil health through addition of higher amounts of organic residues in the soil, (iv) high persistency and regeneration capacity reduce the need for frequent sowing and tillage, v) wider adaptability with capacity to grow under stress conditions, vi) high nutrient and water requirement under intensive cropping vii) multicut nature with capacity to provide regular supply of green fodder for higher income and employment. In addition the leguminous fodders have special significance because of high protein and partial dependence from soil for their nitrogen. To meet the ever increasing demand of fodder can be met through maximization of forage production in time and space. The integration of forage crops in traditional cropping system, identifying remunerative forage/food-forage based systems and finding new niches of forage production are the promising ways of promoting sustainable agriculture. However, the integration of forage crops and livestock into existing cropping systems requires a holistic approach, taking into account the systems ecological, economic and social characteristics and the priority attached to different crop and livestock components by the farmers themselves. Shortages of both land and capital dictate systems that rely as much as possible on multipurpose crops and livestock. Although complex, these integrated systems are also less risk-prone and hence more sustainable not only ecologically but also economically than monocropping systems. Effective utilization of farm wastes and byproducts make to sustain the soil health through forages and dairy based farming systems.

Forage production strategies for irrigated farming systems

Fodders as a group of crops differ from food and commercial crops in several aspects and their package of practices for cultivation vary accordingly. Crop production is concerned with effective utilization of environmental inputs for morphological and physiological response of plant within an agro-climatic environment to produce maximum yield per unit area and time. The development of efficient forage production technologies is the way for sustainability of crop-livestock systems in the country. The most important input for crop production is water especially for fodder crops where the maximum vegetative growth is desired within a short period of time. Provision of irrigation allows the maximum utilization of resources for intensive forage production, which is very important in our country with small land holdings. There are certain fodder crops like berseem, lucerne, turnip, etc. which, can be grown with adequate moisture while in others the production potential can be realized fully with irrigation. A combination of diversified soil types, wide range of climatic conditions (cloudy to sunshine, hot to cold, dry to rainy) and a large group of forage species suited to different agro-ecological conditions and input situations, makes a congenial environment for intensive forage production programme in our country. Major forage crops during *kharif* are guinea grass, napier bajra hybrid, setaria, deenanath (grasses), sorghum, pearlmillet, maize, teosinte (cereals) cowpea, rice bean and guar (legumes); and during rabi season crops like oat (cereals), berseem, lucerne are grown under cultivated condition as a sole or as component crop in cropping systems.

Intensive forage production systems

To meet the fodder needs of dairy animals and livestock keepers intensive forage crop sequences or intercropping systems have been identified in assured irrigated conditions for increasing herbage yield, improving forage quality, reducing fertilizer requirement and ensuring better land utilization. These forage production systems are aimed at producing more green fodder yield with efficient utilization of land, water, nutrients and maintaining soil fertility. Intensive cropping is the only alternative to boost forage yield from irrigated lands. The multicut nature and flexibility in manipulating the duration of forage crops are desirarable characters to increase cropping intensity. The pre-requisites for intensive forage production are (*i*) fodder is required in uniform quantity throughout the year, (*ii*) the fodder crops in the rotation should be high-yielding, (*iii*) the area for production of fodder should be fully irrigated, and (*iv*) other inputs, such as fertilizers and pesticides, should be available in optimum quantity. The different systems of fodder production fall into two categories, viz. the overlapping cropping and the relay-cropping. In the overlapping system, a fodder crop is introduced in the field before the other crop completes its life-cycle. In relay-cropping, the fodder crops are grown in successions, i.e. one after another, the gap between the two crops being very small.

Multiple cropping

Succulent green fodder is produced throughout the year for dairy animals and 3-4 fodder crops are cultivated on the same piece of land in a calendar year. It consists of intercropping, mixed cropping and sequential cropping. These cropping systems are intensified spatially and temporally by growing of berseem in inter rows of hybrid napier grass in spring season and cowpea during summer after final harvest of berseem. Multiple crop sequences like sorghum + cowpea – berseem + mustard – maize and sorghum (multicut) +cowpea-berseem + mustard are promising under irrigated conditions of central India (Table 2). These intensive forage production systems ensures supply of green fodder throughout the year with improved quality, utilization of resources (land, water, nutrients and other inputs) efficiently and provides employment opportunity to the farmers (Table 3). The intensive intensive forage based cropping systems with appropriate soil crop management can produce green fodder 180-300 t/ha/year. Some of the intensive fodder based cropping systems followed in different part of our country. An ideal fodder system is that which gives the maximum outturn of digestible nutrients per hectare, or maximum livestock products from a unit area. It should also ensure the availability of succulent, palatable and nutritive fodder throughout the year. Some of the important intensive fodder-crops rotations and the expected yields are given in Table 3 for different regions.

Table 2. Potential intensive forage crop rotations under irrigated conditions for semi arid region of India

Crop rotations	Green fodder yield (t/ha/year)	Dry matter yield (t/ha/year)	B:C ratio
NB hybrid+ (Cowpea-Berseem+ Mustard)	273.1	44.3	2.41
Sorghum (multicut) – Turnip – oat	190.1	37.4	1.48
Sorghum + cowpea- berseem + Mustard-Maize+Cowpea	180.5	33.3	1.67
Sorghum (multicut)+Cowpea-Berseem+Mustard	172.0	32.3	1.94

(*Source:* Sunil Kumar *et al.*, 2012)

Table 3. Intensive forage crop rotations for different agro-climate zones of India

Zones/ Location* Crop rotation	Green fodder yield(t/ha/ year)
A. North	
Palampur (Himanchal Pradesh) : Sub-temperate, Moist, Red soil	
1. Maize + Cowpea – Lucerne + Oats - Mustard	85
2. NB Hybrid + Velvet bean – Berseem + Mustard	123
Pantnagar (Uttaranchal) : Tarai, Red & yellow soil	
1. Maize + Cowpea – Toria – Oats	177
2. NB Hybrid + Berseem – Cowpea	121
Hisar (Haryana) : Semi-arid, Sandy loam soil	
1. NB hybrid + Berseem	212
2. NB hybrid + Lucerne	176
B. Central and Western	
Jhansi (UP): Semi-arid, Red soil	
1. NB hybrid + Cowpea - Berseem + Mustard	255
2. Sorghum + Cowpea – Berseem + Mustard – Maize+ Cowpea	176
Jabalpur (MP) : Sub-humid, Black soil	
1. NB hybrid + Cowpea – Berseem	176
2. Sorghum+Cowpea–Berseem+ Mustard–Sorghum+Cowpea	169
Anand (Gujrat) : Semi-arid, Black soil	
1. NB hybrid + Cowpea – Lucerne	253
2. Maize + Cowpea – Oats – Maize + Cowpea – Maize + Cowpea	168
C. East	
Ranchi (Kanke): Sub-humid, Red acidic soil	
1. Pearlmillet + Cowpea – Maize + Cowpea - Oats	103
2. Maize + Cowpea – Sorghum + Cowpea – Berseem + Mustard	96
Kalyani (West Bengal) : Sub- humid, Alluvial soil	
1. Maize + Cowpea – Dinanath grass – Oats	131
2. Maize + Ricebean – Berseem + Mustard	112
Jorhat (Assam): Humid, Acidic soil	
1. NB hybrid	106
2. Maize + Cowpea – Maize + Cowpea – Maize + Cowpea	85
D. South	
Coimbatore (Tamil Nadu): Sub humid, Black soil	
1. NB hybrid + Lucerne	225
2. Sorghum + Cowpea – Maize + Cowpea – Maize + Cowpea	111
Vellayani (Kerala): Humid, Red soil	
1. Guinea grass in coconut plantation	135
2. Congosignal grass in coconut plantation	75

(*Source:* Sunil Kumar *et al.*, 2012)

Round the year fodder production

It consists of raising berseem, inter-planted with hybrid Napier in spring and intercropping the inter-row spaces of the grass with cowpea during summer after the final harvest of berseem. This overlapping cropping system is developed at Indian Grassland and Fodder Research Institute (IGFRI) Jhansi to cater the ever increasing green fodder demand of livestock population. In the overlapping system, a fodder crop is introduced in the field before the other crop completes its life-cycle. The suitable combination of these perennial and annual forage species ensure efficient utilization of moisture and nutrients from different soil layers and provide green fodder throughout the year. Under irrigated situations NB hybrid/guinea grass along with seasonal crops like cowpea in kharif/zaid and bersee/lucerne in rabi are found to be suitable for round the year forage production (Table 4). In these cropping systems NB hybrid can be successfully replaced by setaria/guinea grass in humid and sub humid climate. In western and southern part of the country berseem can be replaced by lucern. This NB hybrid based system is having the green fodder production potential up to 270 t/ha. Maize-berseem-pearlmillet fodder system found more productive with high net returns in Punjab (Aulakh *et al.*, 2012). In Punjab, maize-berseem-pearlmillet fodder system found more productive with higher net returns and were almost similar to that obtained with sorghum + clusterbean-oat-cowpea cropping system (Table 5 and 6). The benefit cost ratio was highest in sorghum + clusterbean-oat-cowpea cropping system due to lower cost of cultivation of this system and was lowest in maize-berseem-maize + cowpea system due to its higher cost of cultivation returns (Aulakh *et al.*, 2012).

Table 4. Round the year forage production systems

Crop sequence	Green fodder yield (t/ha/year)
Napier Bajra hybrid + Cowpea – Berseem	260
Maize + Cowpea – MP Chari + Cowpea – Berseem + Mustard	197
Sorghum + Cowpea – Berseem + Mustard	184
Cowpea – MP Chari + Cowpea – Berseem + Mustard	
Napier Bajra hybrid + Cowpea – Berseem – Cowpea	176

Table 5. Total green fodder yield and maize equivalent green fodder yield of different forage based cropping systems in Punjab

Cropping system	Total green fodder yield (t/ha)			Maize equivalent green fodder yield (t/ha)			
	1st year	2nd year	pooled	summer	Kharif	Rabi	Total
Sorghum-berseem	92.5	98.0	95.2		32.2	34.6	66.8
Maize-berseem-pearlmillet	142.1	133.5	137.7	3.6.7	10.8	35.8	83.3
Maize-berseem-maize + cowpea	106.4	108.9	107.8	36.69	9.6	35.0	81.5
Sorghum + clusterbeanoat-Cowpea	108.5	119.4	114.0	18.7	30.3	27.8	76.8
SEm	1.0	1.7	2.4	2.1	0.8	0.9	1.8
CD (P=0.05)	3.5	5.9	8.4	8.3	2.9	3.2	6.1

(*Source:* Aulakh *et al.*, 2012)

Table 6. Economic production of different forage based cropping systems in Punjab

Cropping system	Cost of cultivation (×10³₹/ha)	Gross returns (×10³₹/ha)	Net returns (×10³₹/ha)	B:C ratio
Sorghum-berseem	14.7	46.75	32.06	3.18
Maize-berseem-pearlmillet	18.08	58.34	40.25	3.23
Maize-berseem-maize + cowpea	19.47	57.08	37.61	2.93
Sorghum + clusterbeanoat-Cowpea	14.38	53.77	39.39	3.74

(*Source:* Aulakh *et al.*, 2012)

Food-fodder production systems

In Indian agriculture small and marginal farmers are dominant and owing to land constraints and mounting demand for food and fodder. Introduction of forage species in the existing food based cropping systems will ensure forage availability in these areas without compromising with the yields of food crops. Food-forage-based systems provide a support to small and marginal farmers by adjusting a substantial part of their land exclusively for forage production in grain crop based rotations (Sunil Kumar and Faruqui, 2009). Introduction of legume forages in cereal based systems not only improves productivity, profitability but also soil fertility in the long run. Under irrigated situations, the prominent cropping systems are rice, wheat, sorghum, maize pearl millet based. Inclusion of forages in these systems can meet the needs of fodder demand. In rice based food-forage system rice-berseem sequence found to be promising with average productivity of 67.1 t/ha green fodder yield and 3.35 t/ha rice grain yield. In wheat based systems maize/sorghum (cereals) during kharif as preceding crops and berseem, lucerne and oat as intercrops in association with wheat produces higher green forage yields along with grain yield of wheat. In sorghum/ maize based systems, inclusion of legume fodders (cowpea and clusterbean) as intercrops in these grain crops improves the fodder quality and makes the system more profitable and sustainable (Sunil Kumar et al., 2012). In Bundelkhand region of central India, groundnut - berseem - maize (F) + cowpea (F) was most efficient with respect to forage equivalent yield and net returns followed by sorghum (multi-cut) – berseem and presented in Table (Sunil Kumar and Faruqui, 2009).

Table 7. Forage equivalent yield, economics and system productivity of food-forage based cropping system (mean of 5 years)

Cropping system	Forage equivalent yield (t/ha/ year)	System productivity (t/ha/ year)	Cost of cultivation (×10³₹/ ha/year)	Net returns (×10³₹/ ha/year)	B: C ratio	Employment generation efficiency (man days/ha /day)	Monetary return use efficiency (₹/ha/ year)	Sustainability index
Groundnut - wheat - green gram	115.7	0.32	34.5	39.4	1.14	0.54	107.9	0.82
Sorghum (multi-cut) – berseem	142.0	0.39	32.0	51.3	1.60	0.88	140.5	0.86
Sorghum (single cut) - wheat - green gram	106.8	0.29	30.5	29.6	0.97	0.65	81.0	0.90
Groundnut - 0.96	155.8	0.43	35.0	56.7	1.62		0.78	155.3

berseem - maize(F) + cowpea (F)								
Groundnut - wheat - maize(F) + cowpea (F)	142.7	0.39	35.2	49.9	1.42	0.60	136.8	0.92
Napier bajra hybrid (perennial) + (cowpea - berseem)	130.1	0.30	33.8	44.3	1.31	1.02	121.3	0.88
SEm	3.9	0.02				4.5		
CD (P=0.05)	11.5	0.06				13.3		

Rice based system

Rice-based cropping systems are the most common in India during kharif season because rice is the staple food crop. There is a great opportunity to grow forage crops as substitution or in a sequence. In rice based food-forage system rice-berseem sequence found to be promising with average productivity of 67.1 t/ha green fodder yield and 3.35 t/ha rice grain yield. Berseem requires more number of irrigations therefore other crop sequences like rice – oat may also be adopted (Sunil Kumar *et al.*, 2012).

Wheat based systems

Next to rice, wheat is the most important food-grain of India and is the staple food of millions of Indians, particularly in the northern and north-western parts of the country. Wheat is grown in India in an area of about 30 Million ha with a production of 93 Million tonnes in wide range of soil and climatic conditions. It is sown after the harvesting of kharif season crops and harvested in the month of April. There is ample scope to grow any suitable fodder crops as preceding, succeeding and/or intercrop with wheat. In wheat based cropping system for higher biomass production maize/sorghum can be taken as preceding crop whereas for higher crude protein yield leguminous crops like cluster bean /cowpea may be grown. Berseem, lucerne and oats are the most important fodder crops of winter season. There is a possibility to grow these crops as intercrop in border method of planting in association with wheat. In this system the yield of wheat will not be affected by growing these forage crops as intercrop in border method because these are leguminous dwarf statured crops which are harvested frequently for forage. As such these crops did not cause competition for plant nutrients, soil moisture and solar radiation rather help wheat crop by fixation of nitrogen. In border method, berseem produces higher green forage and dry matter yields over lucerne without the adversely affecting the wheat yield. Growing of lucerne/ berseem in every 4 skipped row of wheat gives bonus yield of green forage (4.6-7.8 t/ha) without causing significant reduction in grain yield of wheat.

Sorghum/Maize based systems

Sorghum and maize being graminaceous widely spaced crops and heavy feeder of nitrogen, provide better prospects for intercropping leguminous forages. In widely spaced crops additive series of intercropping may be better option. Principle involved in parallel intercropping is that the component crops should be complementary and not competitive to each other. The inclusion of legumes in these crops improves the quality of fodder and makes the system more profitable and sustainable. The normal row spacing of 45 to 60 cm is kept for the grain crops and which could be widened by pairing rows and accommodating 2 rows of forage legumes between paired rows. It is observed that growing of cowpea as intercrop did not affect the grain and stover yield of sorghum. Higher green forage, dry matter and crude protein yields of cowpea can be achieved in paired row planting. Inclusion of legume during kharif enhances the yield of succeeding wheat crop. Besides, bonus yield of protein rich cowpea fodder and enhanced wheat yield cropping systems involving leguminous fodder also improves soil health. Baby corn cultivation provides new avenues for crop diversification, value addition and revenue generation besides giving good quality green fodder, which adds enormously to the total economic returns. Baby corn under paired row planting resulted in higher ear yield and monetary returns, but introducing fodder cowpea as an intercrop in paired rows of baby corn was significant in realizing higher ear equivalent yield and total green fodder with good quality without any reduction in ear yield (Kumar and Venkateswarlu, 2013).

Forages in non- food crop production system

The availability of good quality forage is scarce in the regions where oil seeds and cash crops like cotton, sugarcane are grown with assured irrigation facility. Inclusion of forage crops in these existing non-food crops (sugarcane, potato, cotton, etc) play significant role for livestock production, which may prove more remunerative in present entrepreneurial scenario. In north and central India, sugarcane is grown as long duration cash crop. The crop geometry and early growth behaviour offers good opportunity to grow a fodder crop of 70-100 days duration without affecting the cane yield and sugar recovery. The fodder crops grown in inter row spaces provide vegetative mulch against nutrient and moisture loss which favourably affect the physical, chemical and biological condition of the soil. Due to dormant stage of autumn sown sugarcane it gives opportunity to grow other Rabi season crops. Lucerne, berseem and oat has been found suitable under autumn sugarcane based system. The system is beneficial due to efficient utilization of inter row space, improvement in physico-chemical properties of soil, enhanced input use efficiency, additional yield from system, enhanced farm profit. The system may provide green fodder of 40-60 t/ha. Additional farm income with higher B: C ratio of the sugarcane based systems (Sunil Kumar *et al.*, 2012).

Table 8. Green, Dry fodder yield (t ha^{-1}) and economics of baby corn under different planting pattern

Treatment	Green fodder yield (t/ha)			Dry fodder yield (t/ha)			Baby corn ear equivalent yield (kg ha^{-1})	Economics		
	Corn	**Intercrop	Total	Corn	**Intercrop	Total		Gross returns (₹/ha)	Net returns (₹/ha)	B:C ratio
T_1: Fodder corn sole	47.3	-	47.3	10.1	-	10.1	2027	14,190	4,882	0.52
*T_2: Baby corn sole	53.7	-	53.7	9.7	-	9.7	10437	89,120	65,222	2.72
T_3: Fodder corn paired rows	48.3	-	48.3	10.3	-	10.3	2070	14,490	5,182	0.55
*T_4: Baby corn paired rows	54.7	-	54.7	9.9	-	9.9	10848	92,290	68,392	2.86
*T_5: T_4 + Cowpea intercrop	50.5	16.1	66.6	8.8	2.7	11.5	11044	92,430	67,802	2.75
*T_6: T_4 + Clusterbean intercrop	49.8	11.7	61.5	8.5	1.9	10.4	10473	88,200	62,562	2.44
*T_7: T_4 + Pillipesara intercrop	48.8	9.8	58.6	8.4	1.6	10.0	10142	85,620	61,342	2.52
*T_8: T_4 + Fodder corn intercrop	28.6	39.5	68.1	5.1	8.1	13.2	6068	51,020	26,262	1.06
[illegible]	2.62	-	3.06	0.45	-	0.62	320.0	2871	1901	0.07
CD (P = 0.05)	7.96		9.28	1.38		1.90	970.7	8709	5766	0.23

Forage during lean period as catch crops

Under assured irrigated situations of northern India, farmers design to grow their crops under double cropping systems, which offer ample scope for raising short duration fodder crops during lean period. This period falls before the start of rainy season in May-June and during winter season in November-December. Maize, sorghum and pearl millet are grown as grain crops during rainy season followed by wheat in winter season. A gap period exists during May-June which could profitably be utilized for growing fodder mixtures like maize + cowpea, sorghum + cowpea or pearl millet + cluster bean without affecting the production of grain crops in the rotation. Another gap period also falls after harvesting of fodder crops, picking of immature maize cobs and vegetable cowpea, etc. during October-November, where fast growing fodders like mustard and turnip could be grown for forage. Cultivation of forage crops during this gap period provides approximately 30-35 tones of green fodder per hectare. Generally these forage production systems are of immense importance for farmers having small irrigated holdings, whose total area may not be diverted exclusively for fodder production. During winter season, winter maize and late sown teosinte can be successfully grown and leguminous forages like horsegram (*Macrotyloma uniflorum*), moth bean (*Vigna aconitifolia*) and grasspea (*Lathyrus sativus*) are suitable for cultivation on residual moisture after the harvest of late paddy. This forage production systems in lean period has specific benefits of efficient utilization of gap period between two main crops, improvement in soil fertility, additional yield from system and enhanced farm income (Sunil Kumar *et al.*, 2012).

Forage production under horticultural/plantation crops

Horticultural crops are widely grown in different agro-ecological regions in India. Under mono-cropping these crops will not fully utilize the natural resources including space, sunlight, moisture and nutrients owing to its morphological features requiring wider spacing. Forage crops can be grown successfully by utilizing the space between these crops. Guinea grass, signal grass and Congo signal grass are shade loving in nature and are suitable to grow under or near to the horticultural trees and grooves. Combinations of Guinea grass/N-B hybrid + *Desmanthes* (3:1) in between two rows of horticultural/ plantation crops gives 120-150 t/ha green fodder yield in addition to yield from fruit component (Sunil Kumar *et al.*, 2012). This forage integration will enhance system diversification and productivity, improvement in soil fertility and utilize the resources efficiently. Further, to sustain the soil fertility on a holistic perspective, a farming system approach in coconut is developed at Goa (Table . Intercropping of high yielding fodder in the inter spaces of coconut and integration with dairy and biogas unit, with recycling of wastes to the coconut garden through biogas slurry and coconut waste composting. The technology was found to be a suitable intervention to enhance and sustain the returns from coconut garden. (Manjunath *et al.*, 2013)

Table 9. Potential income generation from cocunut based farming system

Technology	Investment (₹/ha)	Production potential	Potential income generation (rs/ha)
High yielding forage intercrops	9,000	111 t/ha	66,600
Coconut	14,000	8,350 Nos	58,450
Integrated dairy (with five milch cows /ha)	87,100	9,470 litres	1,13,640
Biogas unit (4 m3)	3,200		7,200
Total system	1,13,300		2,45,890

(*Source:* Manjunath *et al.*, 2013)

Forage production in non-competitive land areas

Increase in area under forage cultivation is meager due to existing competition between food and commercial crops. Any effort made to increase forage availability in these areas will improve the livestock productivity. Small and marginal land holding farmers with limited forage requirement can effectively grow forages on bunds. Generally field bunds are a common feature of any cultivated lands and occupy 2-10% of the cultivated area. These bunds are prepared and maintained for convenience of all field operations, conserve water and soil. Growing of appropriate forages/bushes stabilizes the bunds through protective cover, soil binding action of roots and also prevents the loss of top soil during heavy rain fall, besides providing considerable amount of forage for feeding animals. Potential non-competitive land uses for forages are field bunds of irrigated and rainfed situation, in hilly areas, terrace risers or bunds, steep uncultivable slopes, thin natural forests near villages, intensive dairy or goat farming on gentle hill slopes and terraced lands. Inclusion of forages in with existing crops/cropping systems offers greater scope for utilizing the land which is otherwise not used for any agricultural purposes. Selected forage species can be grown in non-competitive lands of irrigated farms, field bunds in different agro-climatic conditions.

Conclusion

Forage production strategies for livestock based integrated farming systems under irrigated conditions needs to be propagated to achieve fodder and nutritional security of increasing livestock and human population. Diversification of existing system with change in crops, cropping system, improvement of livestock component and inclusion of forages are essential to improve the on farm income of small and marginal farmers in India. This will also helps in meeting the house hold demand of balanced food, efficient recycling of nutrients and water besides increasing the employment of farm family.

References

Aulakh, CS, Gil MS, Mahey RK, Walia SS and Singh G. 2012. Effect of nutrient sources on productivity of fodder cropping systems in Punjab. Indian Journal of Agronomy **57**(2): 200-205.

B. L. Manjunath, B. K. Swain, Mathala J. Gupta, R. Maruthadurai. 2013 Agricultural Technology Options. Technical Bulletin No. 30, ICAR Research Complex for Goa (Indian Council of Agricultural Research), Old Goa-403 402, Goa, India.

Behera UK, Mahapatra IC. 1999. Income and employment generation of small and marginal farmers through integrated farming systems. *Indian Journal of Agronomy* **44**(3): 431-439.

Jayanthi C, Rangasamy A, Chinnusammy C, Purushothaman S and Planiappan SP. 1994. Integrated farming systems for smallholdings. *Indian Journal of Agronomy*. **39**: 1-7.

Sunil Kumar and Faruqui, SA. 2009. Production potential and economic viability of food- forage based cropping system under irrigated conditions. *Indian Journal of Agronomy* **54**(1): 36-41

Sunil Kumar, Agrawal RK, Dixit AK, Rai Arvind K, Singh JB and Rai SK. 2012. Forage Production Technology for Arable Lands. Indian Grassland and Fodder Research Institute, Jhansi-284003.

Tiwari PN. 1993. Integrated Farming Research for sustaining food production. *Journal of Nuclear Agriculture Biology* **20**: 1-13.

Ramesh Chand, Lakshmi PA, Prasanna and Aruna Singh, 2011. Farm size and productivity: Understanding the strength of small holders and improving their livelihood. Economic and Political weekly **46**(26&27): 5-11.

Gangwar, B and Ravisankar, N. 2014. Diversification of existing farming systems for environmental security and sustainable livelihood of small farm holders. In National symposium on *Agricultural diversification for sustainable livelihood and environmental security and sustainable livelihood of small farm holders* November 18-20, 2014, Ludhiana, Punjab pp 69-73.

Kumar, T.K. and Venkateswarlu, B. 2013. Baby corn (Zea mays L.) performance as vegetable-cum-fodder in intercropping with legume fodders under different planting patterns. Range Management and Agroforestry, **34**(1): 137-141.

Chapter 7

Fodder Production Strategies from Perennial Systems for IFS under Rainfed Conditions

S.N. RAM

ICAR - Indian Grassland and Fodder Research Institute, Jhansi-284003 (UP)

Introduction

In integrated farming systems supply of green fodder to livestock from perennial systems is one of the important component for sustained production of the systems under rainfed conditions. In India, vast area (107.43 m ha) is classified under various types of degraded land (MOA, 1994) where one or more limiting factors render the cultivation of single crops economically unviable. The problem of over grazing is present in the entire country. The increasing pressure on natural vegetation has resulted into acceleration of erosion process and loss of soil fertility. Land degradation has resulted in huge deficits of forage, timber and firewood. In such situation integration of pasture, silvopasture and hortipasture in integrated farming systems not only facilitates the production of vital animals and human needs but also checks lands degradation forces.

Establishment of silvipasture systems

Silvopastures integrate pasture and/ or animals with trees. Woody perennials, preferably of fodder value, are introduced deliberately and systematically and managed scientifically. This system aims at optimizing land productivity, conserving plants, soils and nutrients and producing forage, timber and firewood on a sustainable basis. It involves replantation, substitution or intervention in the existing vegetation by desirable species.

The appropriate silvipastoral systems-combining fodder trees with pastures/fodder crops (with or without animals) with the principles of sustained yield forestry and soil conservation-have the great potential to optimize production and economic returns per unit area (Van Den Beldt 1990).

Trees/shrubs selection

- Ability to thrive and produce acceptable yields on relatively poor soils.
- Grow on inhospitable soils of the given area.
- Develop root systems under conditions of impervious subsoil.
- High palatability of fodder.
- Good coppicing ability to withstand browsing and resilience to drought.
- Resist pests and diseases.
- Yield a favourable cost benefit ratio and generate employment opportunities.
- Improve the local micro-climate and ecological environment.

Suitable species of trees, shrubs, grasses and legumes for different regions

An account of promising forage trees, shrubs, grasses and legumes for various climatic regions of India is given in Table 1.

Table 1. Suitable species of trees, shrubs, grasses and legumes for different regions

Regions	Trees and shrubs	Grasses	Legumes
Humid-regions	*Albizia lebbeck, A. procera, Azadirachta indica, Bauhinia purpurea, Moringa oleifera, Morus alba, L. Leucocephala, Sesbania grandiflora* and *S. sesban*	*Chrysopogon fulvus, Chloris gayana, Dichanthium annulatum, Eremopogon faveolatus, Iseilema laxum, Setaria anceps* and *Sehima nervosum*	*Desmodium uncinatum, Clitoria ternatea, Macroptilium atropurpureum* and *Stylosanthes guianensis*

Semi arid regions	*Acacia nilotica, A. tortilis, Ailanthus excelsa, Albizia lebbeck, A. amara, Azadirachta indica, Dichrostachys cinerea, Hardwickia binata, L. Leucocephala, Parkinsonia aculeata, Prosopis juliflora* and *Sesbania sesban*	*Cenchrus ciliaris, C. setigerus, Dichanthium annulatum, Sehima nervosum, Chrysopogon fulvus Heteropogon contortus* and *Panicum antidotale*	*Desmodium intortum, Macroptilium atropurpureum, Stylosanthes spp.* and *Lablab purpureus*
Arid regions	*Acacia nilotica, A. tortilis, Ailanthus excelsa, Albizia lebbeck, A. amara, Bauhinia variegata, Casuarina equisetifolia, Dichrostachys cinerea, Hardwickia binata* and *Ziziphus mauritiana*	*Cenchrus ciliaris, C. setigerus, Lasiurus sindicus* and *Panicum antidotale*	*Atylosia scarabaeoides, Stylosanthes hamata* and *S. seabrana.*
Temperate and sub temperate regions	*Betula alboides, Celtis australis, Morus serrata, Robinia pseudoacacia* and *Salix spp.*	*Bromus inermis, Dactylis glomerata, Lolium multiflorum, L. Perenne, Phleum pratense* and *Poa pratensis*	*Lespedeza ceraces, Lotus corniculatus, Lupinus augustifolius, Trifolium pratensis, T. repense* and *T. subterraneum*

Planting

The success of plantation depends upon the use of adequate quantity of god quality seeds. The selection of proper seed stand is very important in collection of quality seed. In practice selection of tree for seed collection should be done with care. The seeds should be used as quickly as possible after collection. If, however, they are to be stored, they should be sun dried, put in gunny bags or perforated plastic bags and kept on wooden platforms in well aerated huts. A widespread cause of seed dormancy is the presence of hard seed coat. This may be impermeable to water, impermeable to gases or it may mechanically constrain the embryo. Different kinds of seeds are subjected to different pre-treatment prior to sowing. These includes soaking in cold water and soaking in hot to boiling water. Seeds of several multipurpose trees also require scarification before sowing.

After selection of suitable species of trees planting are done at the onset of monsoon. Pits of 45 cm^3 size at 5x5 m or 6x4 m spacing should be dug during May-June for weathering of soil so as to minimize the attack on seedlings by insect and pests. At the time of plantation of saplings, in each pit a mixture of urea, single super phosphate and potash (125 g in the ratio of 1:4:1) along with 30-40 per cent farm yard manure of the total volume of the pits may be given. Chloropyriphos at the rate of 0.30 % may be applied in soil to prevent termite attack. 5-11 months old tree seedlings may be planted during monsoon in the pits depending on growth of saplings of tree species.

Hortipasture systems

Hortipasture system is a promising land use system in which fruit trees are grown in association with perennial forage grasses and legumes considering both ecological and economic interactions among different components. Animals also form part of hortipastoral system starting after three-year growth of the system. Hortipasture systems supply the protective food (fruit) for human being and fodder for animal.

Establishment of hortipasture systems

Selection of suitable fruit tree

Following are recommended fruit trees and their varieties for different rainfall zones (Table 2).

Table 2. Suitable fruit trees and cultivars for different rainfall zones.

Rainfall (mm) zone	Fruit trees	Cultivars
Less than 350	Ber (*Zizyphus mauritiana*)	Seb, Gola, Mundia
	Gonda (*Cordia myxa*)	Local strain
	Ker (*Capparis decidua*)	Local strain
	Pilu (*Salvadora oleoides*)	Local strain
350-500	Ber (*Zizyphus mauritiana*)	Katha, Maharawali, Bagwadi, Seb, Gola, Mundia
	Aonla (*Emblica officinalis*)	Banarasi, Hathijhool, Chakaiya
	Khirni (*Manilkara hexandra*))	Local strain
	Jamun (*Syzygium cumunii*)	Local strain
	Mulbery (*Morus alba*)	

500-700	Bael (*Aegle marmelos*)	Faizabad selection
	Mango (*Mangifera indica*)	Seedlings
	Custard apple (*Annona spp.)*	Balanagar, Mammoth, Washington
	Sour lime (*Citrus aurantifolia*)	
	Aonla (*Emblica officinalis*)	NA-7, NA-6, Krishna, Kanchan
More than 700	Mango (*Mangifera indica*)	Bombay green, Chausa, Safeda and local cultivar
	Pomegranate (*Punica granatum*)	Jalor seedless, Jodhpur red
	Guava (*Psidium guajava*)	Allahabad Safeda, L-49
	Jackfruit (*Artocarpus heterophyllus*)	Local strain
	Wood apple (*Limonia acidissima*)	Local strain
	Tamarind (*Tamarindus indica*)	Local strain
	Aonla (*Emblica officinalis*)	NA-7, NA-6, Kanchan, Chakaya

Planting

For planting pit with a minimum of 45 cm diameter and 60 cm depth should be dug. The common pit mixture consists of original soil, FYM and small quantity of fertilizers. Chemical fertilizer generally recommended are DAP and single supper phosphate with urea. Chloropyriphos at the rate of 0.30 % may be applied in soil to prevent termite attack.

In dryland planting should be done in early rainy season when the soil profile gets charged with water to a depth of 1 meter. For most part of country the optimum planting time is last week of June till mid September. After correct root placement, good compaction of soil around the seedling must be done to eliminate all air pockets. Weeding more particularly ring weeding twice in first year and once in later years is must so as to improve survival and growth of seedlings.

Establishment of sole pasture or in interspaces of trees

Sowing of grasses and legumes needs a fine seedbed. The land is prepared using disks, harrows or field cultivators as per the need. In sloppy lands, all the tillage practices should be done across the slope. The depth of ploughing should be kept 15-20 cm. Disk plough should be used in soil that have been ploughed and left with rough surfaces, high amount of traces and residues for preparation of a good seedbed.

The best time to work the land will be just before planting time making sure the land is not too wet or too dry. Recommended doses of FYM and fertilizers need to be applied during the land preparation and properly incorporated in the soil.

After preparation of field perennial grasses and legumes are to be established as a sole pasture or in interspaces of trees for forage production. Suitable time for sowing of grasses and legumes is monsoon season. In case of lighter seeds, the seed rate varies from 4-6 kg/ha and slightly heavier seeds such as, *Lablab purpureus, Clitoria ternatea* etc. should be sown at the rate of 15-25 kg/ha. In case of pure stand, grasses are to be sown at the spacing of 50 cm while in mixture with legumes, it should be sown at 100 cm row to row distance and in the interspaces of two rows of grasses, legumes are to be sown.

Grasses can also be established by raising seedlings in nursery. The raising of grass seedlings is best done in the middle of May. After well preparation of land, seed beds (6 x 1.2 m) are prepared. Grass seeds are sown @ 40-50 g/ bed at 0.4-0.8 cm deep in lines at 10 cm apart. Beds are watered using a rose can and covered with wet gunny bags till the seeds are germinated (3-4 days). Gunny bags are removed soon after germination particularly in the evening time. Grass seedlings will be ready for transplanting after 4-6 weeks of sowing.

Seedlings/ rooted slips are transplanted in well-prepared field immediately after the onset of monsoon. Application of 40-60 kg N/ha and 20-30 kg P_2O_5 /ha increased pasture production by 50-100% in majority of grasses besides increasing crude protein content considerably.

Management of pasture, silvipasture and hortipasture systems

Protections from biotic interferences are most vital for proper growth and productivity of the systems. The protection could be offered through barbed wire, chain links supported by angle iron, cement, stone or wooden poles or through fencing by unpalatable bushes. For live-hedges, the suitable species are *Pithecolobium dulce, Lantana camera, Carissa carandas, Agave americana, Opuntia dillenii, Zizyphus nummularia, Jatropha curcas, Saccharum spontaneum and vetiveria zizanoides.*

Heavy infestation of unpalatable bushes in pasture, silvipasture and hortipasture systems not only adversely affects the availability of open space for growing grasses but also the forage production. The standard practice of bush cleaning includes either manual or mechanical felling and removal of stumps or application of selective weedicides on the cut stumps to kill them to stop coppicing.

Contour furrows, contour bunds and staggered contour trenches are found to be effective in soil and moisture conservation in pasture, silvipasture and hortipasture systems. The grasses and legumes are considered best vegetational means for soil and water conservation. The grasses reduce soil and water loss to a great extent. Average soil loss from bare soil is 18 t/ha in alfisols whereas, from the natural grassland it is only 2.5 t/ha from a protected site. Suitable grasses for soil and water conservation for different types of problem soils are *Eulaliopsis binata, Cenchrus ciliaris, Dichanthium annulatum, Chrysopogon, fulvus, Vetiveria zizanoides, Saccharum muja, Pennisetum species* and *Cynodon dactylon.*

The silvopasture systems can be designed and managed to modify cycling in such a way to make more efficient use of nutrients, whether these originate from natural renewal processes or from fertilizer. It may be achieved by reducing the ratio between inputs/outputs and internal cycling. Agricultural ecosystems are highly open with inputs and outputs sometimes as much as 40 per cent of internal cycling: natural forest ecosystems are more closed, inputs and outputs sometimes amounting to less than 10 per cent of internal cycling. There could be several opportunities in modifying this ratio- increase gain from symbiotic nitrogen fixation, enhance uptake of nutrients released by rock weathering, through deep root systems of trees, reduce nutrient fixation on clay minerals and increase availability through release from organic compounds etc.

Studies conducted at Jhansi on the effect of different intensity and interval of cutting on various grasses such as C. *ciliaris*, C. *setigerus*, *Sehima nervosum, Dichanthium annulatum* and *Chrysopogon fulvus* revealed that for getting the maximum forage yield, it should be harvested at 60 days interval and 10-15 cm above ground levels. The larger the cutting interval the higher the biomass production from the pasture, but the crude protein content of the herbage shows reverse trend. Lopping of tree component should be done during lean period.

Production from different perennial systems

It has been possible to increase land productivity from 0.5-1.5 tonnes/ha/year to > 10 tonnes/ha/year (10-year rotation) by developing silvopastures. The additional forage availability through systems is likely to reduce grazing pressure and thus have important environmental implications.

H. binata+*Cenchrus*-Stylo silvopasture produced 2.23 t forage / ha. This is followed by 1.82 t/ha in case of *Chrysopogon-Sehima*-Stylo pasture under the canopy of *H. binata*. *Chrysopogon-Sehima*-Stylo pasture under the canopy of *A. amara* recorded forage production of 1.57 t/ha (Roy and Gupta, 1986).

Highest pasture production was recorded under the trees of *Hardwickia binata* (3.33 t DM/ha followed by *Acacia tortilis* (3.12 t DM/ha) and *Albizia amara* (1.69 t DM/ha). At 25th year, higher bole biomass was recorded in *H. binata* when compared to similar age plantation of *A. amara*. However, because of much higher leaf and pod biomass, higher above ground productivity was recorded in *A. amara* (3.56-5.15 t DM/ha) followed by *H. binata* (2.94-3.87 t DM/ha). *A. tortilis*, at 15th year exhibited above ground productivity in the range of 2.4-3.2 t/ha. Higher production from lopped branches was obtained in *A. amara* (1.68-3.09 t/ha) when compared to *H. binata* (0.28-0.78) and *A. tortilis* (0.23-0.72 t/ha). *A. amara* provided highest average CP (16.3 %) followed by *A. tortilis* (12.2 %) and *H. binata* (8.4 %). (Roy *et al.*, 1999).

During twelfth year of establishment *Dalbergia sissoo* exhibited higher growth (Ht.-8.45 m; CD-16.7 cm; dbh-13.3 cm and canopy diameter -330 cm) under 50 % lopping treatment as compared to other treatments of *D. sissoo* as well as Neem trees. The Pasture production from protected area (managed under and carry system) was higher under Neem silvipasture (3.56 dry t/ha) when compared with *D. sissoo* silvipasture (2.61 dry t/ha). *D. sissoo* trees produced highest leafy fodder (0.18 dry t/ha) at 50 % lopping when compared with other treatment of *D. sissoo* as well as Neem trees during lean period in summer (Rai and Roy, 2007).

High herbage yields was attained in sown pasture on ungrazed areas (4.16 t DM/ha) as compared to grazed areas (3.91 t DM/ha) and natural pasture blocks (3.91 t DM/ha). Highest average annual leaf fodder yield was attained (through two pruning and one lopping) in MPTS + sown pasture stands (0.54-0.65 t DM/ha) followed by MPTS + reseeded pastures stand (0.60-0.65 t DM/ha) and MPTS + natural pasture stand (0.47-0.59 t DM/ha). In all the three systems yield was higher in ungrazed blocks. Among tree/shrub species, highest leaf fodder was derived from *Dichrostachys cinerea* followed by *Dalbergia sissoo, Acacia nilotica and Azadirachta indica* (Roy, 2010).

After five years of establishment *Leucaena leucocephala* (S-24) exhibited better survival (73 %) and growth performance (height -6.6 m, collar diameter-10.28 cm, dbh-8.29 cm and canopy diameter-3.93 m) as compared to *Bauhinia malabarica* (survival- 44 % height -4.32 m, collar diameter-9.78 cm, dbh-6.65 cm and canopy diameter-2.95 m). Pasture combinations of *Cenchrus ciliaris* and *S. hamata* produced 5.49 DMT/ha forage under *Bauhinia* silvipasture and 3.98 DMT/ha under *Leucaena* silvipasture where as *Chrysopogon fulvus* with *S. seabrana* produced 4.65 DMT/ha under *Bauhinia* and 4.83 DMT/ha under *Leucaena* silvipasture. By pruning 0.44 DMT/ha leafy fodder was obtained from *Bauhinia malabarica* trees and 0.52 DMT/ha leafy fodder from *Leuceana* trees in their respective silvopasture systems. *Sesbania sesban* hedge produced 0.73 DMT/ha leafy fodder and *L. leucocephala* hedge produced 1.13 DMT/ha when clipped at 1.0 m height. *Leucaena* silvipasture exhibited more enhancement in organic carbon (8.9 % at 0-15 cm and 20 % at 15-30 cm soil depth) as compared to *Bauhinia* based silvipasture (2 % at 0-15 cm and 16.37 % at 15-30 cm soil depth). Similar trends were recorded with respect to available N and K at both of the soil depth where as available P was higher at 0-15 and lower at 15-30 cm soil depth (Rai and Gupta, 2012).

Intercropping of Guinea grass with *S. hamata* in association with *Hardwickia binata* produced significantly higher dry forage yield (8.26 t/ha) as compared to intercropping of *Chrysopogon fulvus* (6.27 t/ha) and *Cenchrus ciliaris* with *S. hamata* (7.52 t/ha) during 4th year of establishment. *H. binata* attained mean height 4.05 m, collar diameter 8.75 cm, dbh 6.64 cm and canopy spread 2.11 m in fifth year of establishment (Ram *et al.*, 2012).

In the third year after establishment, *Ficus infectoria, Madhuca latifolia, Morus alba* and *Acacia nilotica* recorded about 90 % survival. Maximum plant height and collar diameter was recorded with *Acacia nilotica* (317,09 and 6.25 cm) followed by *Morus alba* (256.38 and 5.10 cm) and *Ficus infectoria* (133.80 and 3.62 cm). Maximum pruned biomass was obtained from *M. alba* (2.83 kg/plant) followed by *A. nilotica* (2.23 kg/plant) and *F. infectoria* (0.98 kg/plant). Maximum green fodder production (290-520 q/ha) was obtained from *Panicum maximum* followed by *Chrysopogon fulvus* (180-290 q/ha), *Cenchrus ciliaris* (70-120 q/ha), *Clitoria ternatea* (100-210 q/ha) and *Stylosanthes seabrana* (90-140 q/ha) with different tree combinations. Forage production of grass and legumes with different alley combinations was maximum with *P. maximum* (290-450 q/ha) followed by *C. ciliaris* (100-350 q/ha) and *C. ternatea* (70-77 q/ha) (Kumar *et al.*, 2014).

In Aonla based hortipasture system, maximum fruit yield (20.9 t/ha) was recorded under intercropping with *S. hamata*. The forage production was higher in association with tree (3.9 t DM/ha) as compared to sole pasture (3.6 t DM/ha) (Sunil Kumar and Shukla, 2010).

In ber based hortipasture system, pasture production under pruning of secondary branches of ber tree at 20 cm length from base was significantly higher (4.05 t DM/ha) as compared to pruning of secondary branches of ber tree at 60 cm length from base (3.77 t DM/ ha). Among pasture, Guinea grass + *S. hamata* produced maximum yield (6.29 t DM/ha). Pruning of secondary branches of ber tree at 40 cm length from base recorded significantly higher fruit yield (26.28 kg/ tree) as compared to pruning at 20 cm length from base (18.31 kg/tree) (Sunil Kumar and Ram, 2009). Six year old Guava based hortipasture system gave 6.75 t/ha dry forage yield and 6.25 t/ha fruit yield (Sunil Kumar and Shukla, 2013).

Intercropping of *S seabrana* with Guinea grass gave maximum dry forage yield 6.68 t/ha followed by *S. hamata* (6.29 t/ha), *M. atropurpureum* (5.60 t/ha) and *C. ternatea* 5.41 t/ha (Ram and Trivedi 2012). Harvesting of Guinea grass at 60 days interval along with application of 80 kg N/ha in combination with 10 t FYM/ha gave 6.31, 11.59 and 8.17 tonnes dry forage yield/ ha during first, second and third years of establishment under semiarid rainfed conditions (Ram and Trivedi, 2013).

Resource conservation

Sustainable management of natural resources of land, water and vegetation is essential in providing livelihood and environmental security. Ever-increasing demographic pressures coupled with developmental activities are causing tremendous pressure in the utilization of these resources, leading to various kinds of ecological problems, such as droughts, floods, landslides, mine spoils, siltation of reservoirs, deterioration of water bodies, loss of biodiversity etc. The increasing frequency of these problems is the outcome of excessive biotic and abiotic interferences that have resulted into considerable degradation of our natural resources.

Vegetation is one of the principal biological methods of conservation of soil and water resources. Tree play a vital role in soil and water conservation, regulation of water yield and rate of siltation in reservoirs. Thick canopy of trees and grass can intercept large amount of water and thick layer of humus and leaf litter on land is mainly responsible for higher infiltration rate low run off (Singh *et al.*, 1983; Singh and Gupta, 1987; Mathur *et al.*, 1982).

The erosion from sloping land can be drastically reduced with the establishment of closely spaced contour hedge rows. The ameliorative effect of alley of *Leuceana* on runoff (13 mm) and soil erosion (0.10 t/ha/year) was significant (Lal, 1987).

Comparative study of silvipasture systems models for eroded lands indicated that in Shvaliks, *Eucalyptus-Eulaliopsis binata* (bhabar grass) system was most effective in reducing the soil loss (0.07 t/ha) followed by *Acacia catechu*- Napier grass (0.24 t/ha), *Leucaena*-Napier grass (0.28 t/ha), Teak – *Leucaena*- bhabar (0.43 t/ha), *Eucalyptus- Leucaena*-turmeric (0.59 t/ha), Poplar- *Leucaena*-bhabar (1.54 t/ha) as compared to arable crops (2.69 t/ha) and cultivation fallow 5.65 t/ha (Grewal, 1993).

In another study at Shiwaliks in Chandigarh, interception of rainfall by crows of *Dalbergia sisso* and *Eucalyptus* hybrid was studied for 4 years (1969-72). Observations indicated that there was a gradual increase in interception with the increase in age of *D. sissoo* whereas; no such change was seen in E. hybrid. It was observed that *D. sissoo* intercepted more rain water than E. hybrid (Mishra *et al.*, 1972).

The findings on silvopasture study (*D. sissoo + C. fulvus; A. catechu + E. binata; E. hybrid + C. fulvus; and L. leucocephala + Napier*) over a period of 14 years suggested that gravely and bouldery lands could be effectively utilized by putting them under perennial vegetation for improving soil fertility and biomass production (Samra *et al.*, 1999).

Thus, for development of pasture, silvipasture and hortipasture systems in integrated farming systems, protection of area, adaption of soil and water conservation measures and selection of suitable grass, legume and tree species for different regions–humid, semi arid, arid regions and temperate and sub temperate are very much essential. Development of perennial systems of forage production in integrated farming systems could easily meet the deficit of forage in the country and also conserve natural resources.

References

Grewal S S. 1993. Agroforestry systems for soil and water conservation in Shivaliks. In: Proceedings Agroforestry in 2000 AD for semiarid and Arid Tropics (eds. A.S. Gill, R. Deb Roy and A.K. Bisaria). NRCAF, Jhansi:82-85.

Kumar R V, Roy A K, Singh H V, Kumar S and Ghosh P K. 2014. Silvipasture system for round the year top feed and fodder availability. *Annual Report, IGFRI*. p 36-37.

Lal R. 1987. Managing soils of sub-Saharan. *Africa Science* **236**: 1069-1076.

Mathur H N, Singh R P and Gupta M K. 1982. Comparative study of infiltration in soils under forest cover and agriculture in temperate climate. *Indian For.* **108**(10): 648-653.

Ministry of Agriculture, 1994. Draft Report of Land Degradation in India, Soil and Water Conservation Division, Department of Agriculture and Cooperation, MOA, New Delhi.

Mishra P R, J P Singh and Agnihotri Y. 1972. Forest interception and retention study-Chandigarh Rep. CSWCRTI, Dehradun.

Samra J S, M K Vishwanatham and Sharma A R. 1999. Biomass production of tree and grasses in a silvipasture systems on marginal lands of Doon vally of North West India. 2. Performance of grass species. *Agroforestry Systems* **46**: 197-212.

Singh R P and Gupta M K. 1987. Rainfall interception by Pinus wallichiana plantations in temperate region of Himachal Pradesh, India. *Indian For.* **113**(8): 559-566.

Singh R P, Sharma K C., Mathur H N, Gupta M K and Gupta A K. 1983. Interception studies in Cedrus deodara plantation in Himachal Pradesh. *Indian For.* **109**(5): 261-266.

Sunil Kumar and Ram S N. 2009. Effect of pruning intensities and psture combinations on productivity of ber based hortipastoral system. *Annals of Arid Zone* **48**(1): 63-66.

Sunil Kumar and Shukla, A. K. 2010. Productivity, economics and soil nutrient buildup in grownup aonla based hortipastoral system under semiarid condition. *Range Mgmt. & Agroforestry Symposium Issue* (A): 29-30.

Sunil Kumar and Shukla A K. 2013. Evaluation of Guava based hortipasture system under moisture consevation practices. *Annual Report. IGFRI.*

Ram, S N and Trivedi, B K. 2013. Response of Guinea grass (*Panicum maximum*) to fertility levels and harvest intervals. *Range Mgmt. & Agroforestry* **34**(1): 69-73.

Ram S N, Roy A K and Shukla A K. 2013. Evaluation of *Hardwickia binata* based silvopasture systems under moisture consevation practices. *Annual Report. IGFRI.*

Rai M P and Roy M M. 2007. Herbage dynamics and utilization of forage from grown up Neem and Sissoo silvipasture. *Annual Report, IGFRI.* p 53-54.

Roy R Deb and Gupta S K. 1986. Studies on establishment and growth of plantation species under various silvopastoral combinations. *Annual Report, IGFRI.* p 86-87.

Roy M M. 2010. Post grazing regeneration in pasture/silvopasture stands in semiarid region. *Annual Report, IGFRI.* p 37-38.

Van Den Beldt R J. 1990. Agroforestry for semiarid tropics. In: Agroforestry: Classification and management (ed. K. G. Mc Dicken & N. T. Vergara). John Willey and Sons, New York: 150-194.

❑❑❑

Chapter 8

Forage Production Strategies in IFS to Match with the Farmer's Resource Base

INDER DEV, ASHA RAM, MADHULIKA SRIVASTAVA, MAHENDRA SINGH AND RUPALI TIWARI

ICAR - Central Agroforestry Research Institute (CAFRI), Jhansi-284003 (UP)

Introduction

India's economy is agro-based and about 69% of the total population depends on it for their livelihood (GoI, 2013a). Agriculture contributes 14% to the GDP and employs 49% of the total workforce (GoI, 2013b). Back in the 60s, world was going through severe food crises. At that time we needed to modernize our agriculture and we needed new technologies and then Green Revolution ushered in. At present productivity is declining, resources (soil and water) are degrading, pesticides are poisoning our food systems and land holding size is decreasing. Livestock is the integral component of Indian agriculture since time immemorial. Its contribution to national economy through milk, meat, wool as well as farmyard manure is enormous. The productivity of livestock and growth of animal husbandry are closely linked with the biomass and quality of forages. Forages covers vast array of species in cultivated and range grasses and legumes category. Currently there has been radical change in realising the importance of forages in integrated farming system, crop diversification, watershed management, restorationof degraded lands and climate resilient agriculture. India has the largest number of livestock, representing over 17% of the world. Among four important species of livestock, cattle represent over 43% of the population followed by buffaloes (19%), goats (26%) and sheep (10%). The share of livestock rearedis highest in marginal followed by small and semi-medium by operational holders. However, the higher number of livestock per holding possessed by large holders related to marginal holders implying that small land holders are playing seminal role in development of livestock sector in country. Furthermore, population growth, urbanization and income growth are fuelling a substantial increase in the demand for food of animal origin. Keeping these issues in mind, scientists are looking for an alternate agriculture system so that demands can be met in sustainable manner. Fodder production is a component of the farming system and efforts need to be made for increasing the forage production in a farming system approach.The livestock sector provides livelihoods to 1.3 billion people, mostly smallholder farmers and pastoralists in the developing world. Rapidly increasing demand for livestock products means that the production footprint of livestock must significantly decrease. There are both intensive and extensive systems that can contribute to production and environmental goals. Due to fast agricultural and industrial development in India, the agro-ecosystem of country been depleted very deleteriously. Heavy deterioration of soil health, lowering of water table and high rate of environmental pollution are the matter of great concern for the future of the India. Grasslands represent some 70% of global agriculture land area; unfortunately as much as 35% of the grasslands are degraded. Forest cover of the many North Western states like Punjab, Haryana has declined sharply due to heavy pressure of agriculture. Furthermore, the excessive use of the agro-chemicals in the soil and crops has further worsened the condition of human, animal, environment and soil health. Farmers are not willing to spare their productive land for sole cultivation of trees. More specifically, In Indian condition, monoculture of the rice-wheat rotation has depleted the soil health and water resources in Indo-Gangeticplains. There is a dire need to rehabilitate the soil health and maintain water resources to develop a suitable sustainable land use like Agroforestry based production systems.

1. Integrated Farming System

Integrated Farming System provides opportunity which would considerably provide varieties of farm produce and improve the economic condition of the marginal and small farmers. Integration of different agricultural allied enterprises with crop activity as base would provide ways to reuse and recycle produce/waste material of one component as input in the other linked component and to reduce the cost of production of the economic produce of the component two and finally to enhance the net-income of the farm as a whole. Integrated farming system (or integrated agriculture) is a commonly and broadly used word to explain a more integrated approach to farming as compared to monoculture approaches. Integrated farming is a system which tries to imitate the nature's principle, where not only crops but, varied types of plants, animals, birds, fish and other aquatic flora and fauna are utilized for production. These are combined in such a way and proportion that each element helps the other; the waste of one is recycled as resource for the other may sometimes be known as Integrated Biosystems. In this system an inter-related set of enterprises used so that the "waste" from one component becomes an input for another part of the system, which reduces cost and improves production and/or income IFS works as a system of systems. In mixed farming situations, livestock are important sector to support the small and marginal farmers by providing employment throughout the year. IFS ensure that wastes from one form of agriculture become a resource for another form. Since it utilizes wastes as resources, we not only eliminate wastes but we also ensure overall increase in productivity for the whole agricultural systems. The basic principle is to enhance the ecological diversity - by choosing the appropriate cropping methodology with mixed cropping, crop rotation, crop combination and inter cropping so that there is less competition for water, nutrition and space and adopting eco-friendly practices; by following multistorey arrangement so that the total available area is effectively used and there is a high level of interaction among biotic and abiotic components; by integrating subsystems by which the various components interact positively, so that the whole farm productivity is increased. IFS is a labour intensive system, thereby engaging the farmer family productively on their own farms, throughout the year. IFS will lead to collective efforts among the farmers like collective purchase of inputs and collective marketing of produce, thus reducing their costs of production (Das, 2013). All over the world, farmers work hard but do not make money, especially small farmers because there is very little left after they pay for all inputs (seeds, livestock breeds, fertilizers, pesticides, energy, feed, labour, etc.). The emergence of Integrated Farming Systems (IFS) has enabled us to develop a framework for an alternative development model to improve the feasibility of small sized farming operations in relation to larger ones. (Anonymous, 2010). It takes 3-4 years to establish a good integrated farm with market linkages to ensure nutrition and livelihood of a family. When we have many such farms in the village, there will be enormous scope for employment and business opportunities, especially for the youth for selling seeds, seedlings, manures etc. (Das, 2013).

2. Farmers' Resource base

In the Integrated Farming System the operational land holding and livestock are the two prime resource base owned by the farmers. The distribution of operational land holdings and average number of livestock owned by the farmers are presented in Table 1 and are depicted in Fig. 1 and 2.It is evident from the data and figures that the share of marginal, small and semi-medium operation holdings taken together constituted 95, 68 and 88% in terms of number of holding, area operated and livestock possessed, respectively.

Amongst allied enterprises in Integrated Farming System, rearing livestock is the most important and complementary to all other enterprises. Livestock is the backbone of Indian agriculture and plays a key role in providing employment especially in rural areas. Livestock are an important asset and livelihood option for poor people in our country, but poor feed quality and dry season feed shortages are a serious limitation for sustainable livestock production. There is tremendous pressure of livestock on available feed and fodder, as land available for fodder production has been decreasing. As per Draft Report of Working Group on Animal Husbandry and Dairying for Five-Year Plan (2002-2007, Govt. of India, Planning Commission, August-2001) the estimates for 2015 reveals that the deficit in green (63.50%) and dry fodder (23.56%) is increasing every year, while for concentrates, the gap is almost static. However, this gap is critical and is going to determine the type of animals and husbandry practices to be followed. To achieve targeted growth rate of 6% in livestock output, progress in livestock infrastructure, institutional efforts and availability of livestock feed is required to be accelerated by about 50% (Chand and Raju, 2008). Scarcity of feed and fodder resources (both quantity and quality), low production potential of animals, non-availability of critical inputs or services in time along with access to capital and markets, are primary reasons for low productivity of dairy animals (Mishra *et al.*, 2009).

Table 1. Distribution of operational holdings and average number of livestock owned by category of land holdings, 2010-11(Number of holding: million; Operated area: million ha: and Average size of holding: ha)

S. No.	Category of land holdings	Number of holdings	Operated area	Average area per holdings	Distribution of livestock(%)	Livestock per holding Cattle & Buffalo	Sheep & Goat
1	Marginal (< 1.00 ha)	92.83	35.91	0.39	50.1	1.34	0.74
2	Small (1.00 to 1.99 ha)	24.78	35.24	1.42	21.6	2.23	1.39
3	Semi-medium (2.00-3.99 ha)	13.90	37.71	2.71	15.9	3.05	1.93
	Sub Total	**131.50**	**108.86**	**0.83**	**87.6**	**1.67**	**0.99**
4	Medium (4.00 to 9.99 ha)	5.88	33.83	5.76	9.6	4.43	2.91
5	Large (10.00 ha & above)	0.97	16.91	17.38	2.8	6.06	4.98
	Sub Total	**6.85**	**50.74**	**7.41**	**12.4**	**4.67**	**3.21**
	Total (all size of holdings)	**138.35**	**159.59**	**1.15**	**100**	**1.82**	**1.10**

Source: Author's calculation with data from Report on 'Agriculture Census-2010-11 (Phase-I), & Input Survey -2006-07, Agriculture Census Division, Department of Agriculture and Cooperation, Ministry of Agriculture, Government of India.

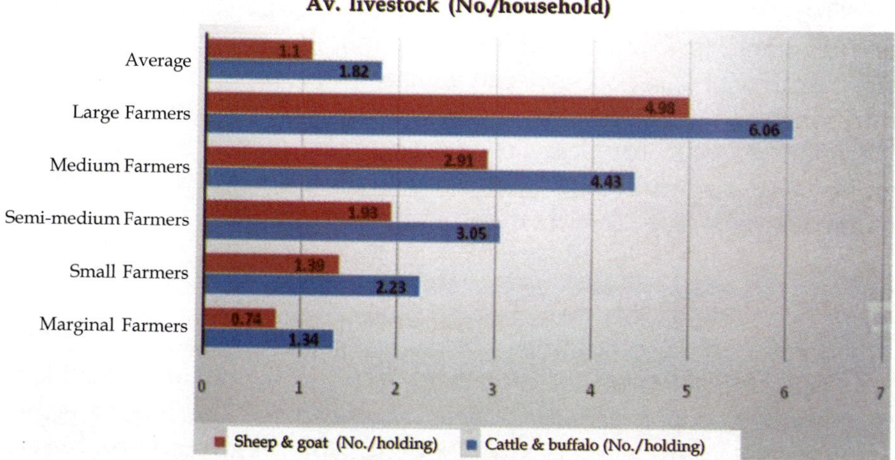

Fig. 1. Distribution of operational land holding

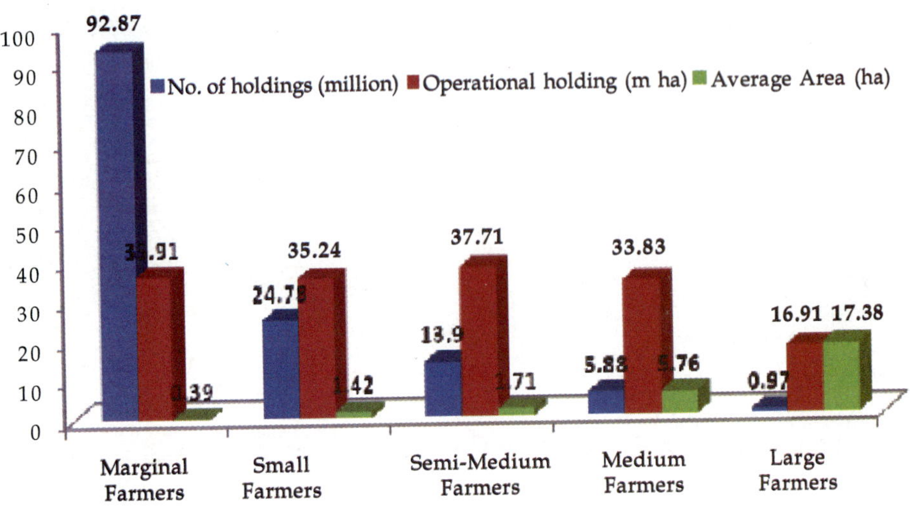

Fig 2. Average number of livestock per household

The availability of nutritious green fodder in adequate quantity plays an important role in minimizing the cost of milk production, because feeding cost alone constitute about 70% of total production cost. Better feeding could be achieved by ensuring the adequate supply of good quality forage from better varieties and improved cultural practices. Intensive methods of forage production may not be viable option for integrated farming systems in various agro-ecosystems.

3. Constraints in forage production w.r.t the farmers' resources

There is heavy stress on account of a large-scale exploitation for fuel wood, timber and fodder, mismanagement of forest resources and frequent fires. There is acute shortage of fodder especially green nutritious fodder, which is major cause of low productivity of livestock. The main reasons for low productivity is insufficient and low quality fodder and feed including grazing facilities. Some of the main constraints are described hereunder (Dev, 2001).

3.1 Less area under fodder crops

At present land holdings are very small and farmers are always biased in choice of the crops.Due to the marginal and small land holdings the farmers are deficient in green nutritive fodder as they cannot spare even a small portion of their land for fodder cultivation.

3.2 Uncontrolled grazing

Uncontrolled grazing in community property resource (CPR) and pasturelands has led to a decline in biomass availability. The grazing pattern has created manifold problems in these pastures. Obnoxious weeds have invaded the pastures. Excessive and continuous grazing has severely damaged these lands.

3.3 Poor management practices of grasslands

Livestock keepers (of all the categories) have access to the CPR and other grazing lands. Presence of inferior and unproductive grass species, lack of fertilizer application, and absence of legume component, impropercutting and indiscriminate grazing are some of the important factors responsible for poor productivity of grasslands.

3.4 Heavy livestock population

Huge population and poor fodder availability has widened the gap between demand and supply of forage crops. It is a fact that considerable fodder resources are wasted on maintenance of an excessive number of poorly fed and low yielding animals, which contributed to process of pasture destruction.

3.5 Over exploitation and unscientific management of fodder trees

Tree leaf fodder is the major feed resource during lean periods, particularly during the winters. The tree leaf fodder provides 50-90% of the forage demand during lean periods (Negi, 1977). The over exploitation and unscientific management of fodder trees has depleted this resource at huge environmental cost.

4. Strategies to increase fodder production in integrated farming system

Keeping in view the constraints in fodder production and in order to overcome the gap, the emphasis need to be given on several steps for augmenting the fodder production. Existing resource utilization pattern needs to be studied in totality according to a system approach. The holistic approach of integrated resource management will be based on maintaining the fragile balance between productivity functions and conservation practices for ecological sustainability. Some of the scientific interventions, which could help in improving the productivity of forages, are described here.

4.1 Agronomic management

The herbage production in integrated farming system can be enhanced with the adoption of improved technology. Some of the important components of this technology are: (a) Selection of suitable fodder crop with improved varieties; (b) Land preparation, seed rate and time of sowing (c) Control of bushes and weeds in crop field (d) Introduction of fodder legumes/grasses in food crops as intercrop (e) Nutrient management(f) Cutting and grazing management (Dev, 2001).

4.1.1 Selection of suitable fodder crop with improved varieties

It is the one of the most important management practice for improving fodder production. The suitable crop and variety should be selected according to farmers' need and available resources. A large number of improved varieties in different forage crops suitable for different agro-ecological zones has been developed and released by ICAR and SAUs. These varieties resulted in substantial increase in the productivity and production of forages in the country. The improved varieties released/notified during the past three decades are given in ICAR Handbook.

4.1.2 Control of bushes and weeds in crop field

The bushes and obnoxious weeds and poor quality grasses may offer severe competition for light and nutrients. Several weeds are very detrimental to health of animals and human beings. So proper integrated weed management practices should be followed for better fodder production. Hand weeding, mechanical weeding, biological control and chemical weed control are some of the weed management practices, which could be adopted for weed management.

4.1.3 Pasture establishment

The successful establishment of a pasture requires more skill and care, as compared to other crops. The method of introduction of improved grasses and legumes in the natural grasslands should be cost effective with minimum soil working. Some of the methods of establishment such as i) Scratching or pitting (Singh,1995) and ii) Hoof and teeth method (where the pasture could be heavily grazed; followed by throwing seeds, then allowing the animals to trample the area when soil is wet).

4.1.4 Introduction of legumes/grasses

Forage legumes enrich the N content of the soil and have a high nutritive value. Legumes can be grown in mixtures with grasses. By introduction of legumes the quantity as well as quality of herbage production can be substantially increased. Among the legumes, Siratro (*Macropteliumatropurpureum*), *Stylosantheshamata, S. scabra, Glycine javanica, Dolichosauxilaris, Desmodium* spp and *Centrosemapubescens etc.* have shown good performance (Melkania, 1995). Legumes and grass species can be introduced during July by seeding and tussock planting, respectively. The herbage yield and nutritive value of the hay from grasses-legume mixtures were found five and two times higher, respectively than the hay of local species. It is essential that during the first year of seeding/tussock planting, grazing is restricted in treated sites and the grass cutting is done carefully to help the establishment of introduced fodder species. Some of the grasses; *Cenchrusciliaris, Dactylisglomerata, Dicanthiumannulatum, Festuca* sp., *Lolium* sp., *Pennisetumpedicellatum, etc.* and legumes; *Desmodiumintortum, Dolichos lablab, Phaseolus artopurpureus, Stylosantheshumilis, Trifolium* sp. *etc.* have been found adapted to different agro-climatic regions of India (Shastry and Patnaik, 1990).Legumes introduced in the pastures generally do not establish well due to ineffective nodulation. *Rhizobium* inoculation of the pasture legumes provides synergistic effect for better establishment.

4.1.5 Fertilizer management

The present poor production potential of pastures could also be attributed to poor fertility of soils. To raise the fertility status and rectify the deficiencies, soil testing coupled with field trials need to be conducted to work out the fertilizer requirement of different pastures. Generally, no fertilizer is added to rangelands except the dropped excreta by animals. Judicious use of fertilizer can boost the vegetative growth and is also economically feasible. Application of nitrogen fertilizer must be given in split doses for better utilization, whereas phosphorus and potash should be supplied as basal dose in case of grasses.

4.1.6 Cutting and grazing management

The response to cutting of a forage plant depends upon its seasonal yield of carbohydrate storage, its growth habit and extent of inflorescence development. Frequency of cutting also significantly influences the yield and quality of herbage produced. The areas with high temperatures may require larger interval and low intensity of cutting to build up sufficient carbohydrate storage for regrowth. The Himalayan grasslands experience intense grazing pressure on account of being the prime source of forage. Grazing contributes more than 50% of the herbage requirement for sedentary and semi-migratory flocks, while for migratory flocks 100% herbage is provided by grazing. Controlling the time, duration and intensity of grazing appears to be the key factors in grazing management. Periods of rest allow grazed perennials to replenish leaf area, seed set and store food reserves in their roots (Merrill, 1983; Adams *et al.*, 1991). Continuous or too frequent access by large numbers of cattle to the same range impedes the ability of new growth to store food. The grazing can lead to the disappearance of nutritive species and infestation by less palatable species and weeds. Deferred rotational grazing system was found superior in *Sehima* dominated grasslands, resulting in greater number of animal days as compared to continuous approach. Rotational grazing has steadily gained the popularity in last two decades, because it offers better control over livestock distribution and feeding pattern with goals of periodically resting vegetation (Adams *et al.*, 1991). Recently grassland management systems have been developed that not only reduce impacts but also use livestock as a tool to enhance productivity and ecosystem function. It is not the presence or absence of livestock but the choice of grazing management system, and its suitability for the landuse, that leads to positive or negative effects. Improved grazing management schemes for native or synthesised silvi-pasture lands is described as under:

Controlled grazing	Based on the availability of forage from ground vegetation as well as trees/shrubs; mixed herd grazing may be introduced based on the concept of land's carrying capacity.
Rotational grazing	The grazing area is divided into 4-6 paddocks and livestock species are allowed to graze in rotation for 7-15 days. By the time last paddock is grazed, the first one is ready again.
Deferred rotational grazing	Some of the rotational paddocks are left un-grazed for use in dry season and for proper seed formation. These paddocks should be changed every year to allow good seed bank in the soil in the entire area.

Strip grazing	Grazing animals are confined to smaller areas of pasture paddocks rather than grazing randomly. In this scheme the advantage is that the area is fully utilized and the animal gets access to just right amount of feed requirement.

Source: Patnayak, (1982)

4.2 Growing of fodder crops and fodder trees

For augmenting fodder availability, emphasis needs to be given to cultivated fodder crops on larger areas. Foliage of fodder trees could be fed to the livestock in mixture with crop residues and hay. Mixing of tree foliage with dry roughage improves their palatability and nutritive value (Dev, 2001).

4.3 Silvipastoral System

Silvipastoral systems are the most important for increasing fodder production from the marginal, sub-marginal and other wastelands, which comprise about 50% of the total land area. It involves planting of multipurpose trees in the existing pastures/grazing lands or planting such trees on wasteland/denuded lands followed by sowing /planting of grasses and or legumes in between the inter-spaces of trees. The trees and shrubs may be used primarily to produce fodder for livestock or they may be grown for timber, fuel wood, and fruit or to improve the soil. This system is classified in to three categories i) Protein bank ii) Livefence of fodder trees and hedges iii) Trees and shrubs on pasture.

4.4 Agrisilvipastoral system

Under the agrisilvicultural system multipurpose trees (MPTs) including fodder cum fuel trees can be grown in association with crops. Trees are pruned annually, yielding fodder as well as fuelwood. In addition to annual pruning, few trees are also cut down in order to allow light penetration and minimization of competition with the crops. Under alley cropping system MPTs like *Leucaenaleucocephala*and even perennial pigeon pea *etc.* are pruned frequently to provide leaf fodder to get better crop production.

4.5 Agri-horti-silvicultural system

Under this system besides growing fruit trees and fodder crops, fast growing NFTs like *Leucaenaleucocephala*can be lopped two to three times in a year to provide fodder (2.5-3.0 tha^{-1}) and fuelwood (1.8-2.5 tha^{-1}). These fodder trees also provide some protection to the fruit trees during summer and cold winters.

4.6 Hortipastoral system

In this system forage are grown in wide inter-row spaces of fruit trees for economic utilization of orchard lands. Forage from hortipasture is consumed fresh and is also conserved as hay for winters. There is considerable area under orchards and inter spaces between fruit trees could be utilized for the production of fodder by growing perennial grasses and legumes.

4.7 Forage production on terrace risers or bunds

A non- competitive land use systems for forage production in the hills is to grow forage on terrace bunds and risers. Forage grasses/ legumes/fodder trees grown on terrrace risers and bunds arrest the nutrient loss in runoff water under high rainfall conditions of this region. This gives an added advantage to produce forage without any fertilizer or manure.

4.8 Efficient Use of Crop Residues

Although about 54% of the fodder needs are met from various crop residues, no serious efforts are presently made to either increase the yield or quality of this fodder. Farmers in many regions have been wasting the crop residues, either by feeding the stalk without processing or by burning. Subsequently, with the development of dairy husbandry particularly in peri-urban areas, crop residues are now in good demand. During the last 5 years, the retail price of sorghum stalk has been almost 50% of the value of sorghum grain. With such demand, farmers have started shifting back to the old varieties with higher stalk yield. Similar demand for high stalk yielding varieties has now set a new direction for breeding and selection of new varieties, which have higher fodder quality and yield, without any reduction in grain yield.

Timely harvesting of crop residues, proper processing and storage can also enhance the quality of the forage and prevent wastage. Harvesting of stalk before it turns fibrous for direct feeding or converting into silage, can keep the nutritive value high while reducing methane generation by the ruminants. There are various methods of treating the crop residues before feeding, to improve its nutritional value. It has been reported that even chaffing of stalk before feeding, can reduce the emission of methane by 10% while saving the wastage by 25-30%. Further treatment of crop residues by way of soaking in water and treating with steam under pressure, can also improve the nutritive value and palatability. There are other methods like urea treatment in addition to molasses and physico-chemical methods like urea ammonisation, by storing the urea treated straw in anaerobic condition which can further improve the quality.

It is necessary to set up fodder banks in fodder surplus areas and process into compact feed blocks either directly or in addition to concentrates and minerals. Such blocks can easily be transported to different parts of the country which are facing fodder shortage. Shortage of fodder is often a temporary problem which is not confined to any particular region. The shortage arises either due to crop failures or due to natural calamities like floods and droughts which are unpredictable and not location-specific. Hence fodder banks established in fodder surplus areas need to keep a watch and look out for opportunities to supply to regions facing seasonal shortage.

4.9 Fodder crops for wastelands

Considering the limitations of traditionally cultivated fodder crops, it is necessary to introduce various non-traditional fodder crops for growing on marginally productive farms and denuded community lands. There are many hardy grasses and legumes like stylo, seratro, hedge lucerne, etc. which can be grown on wastelands without irrigation. There are many fast growing shrubs and trees which can be lopped regularly as fodder. Such tree species can be established on field bunds, home gardens and along farm boundaries.

4.10 Pastures on CPR

Development of community pastures is another excellent opportunity. Generally, about 5 to 10% of the land area in every village is reserved for community pastures. However, a part of this land is encroached or diverted by the local government for other purposes. Nevertheless, a significant portion is still available for common grazing. Over the years, in the absence of controlled grazing and care, the productivity of these community pastures has been severely eroded. Such lands can be brought under silvipasture development involving local people. There is a need to develop suitable fodder shrubs, trees and grasses for development of pasture lands. Fodder species for introduction under agroforestry need to be shade tolerant and resistant to pests and diseases. Establishment of leguminous shrubs particularly in fruit orchards, can even enhance the fruit production. Seed production of elite genotypes should also be taken up on a large scale to meet the demand.

5. Conclusion and policy implications

By and large most of the livestock keepers especially the marginal and small farmers are by dependent for livestock rearing on the common grazing lands and pasturelands besides their own resources. Therefore, it is suggested that besides improving the forage production from their own lands, it is imperative to improve the common grazing resources for improved productivity of the livestock.

References

Adams, B.W., Ehlert, G. and Robertson, A. 1991. Grazing systems for public grazing lands. Range notes No. 10 (Alberta forestry lands and wildlife, public lands division, Leth-bridge, Alberta). pp. 1-8.

Anonymous, 2010. A Manual on Integrated Farming Systems (IFS) by Caribbean Agricultural Research and Development Institute (CARDI), July, 2010. Ministry Of Economic Development, Belize 9th European Development Fund.

Das, Anshuman. 2013. Agricultures, Integrated Farming: An approach to boost up family farming LEISA India, 15 no. 4 December.

Dev, I. 2001. Problems and prospects of forage production and utilisation of Indian Himalaya. ENVIS Bulletin **9**(2): 11-18.

GoI, 2013a. Agricultural Statistics at a Glance. Directorate of Economics and Statistics, Ministry of Agriculture, Govt. of India.

GoI, 2013b. National Sample Survey Office. Ministry of Statistics and Programme Implementation, Govt. of India.

Melkania, N.P. 1995. Forage resources and forage production systems in Central Himalaya. In: New Vistas in Forage production (ed. Hazra, C.R and Misri, Bimal). AICRPF (IGFRI). Publication Information Directorate, New Delhi. 203–211.

Merrill, J. 1983. "The XXX Ranch: Managing Range for Ecology and Economy", Publication No. ys-83-1, Chapter offprint from 1983. Year Book of Agriculture (U.S. Deptt. of Agriculture, Washington, D.C., pp. 87-94).

Mishra, A. K, Rama Rao, C.A, Subrahmanyam, K.V. Ramakrishna, Y.S. 2009. Improving dairy production in India's rain-fed agro ecosystem: constraints and strategies, *Outlook on Agriculture,* **38**(3): 284-292.

Negi, S.S. 1977. Fodder Trees in Himachal Pradesh. Indian Forester. 103: 616-622.

Patnayak, B. C. 1982. Problems and prospects in nutrition of arid zone livestock: In: aridZone research and development. (H. S. Man, Ed.) Scientific Pub. Jodhpur, India. pp. 369-379.

Raju, S.S. and Chand, R. 2008. Agricultural Insurance in India Problems and Prospects. National Centre for Agricultural Economics and Policy Research (Indian Council of Agricultural Research), New Delhi. Pp 78.

Shastry, G and Patnaik, U.S. 1990. Grassland management. In: Watershed Managent (ed. V.V. Dhruranarayana, G. Shastry and U.S Patnaik). Central Soil and Water Conservation Research & Training Institute (CSWCR & TI), Dehradun, ICAR Publication. pp. 68-73

Singh, L.N. 1995. Temperate Pastures and their Management (in press), IGFRI, Jhansi.

❑❑❑

Chapter 9

Integrated Dairy Farming System

B P KUSHWAHA & S B MAITY

ICAR - Indian Grassland Fodder Research Institute, Jhansi-284003 (UP)

Introduction

Dairy farming from being a traditional family run business today has grown hugely to an organised dairy industry with technological specialization in every part of the process. We have seen a tremendous growth in dairy farming equipment that help modern dairy farms to manage thousands of dairy cows and buffaloes. This huge boost in the industry has created a lot of farming jobs for the people. But many of the dairy farms still manage and run organic dairy farms in villages and supply the milk to get processed by large companies and finally sell to the retail outlets. In India milk production has reached to 132 million tons in 2012-13 from 80.60 million tons in 2000-01. Further, the production is estimated at 139 million tonnes in 2013-14. This 5 percent increase over the previous year is on account of good monsoon, low feed cost and high milk prices. Going back to 70s, India was a milk deficient nation. To tide over these crises, the National Dairy Development Board (NDDB) under the leadership of late Dr Verghese Kurien launched operation flood. The cooperative model of production and marketing promoted by operation flood encouraged farmers to produce more and as a result India surpassed the US as the largest producer of milk in 1998. Operation Flood operated in three phases from 1970 to 1996.

However, India still has an issues related to low productivity. Productivity of the cows and buffaloes in India is one among the lowest in the world. As compared to more than 7500 kg per cow of dairy animals yield annually in the US, Denmark, Sweden, Netherlands, Finland etc. Indian dairy animals yield only 1169 kg in a year. This is lower than world average of 2200 kg per year. Lower productivity has become a growing concern for India's dairy industry and policy makers as well. The economic survey 2011-12, revealed that India's demand for milk is growing at twice the rate of growth in supply and if India wants to remain self-sufficient in milk and prevent import, it will have to increase its annual milk production by at least 5-6 per cent annually (*Back ground notes, India's Premier Biennial, Agro technology Fair, 22-25 November 2014, Chandigarh, India*).

On the production side, milk production in the country has grown 3.5 to 4 times higher than the world's average growth. There are huge regional variations in terms of milk production. According to the Department of Animal Husbandry, Dairying and Fisheries, north and north central region contributes more than 50 per cent of total production in the country. In spite of all these, that there exists a demand supply gap. Buffalo contribute more than 55 percent of country's total milk production, cattle contribute 40 percent. Around 4 percent of total milk comes from goat. Major constraints in the dairy farming are as follows:

1. Low producing animal
2. Inadequate breeding facilities
3. Poor outreach of veterinary health services
4. High cost of milk production and low milk price to the producer
5. Shortage of quality feed and fodder: The area under fodder crops has stagnated around 8.5 to 9.0 million hectare (4.6 percent of total cultivated area).

Integrated dairy farming system (IDFS)

Dairying is one of the important money generating components of the integrated farming system, especially in smallholding system in rural areas where production resources have to be used to maximum advantage to generate immediate farm income. Cattle are identified as a dairy animal throughout the world. However in India buffalo makes a most important contribution to both rural and the national economy. Indian farmers, especially in north and central India, often use both cattle and buffalo together for milk production, in order to fetch a higher price for high fat content milk.

Most of the dairy farmers concentrate mainly on milk production which is subjected to a high degree of uncertainty in income and employment to the farmers. The income from the livestock remained fluctuating and also uncertain. Infertility in cows and buffaloes and large scale mortality in poultry and goats are some of the major problems identified for such uncertainty. Integration of various enterprises *viz.*, cropping, animal rearing, fishery, poultry, forestry etc. have great potentialities to boost up the farm income. These enterprises not only supplement the income of the farmers but also help in increasing the family labour employment. The IFS approach introduces a change in the farming techniques for maximum production in the livestock and crops and takes care of optimal utilization of resources. The farm wastes are also better recycled for productive purposes in the integrated system. A judicious mix of enterprises like livestock, fishery, poultry, sericulture etc. suited to the given agro-climatic conditions and socio-economic status of the farmers is expected to bring prosperity in the farming. The major components of any IFS are crops, animals and trees. The needs and resource base of the farmer, demand and supply of the commodity, climate, soil type, rainfall and growing season are some of the major factors which decides the selection of IFS components in any farm. C Devendra (*www.ilri.org/InfoServ/Webpub/fulldocs/SmHDairy/chap15.html) have described the advantages of integrated dairy farming system as follows:*

- Diversification in the use of production resources
- Reduction in, and spread of, socio-economic risks
- Involvement of a preponderance of small farms and small farmers
- Use of large populations of ruminants (buffalo, cattle, goats and sheep)
- Integration of crops (annual and perennial) and animals
- Use of animals and crops play multipurpose role
- Low input use and traditional systems
- Involvement of the three main agro-ecosystems (highlands, semi-arid and arid tropics, and subhumid/humid tropics).

Mainly there are two types of integrated dairy farming systems being followed. System one combing annual crops and animals are more common in irrigated and lowland areas where water is available and enables intensive crop production. System two combining animals and perennial crops are more common in upland and rain fed areas. Between the two systems, dairying involving mainly cattle and buffalo is much more common in the first system involving annual cropping. These are the areas which are usually irrigated where intensive cereal cropping is common, and a distinctive market pull exists because of the peri-urban demand for milk. Thus, dairy production is an expanding feature in these situations in which it is integrated with crop cultivation and provides an economic motivation for farming systems. (C Devendra (*www.ilri.org/InfoServ/Webpub/fulldocs/SmHDairy/chap15.html*). Integrated dairy farming system prevails throughout the country in one or the other form, some examples are as follows:

- *North* and *Central India:* wheat, paddy, pulses, sorghum, berseem, lucern, maize, bazra and cattle and buffalo production.
- *South India:* paddy, tree crops, root crops, spices, vegetable, cattle, sheep, goats and poultry.
- *North east:* rice, maize, vegetable, cattle, pig, goat and sheep etc.
- *Western India:* Wheat, sorghum, bazra, lucern, cattle, sheep and goat.

Socio-economic benefits of integrated systems

Dairying in integrated system has significantly contributed to the development of peri-urban dairy system in our country through supply of animals. Pregnant cows and buffaloes are often brought from the rural areas to the cities and peri urban areas for milk production. Some case studies have been presented below, which give an indication of the contribution of the dairying in integrated farming system. Economic analysis of different farming systems (one hectare of irrigated land or 1.5 ha of un-irrigated land) indicated that under irrigated conditions, mixed farming with crossbred cows yielded the highest net profit, followed by mixed farming with buffalo, and arable farming. However, mixed farming with Hariana cows made a loss (Singh *et al.*, 1993). Comparative productivity and economies of dairy enterprises (mixed farming with three crossbred cows on one hectare of canal irrigated land versus mixed farming with three Murrah buffalo) indicated that mixed farming with crossbred cows under canal-irrigated conditions was more efficient for the utilisation of land, capital, inputs and the labour resources of the farmer (Kumar *et al.*, 1994).

Baseline surveys in Gujarat India, indicated that around 75 per cent of rural households kept cattle in the face of under-employment.

More particularly, the farm surveys showed that cattle kept mainly for milk, contributed 32 per cent and 20 per cent for tribal and non-tribal ethnic groups respectively (Patil and Udo 1997).

The results of a study (Radha et al; 2000), conducted on survey based, in Telangana zone of Andhra Pradesh, with three agricultural and livestock based farming systems viz., dairy, poultry and sheep rearing clearly revealed that all the farming system generated more than 3 times additional employment over arable farming. The net returns were higher in agriculture + dairy (Rs. 35293) followed by agriculture + poultry (Rs. 26830) and agriculture + sheep rearing (Rs. 14665). Among different farming systems, the agriculture + dairy was proved to be more promising than others.

IFS studies conducted on farmers fields in Punjab conditions, gross profit was found to increase from Rs.81200/ha/annum in cropping (Rice-wheat) alone to Rs.154000/ha/annum in crop+dairy and Rs.113200/- in fish+piggery system of farming (Gill,M.S.,2004). Singh *et al.* (2011) studied a Integrated Farming System model at Modipuram in Western Uttar Pradesh. They developed IFS model on 1.5 hectare area comprises of crops (0.72 ha), dairy (0.32 ha), horticulture (0.22 ha), fishery (0.10 ha) and miscellaneous (0.14 ha) which was used for goat, apiary, vermicompost, threshing floor and farm building etc. The net returns were higher in dairy (Rs. 87020) followed by crops (Rs. 74430), horticulture (Rs. 10260), fisheries (Rs. 4940) and Apiary (Rs. 4200). Results revealed that farmers can earn 82.47% more profits (Rs.135.82 thousands/ha/year) under IFS system than crops alone (Rs.74,435/ha/year).

Conclusion

The preliminary research conducted under integrated dairy farming system clearly indicated the importance of dairying in the system. For small farmers dairying alone cannot be a profitable business until it is integrated with the other agricultural activities. The integration is to be made in such a way that product of one component should be the input for other enterprises with high degree of complimentary effects on each other. The fodder fed to the cattle produces milk, the dung; urine and litter produce farmyard manure and energy used for crops and other farm activities. Thus integrated farming system is a complex enterprise. Potential improvements and increased productivity from this enterprise can only come from better understanding of the nature and extent of the interaction with other sub-sectors, like crops and natural resources, economic benefit, as well as the impact on the livelihoods of small farmers and environment.

References

Gill, M.S. (2004). Methodologies for farming system approach- A case study. In: Proceedings of National Symposium on Alternative Farming Systems held at PDCSR, Modipuram, 16-18 September, 2004, pp. 95-108.

Kumar H., Singh J.N., Kadian V.S., Singh K.P., Saxena K.K. and Kumar H. 1994. Comparative productivity and economics of dairy enterprises under mixed farming systems. *Farming Systems* **10**: 36-44.

Patil B.R. and Udo H.M.J. 1997. The impact of crossbred cows in mixed farming systems in Gujarat, India: Milk production and feeding practices. *Asian-Australasian Journal of Animal Sciences* **10**: 253-259.

Radha,Y., Eshwaraprasad,Y. and Vijayabhinandana, B.(2000). Study on income and employment generation on agricultural based livestock farming system. Paper presented at VIII Annual Conference of AERA at TNUASU, Chennai, 28-29 Dec., 2000.

Singh K.P., Singh S.N., Kumar H., Kadian V.S. and Saxena K.K. 1993. Economic analysis of different farming systems followed on small and marginal land holdings in Haryana. Haryana *Journal of Agronomy* **9**: 122-125.

Singh, J.P., Gangwar, B., Pandey, D.K. and Kochewad, S.A. (2011). Integrated Farming System Model for Small Farm Holders of Western Plain Zone of Uttar Pradesh. PDFSR Bulletin No. 05, pp. 58. Project Directorate for Farming Systems Research, Modipuram, Meerut, India.

□□□

Chapter 10

Animal Health and Infertility Management Issues and Strategies under Integrated Farming System

K.K. SINGH AND S.B. MAITY

ICAR - Indian Grassland and Fodder Research Institute, Jhansi-284003 (UP)

Introduction

Infertility in cattle accounts for major economic losses in dairy farming and dairy industry in India. In cattle, nearly 10-30 per cent of lactations may be affected by infertility and reproductive disorders. To attain good fertility or high calving rate both the male and female animals should be well fed and free from diseases. Artificial breeding program depends on the inherent fertility, health, nutrition and management of cow as well as the ability of handlers to preserve store and deliver healthy semen dose to cow/buffalo at proper time during the estrus. Due to infertility if open span is increased by one estrous cycle, this accounts loss of production of 20-25 days in addition to cost of treatment and vacant days, labor, etc.

Infertility in farm animals is due to various causes such as nutrition, physiological disturbances and infectious causes, which may work separately or in combination. For this reason each cause must be accurately diagnosed and treated according to conditions found in examination. Recent studies showed that some metabolites and nutritional factors are responsible for infertility. The energy shortage in postpartum period led to decrease in luteinizing hormone secretion frequency and dynamic follicle diameter, however, it increases calving interval time.

Absence of periodic manifestation of estrous without pregnancy is anestrus. In this condition there is insufficient stimulus from pituitary for secretion of FSH or LH resulting in failure of maturation and rupture of Graafian follicle. Malnutrition due to non-availability of feeds or physiological problems in assimilation may lead to anestrus.

Infectious infertility is necessarily caused by venereal infections and mostly through sexual passage i.e. natural or artificial breeding. Spread of these infections is due to movement of livestock from different regions of the country. Cystic ovarian degeneration is multi-glandular degeneration showing nymphomania, irregular estrus cycle and it is mostly permanent degeneration.

It is possible to detect infertile, sub-fertile and sterile animals by thorough investigation of management and health. There are different aspects such as nutrition, management, infectious agents that affect the infertility. Nutritional aspects include green fodder, balanced feeding and mineral supplementation. Management aspects include breeding strategies, timely detection of estrus care of pregnant animals, care of post parturient dams, vaccination schedule for important diseases. Infections include bacterial, viral, protozoal and fungal infections affecting genital tracts.

Cause of Infertility

The causes of infertility are many and can be complex (Arthur, 1982). They relate to Graafian follicle development and maturation, oestrus onset, successful coitus, ovulation, fertilization, implantation, and the development and delivery of the foetus and its membranes. Anything interfering with these routines, such as diseases, poor nutrition, inadequate herd management, hereditary and congenital factors, hormonal disturbances or environmental changes, makes the animal infertile, if only temporarily (Osmanu, 1979).

1. Congenital Causes

Congenital causes of infertility are often inherited. They include developmental abnormalities of the ovaries, oviducts, uterus, cervix, vagina and vulva. Some are lethal, a few have a morphological significance and others a functional significance. Common morphological conditions include ovarian (gonadal) hypoplasia and aplasia, anomalies of the tubular genitalia, hermaphroditism, freemartinism, arrested development of the Mullerian ducts (White heifer disease) and double cervix (Lagerlof, 1963).

Bovine gonadal hypoplasia is not easy to diagnose (Lagerlof, 1963) and in cases of bilateral ovarian hypoplasia heifers do not develop secondary sexual characteristics. They are anoestrus and infertile. Where the condition is unilateral, normal sexual organs and oestrous activity may be observed. Such animals are fertile, although less so than normal. The condition is potentiated by an autosomal recessive gene with incomplete penetrance, and therefore the incidence of gonadal hypoplasia can be reduced by using only animals (both male and female) with normally developed sexual organs as breeding stock.

2. Functional cause

In a sample of 510 Gir cattle in India, functional abnormalities were responsible for 59.4% of the cases of reproductive disturbance, compared with 23.3% attributed to pathological causes, 8.8% to anatomical factors and 4 8% to old age (senility) (Kodagali, 1974). The causes of functional infertility included cystic and inactive ovaries with anoestrus, early embryonic mortality with repeat breeding, and prolonged gestation. Anoestrus often reflects a hormonal disturbance and accounted for 47.8% of the cases. Repeat breeding, where cows require three or more services to conceive, accounted for 11.5% of cases. Singh et al (1981) also found functional infertility to be more common than infertility due to infectious diseases (76 vs 24%).

Cystic ovaries and retained (persistent) corpora lutea

Cystic ovaries contain one or more persistent fluid-filled cavities larger than a ripe follicle (Arthur, 1964). This is sometimes referred to as cystic ovarian disease (Chauhan *et al.,* 1984). Ovarian cysts can be classified as follicular and luteal. They may vary in size from that of a ripe follicle to that of an orange. Their effects also vary according to their number and degree of luteinisation. Many unluteinised follicles tend to lead to nymphomania with frequent, irregular heats, whereas a cow with a few extensively luteinised cysts may become anoestrous. Cows with long-term cysts may show virilism. In addition to the pathologic follicular and luteal cysts, there are the non-pathologic cystic corpora lutea. These are normal structures that follow a normal ovulation but have a fluid-filled central cavity 7-10 mm in diameter. On rectal palpation they feel like normal corpora lutea but more fluctuant and soft. They do not alter the oestrous cycle duration and when conception occurs, it can be maintained to term (Roberts, 1971).

The term "cystic ovaries" is therefore usually applied to pathological follicular and luteal cysts. Cystic ovaries are conventionally diagnosed by rectal palpation, but it may be difficult to differentiate between follicular and luteal cysts. Although both tend to be smooth and convex, follicular cysts are more tense and thinner-walled.

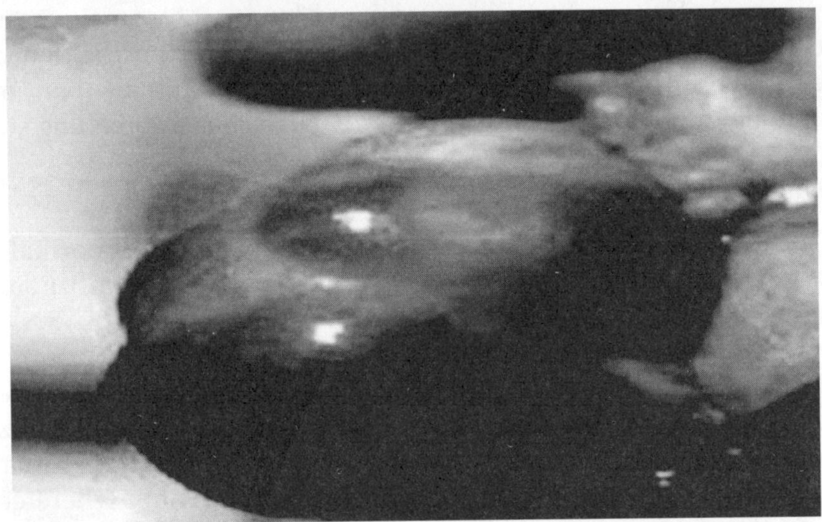

Picture 1. Ovarian cyst

Forage legumes are often advocated as a means of improving animal nutrition in the tropics. Some, however, contain substances, such as phytoestrogens, that may reduce fertility. Feeding cows for prolonged periods on clover, lucerne (alfalfa), or other plants rich in phytoestrogens may lead to cystic ovaries. Little (1976) assessed several pasture species, particularly tropical legumes, for oestrogenic activity and found that lucerne (*Medicago sativa*) had a slight oestrogenic potential. More evidence is needed on the oestrogenic effects of forage legumes, especially where they form an important component of natural pastures.

Cows with cystic ovaries can show either nymphomania or anoestrus. Nymphomaniac cows exhibit prolonged periods of frequent oestruses. They mount others and stand for mounting, even by a bull. In contrast, cows with virilism mount others but do not stand for mounting. Nymphomaniac cows have been called "bullers" because of these homosexual tendencies. They develop oedema of the vulva, copious mucous discharges may be seen and the sacro-sciatic ligament frequently sinks, causing the coccyx to tilt up.

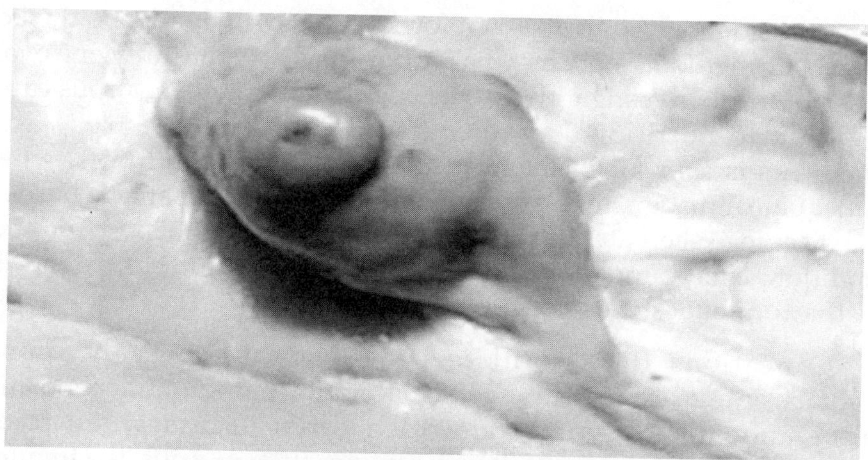

Picture 2. Persistent corpus luteum

Diagnosis of persistent corpus luteum is based on the presence of a large, persistent corpus luteum on the ovary, anoestrus and persistently high levels of progesterone. Frequent visits and rectal palpations of affected cows, with accurate records of the rectal findings at each visit, are required before a diagnosis of persistent corpus luteum can be confirmed. Once confirmed, the persistent corpus luteum can be enucleated or lysed by prostaglandins. These continue to produce enough progesterone to prevent further follicular development, ovulation and oestrus. Severe endometritis may be associated with a persistent corpus luteum due to toxic damage to the endometrium, which prevents proper secretion of luteolytic prostaglandins. This can also occur with pyometra, foetal mummification and maceration, i.e. conditions that simulate pregnancy (Boyd, 1977). Lamming (1977) reported persistent corpus luteum to be rare (under 2%) in cows with normal uteri.

Other causes

Anestrous

Anestrous is considered as a problem when cows are not seen in heat. Failure to observe heat and heat detection must always be ruled out as the primary problem. Review the heat detection procedures and heat detection efficiency with the producer. Pregnancy can be a prominent cause of anestrous and must always be ruled out before anything else. Other reasons for anestrous are ovarian aplasia, which is a rare condition when one or both ovaries are absent. It is caused when the gonadal ridge does not form correctly. Palpation is the best diagnostic method. There may be partial aplasia, when 'only one ovary does not form. Investigations in the country revealed anestrous cases, 73% due to nutritional problems and 27% true anestrus cases. Anoestrus is a major problem in the tropics and subtropics, where inadequate nutrition, high ambient temperature, high parasite burdens and disease exacerbate the problem. Low body weight and poor body condition, compounded with lactation stress, can further extend the postpartum anoestrous period.

Conditions that simulate pregnancy, such as pyometra (pyometron), severe metritis, foetal maceration or mummification, may cause anoestrus. Pyometra, which often occurs when foetal membranes are retained or following postpartum metritis, is a frequent cause of anoestrus. These conditions damage the endometrial lining of the uterus and reduce secretion of luteolytic prostaglandin. The cyclic activity of the ovary is thus interrupted in the luteal phase and the cow or heifer is anoestrous until the condition is corrected. Without regular veterinary care and palpation, these conditions may remain undiagnosed and the cow may be believed to be pregnant.

Although anoestrus can, to some degree, be overcome by treatment, it is more practical to ensure that animals are well managed and are fed to maintain good condition during critical periods, i.e. prior to mating and during lactation, in order to avoid anoestrus.

Anoestrus is normal in pregnant, prepubertal and recently-calved animals. Pregnancy is a common cause of anoestrus, but this is often overlooked where service records are poor. Since many treatments for anoestrus terminate pregnancy, the possibility that cows presented as anoestrous are pregnant should be eliminated before treatment begins.

Ovarian Hypoplasia

Where one or both ovaries are smaller than normal. This is a recessive trait in the Swedish Highland breed and is caused by a single autosomal recessive gene with incomplete penetration. True anestrus must be ruled out. There may be partial hypoplasia with this condition.

Suckling by the calf inhibits cyclic activity in the cow by decreasing LH release. Summer heat tress occurs, when the combined effects of heat and humidity produce a high "apparent temperature". If you compare estrus duration at 33.5°C vs 18.2°C, then the estrous cycle is longer and the estrus duration is shorter and estrus intensity is less.

Ovarian Atrophy: Ovarian atrophy is caused from nutritional problems and is most often seen in dairy cows with high production.

Mummification: A mummy in the uterus simulate a pregnancy and prevent the normal luteolytic mechanism. This results in anestrous. Treatment for this condition is administration of prostaglandin to lyse the corpus luteum.

Repeat breeders

Repeat breeding can be caused by a number of factors, including subfertile bulls, endocrine problems, malnutrition, reproductive tract infections and poor management. Vandeplassche (1982) cited previous work in which adhesions of the ovarian bursa, salpingitis, cystic ovaries and endometritis were found in repeat-breeding cows.

Repeat breeding can be treated by enucleating the corpus luteum or causing its lysis by prostaglandins, uterine massage or manual stimulation of the clitoris after artificial insemination or infusion of the uterus with 50-200 ml of 1 to 5% Lugol's iodine, which has a stimulating effect on the uterus. Enucleation of the corpus luteum may cause adhesions between the ovary, bursa and fimbria or haemorrhage in the ovary. Pharmacological enucleation with prostaglandins avoids this problem. Intra-uterine infusion of antibiotics has been used but there is little evidence that this increases fertility during the following cycle. However, fertility tends to increase by the second or third heat after treatment. Human chorionic gonadotrophin can be administered at the time of AI to promote ovulation or progesterone can be injected 4 to 5 days after service. Details are given in the following sections on how pathological causes can be handled.

Out of cases examined in Goa, 63% cases were repeat breeding, 29%were anestrous and others 8%. A repeat breeder is defined as a cow that has calved before, is less than 10 years old, has normal heat cycles, has no palpable abnormalities has been bred 3 or more times and is not pregnant. We normally assume a problem exists when the incidence is 10-15 %. In case of repeat breeding cases 42% cases were infectious, 36% were due to management problems and 22% were due to hormonal imbalance.

1. Non infectious cause of fertility

a) Managemental causes

Breeding health can be judged from the conception rate within the herd. High conception rates within the herd can be maintained to certain extent by making adequate provisions for timely services from high fertile bulls over a suitably long period of timeto give enough chance to the cows toexpress their fertility at proper time.The conception rate of a herd shouldbe estimated as the percentage of females becoming pregnant to the firstservice. The conception ability of theindividual animals and of the herd thuscan be known and is not vitiated by thepresence of individuals havingsporadic type of transient forms ofinfertility. It is possible to ascertain theaccuracy of the conception rate bysubjecting all females to pregnancydiagnosis within 6 to 8 weeks after service. A satisfactory conception rate in a herd can only be maintained with the high degree of breeding fitness in majority of animals in the herd. It is usually observed that draught type breeds have better conception rates than the dairy type.

Conditions of overstocking, lack of spacious accommodation and poor hygiene predispose the stock to known and unknown infections. In the assessment of herd infertility a detailed history should be obtained from the owner, which should include 1) size of the herd. 2) name of the breed 3) the management practices 4) feeding standards 5) data on all calving, service dates and 6) reproductive disorders - abortions, retained placenta, abnormal discharges and infertility treatments. Cognizance should also be taken of all the fresh introductions to the herd. The clinician should make a comprehensive study and then examine each cow irrespective of the symptoms of breeding inefficiency. During the recent years, large number of animals are being bred by artificial insemination to accrue thebenefits of superior germplasm. In such instances, difficulties have been experienced in the assessment of herd infertility since the bull's ability to fertilize is not clearly known and the infertility investigations are mainly focused on the females without reference to male relationship. Artificial insemination has practically eliminated the fear of transmission of coital infections. One should not ignore the likely possibility of transmission of other types of infections through semen when large number of cows are involved. The presence of mixed infections is also a likely factor.

Anestrus, or failure to show signs of estrus is in most cases a failure to detect estrus. In one study of cows reported to be anestrous, 90%

were determined to be cyclic but not detected in estrus and only 10% were truly anestrus (no ovarian activity). Undetected estrous signs in cows with normal ovarian activity resulting from:

- Inadequate estrous detection since 66% of estrous signs are shown between 6 p.m. and 6 a.m. Cows with short estrus (less than 12 hours in length) may be missed even with twice-a-day estrous detection. This is particularly true when cows are observed for estrus during a time when they are unlikely to exhibit standing behavior.
- Inadequate animal identification and/or inadequate records.

b) Nutritional infertility

Post partum nutrition is most important for fertility. If TDN is low both pre partum and postpartum, fertility suffers. Vitamin A had no effect on fertility, but may cause irregular cycles. Vitamin D deficiency suppresses signs of estrus and delays ovulation. Vitamin E deficiency may cause reproduction problems.Vitamins and minerals are often suspected in infertility and anestrus but little hard evidence supports these claims.By feeding cows after they calve,you cannot expect to improve thenumber of cycles. Feeding in pre-partum is the best way to assure early return to cycle in cows. By providing good pre-partum nutrition, it is easy tomaintain adequate pre-partumcondition, so the stress of postpartumlactation produces a shorter durationof negative energy balance.

c) Hormonal infertility

Silent Heat

Silent heat is generally not aproblem and usually is manifested byunobserved heats by farmer. However,the first postpartum heat is normallysilent, because there are no estrogenreceptors. This is a result of the lowpostpartum progesterone.

Delayed Ovulation

If a cow ovulates more than 18hours after end of heat, then ovulationis said to be delayed. This may bediagnosed by palpation and can betreated with GnRH. This only occurs in < 2% of cows and is more likely to be aheat detection problem.

2. Infectious infertility

Diseases resulting in bovine reproductive failure can be an infectionwith a bacterium, virus or parasite. Symptoms are usually similar and diagnosis requires the services of a trained veterinarian and often the veterinary laboratory. The symptoms of infections in most animals are similar regardless of the cause of infection. Cows may return to heat regularly after breeding or may miss a few heats onlyto start cycling again early in the season. Aborted fetuses are usually not found, but cows may have acreamy white discharge from the vulva. Some farmers may not even realize that there is a problem until anunusually large number of cows are diagnosed "empty" at the time ofpregnancy checking. Some of the important causes are discussed below.

Fertilization failure

In normal heifers 100 %fertilization has been found one dayafter breeding. This drops to 85 % inbreeders. Therefore, repeat breeders seem to have more of a fertilizationfailure. If embryos are fertilized and transferred, one can have normalpregnancy rates. Release of PGF from inflammatory conditions such as mastitis can causeluteolysis and pregnancy loss. Inabilityto prevent PGF release (shownexperimentally in a repeat breeder inresponse to oxytocin administration) causes return to estrus. Fertilization rates are normal inheat stressed cows, but day 1 to 2 embryos are affected most by the heat. As blood flow to uterus decreases toshunt it to the rest of the body forcooling, the uterine temperature rises, nutrients decrease, and wasteproducts increase. Salpingitis, metritis and cervicitis can cause achange in the uterine environment that leads to infertility.

Metritis

Inflammation of the uterus is known as metritis. Cows normally havea red-to-brown discharge during thefirst two weeks after calving. Ifdischarge persists beyond 2 weeks orif the discharge is foul-smelling, this is evidence of metritis. Possible factors involved are retained placenta, injuryto the reproductive tract can occur dueto a difficult calving or excessive forceused to assist at calving. Injuries canalso occur at the time of breeding oruterine treatment. Post breedinginfusion in these cows is not routinely helpful.

Pyometra

A pyometra is a uterus filled with pus that has a closed cervix and a corpus luteum on the ovary. The pus prevents the normal luteolytic mechanism from happening. This results in anestrus. The fluid in the uterus mimics a pregnancy, so the cow do not return to heat. Treatment for this condition is administration of prostaglandin to lyse the corpus luteum.

Abortion

Abortion is defined as fetal death and expulsion between 42 (an estimated time of attachment) and 260 days (the age at which a fetus is capable of surviving outside the uterus) of gestation. The condition does not include fetal maceration and mummification. Pregnancies lost before 42 days are usually referred to as early embryonic deaths, whereas a calf that is born dead etween 260 days and full term is defined a still birth. A rate of 3 to 5 abortions per 100 pregnancies per year is of ten considered "normal."However, the loss of any pregnancy can represent a significant loss of (potential) income to the producer and appropriate action should therefore be taken to prevent abortions and to investigate the cause of abortions that may occur. Abortion is the most important condition that limits cow's ability to produce a calf and considerably erodes the profit. The greatest risk of fetal loss is during the first trimester of gestation and then progressively decreases as gestation advances with a slight increase in the risk toward the last month of gestation.

Either infectious or non-infectious agents may cause abortion. The infectious causes include bacterial, mycotic, viral, and protozoal. Historically, it has been suggested that 50-65%,. 20-25%, and 15-25% of infectious abortions were caused by bacterial, fungal, and viral causesrespectively. The non-infectious causes include nutritional factors, chemicals, drugs, toxins, poisonous plants, and hormonal agents. Under modern systems of dairy management cows continue to experience a seemingly high risk of abortion, despite intensive efforts toward immunizationagainst infective agents. It should beremembered that because of theendemic nature of abortion occurring invaccinated cattle, factors other thanthe infectious agents might contributeto the risk of abortion. Many abortions go unnoticed orundiagnosed and we always suspect fertility problems when cows are foundopen rather than embryo/fetal loss.

Infectious causes of abortion

Bacterial

Bacterial abortions result from brucellosis, leptospirosis, campylobacteriosis (vibriosis), listeriosis, *Haemophilus somnus*complex, and ureaplasmosis. Bacteria like *Salmonella, Actinomyces, Escherichia coli, Streptococcus, Staphylococcus, Bacillus, Pseudomonas, Proteus, Pasteurella, Nocardia,* and chlamydia species, as determined by the microbiological findings, can cause abortion. All these organisms and few others that are not listed have been isolated from sporadic cases of abortion. These are secondary to either a septicemia in the dam or ascending infection through the vagina and cervix or due to persistent endometritis.

Brucellosis

Bovine brucellosis is the well known and most controversial infection of the bovine reproductive system. Brucellosis generally has been thought of as a cattle disease, but it is also seen in swine, sheep, goats, dogs, horses, and wildlife, and can be readily transmitted to humans. The disease represents a real occupational hazard for veterinarians, slaughter men, and cattle producers. Brucellosis is caused by the bacterium *Brucellaabortus.* The organism has an affinity for certain body tissues such as the udder, uterus, lymph nodes, testicles, and accessory sex glands. Because of its affinity for the uterus, abortion is the usual sign of the disease.Brucellosis is a particularly difficult disease as there is no sure way to identify infected cattle by their appearance, all infected cattle do not abort. In addition, the incubation period for brucellosis is variable. Brucellosis is primarily transmitted to susceptible animals by direct contact with infected animals. Essentially, the only time an infected cow transmits the organism is at or around calving or abortion.

Leptospirosis

Leptospirosis is a contagious, bacterial disease of animals and humans. In cattle, horses, pigs, sheep, goats, and dogs, it has been characterized by a wide variety of conditions including ever, icterus (jaundice), hemoglobinuria (bloody urine), abortion, and death. However, the concept of this disease has recently changed. It is used to be considered a highly fatal disease, but is now thought to be a widespread, mostly subclinical infection of many species of wild and domestic animals. Signs of leptospirosis in cattle range from mild, unapparent infections to acute infections that end in death.

Campylobacteriosis, Vibriosis

Campylobacteriosis (vibriosis) is a venereal disease of cattle caused by the organism *ampy/obacterfetus*subspecies fetus. Before 1973, this organism was known as *Vibrio fetus*subspecies venerealis, Campylobacteriosis is characterized by infertility with an increased number of services necessary for conception. Early embryonic deaths are common. In a herd that has never been exposed, and where no immunity exists, an acute type of infertility problemdevelops. In this case, infertility caused by endometritis results in early embryonic death and a prolonged period (up to 120+ days) passes before successful conception occurs. Spread of the organism to the male is primarilyby way of copulation with an infected female.

Viral infections

Several viral diseases, including infectious bovine rhinotracheitis (IBR), bovine virus diarrhoea/mucosal disease (BVD/MD), pare-influenza-3, G-up, infectious (contagious) bovine epididymitis and vaginitis complex (Epivag), transmissible fibropapilloma and epizootic bovine abortion have been associated with cattle infertility. Some viral causes are as follows.

Infectious bovine rhinotracheitis

Infectious bovine rhinotracheitis (IBR) is caused by a herpes virus). IBR can affect the respiratory, reproductive, nervous and digestive systems of cattle. The disease can be transmitted sexually or by droplet inhalation; in its venereal form it has been associated with infertility. Cows and heifers served by an infected bull develop a pustular vulvo-vaginitis 2 to 3 days later. This may clear up after 2 weeks. Some develop superficial ulcers on the mucosa of the vestibulum and may discharge yellowish pus. Infection rarely extends directly to the cervix or uterus, and therefore pregnancy is rarely terminated. However, if cows are artificially inseminated with semen from an infected bull, the virus is deposited directly into the uterus and induces endometritis. Infection of the uterus disrupts the cow's oestrous cycles and reduces its fertility. The virus may also invade the uterus systemically in cows suffering the respiratory form of the disease. After invading the uterus the virus may remain dormant for several months. Abortion occurs after 4 to 7 months. The foetus dies soon after being invaded by the virus. The dead foetus may be retained for several days and appear mummified when finally expelled.

Bovine virus diarrhea

Bovine virus diarrhoea/mucosal disease (BVD/MD) is widespread. It is caused by a Toga-virus. Infection may be acute, mild or chronic. When the virus infects a pregnant cow it may also infect the foetus and kill it. Calves born alive may be stunted, with cerebellar hypoplasia, brain cavitation and mucosal ulceration. These signs aid diagnosis. Confirmation is by the demonstration of antibodies in foetal blood prior to ingestion of colostrum. A double sample with dam's blood is very helpful. A herd test will indicate if BVD/MD was prevalent at the time of abortion.

Infectious (contagious) bovine epididymitis and vaginitis complex

Infectious bovine epididymitis and vaginitis complex (Epivag) appears to be confined to East and southern Africa. It was first described in Kenya and subsequently in South Africa (Arthur, 1964). Epivag is a venereal disease. In cows the main symptom of Epivag is muco-purulent discharges from the vagina Epivag can cause permanent lesions on the Fallopian tubes. In the bull, it causes the epididymis to swell, sometimes to the size of a golf ball. The disease can be controlled by slaughtering infected bulls and by using artificial insemination.

Mycoplasmas

Mycoplasmas are infective agents distinct from both bacteria and viruses. Several species of *Mycoplasma* cause disease in cattle. They have been associated with infertility, but their exact aetiological role is difficult to ascertain because they are present in the tracts of healthy animals.

In a study of cattle in South Africa, *Mycoplasma bovigenitalium* was found in 43% of males, 47% of females, 25% of foetuses and 11% of placentas (Trichard and Jacobsz, 1985). *Mycoplasma ergini, M. alkalescens, M. laidawii, M. bovis, M. bovirhinis, M. verecurdum* and *M. conjuctivae* were also isolated.

Mycoplasmas can be transmitted by discharges from the respiratory and reproductive tracts, and by milk, of infected animals. Infected cattle develop antibodies, but these do not give complete protection.

Symptoms associated with mycoplasmoses can also be observed with other conditions. Diagnosis can only be confirmed by isolating the organisms from nasal or reproductive-tract mucus and discharges, milk, arthral fluids, foetal tissues or the placenta.

Non-infectious causes of abortion

Nutritional

Starvation may result in placental insufficiency and abortion, however, it rarely occurs in a modern dairy practice. Vitamin A deficiency has been suggested to result in thickening anddegeneration of placenta and abortion in late gestation. Iodine deficiency has also been uggested as a cause of abortion.

Chemicals, drugs, and toxins

Toxic agents may also cause abortions or early embryonic deaths. Cattle are susceptible to fertilizer nitrites and nitrates orthe nitrates found in plants under certain conditions (e.g. drought-stress). If a cow is exposed to sufficiently high levels of nitrates/ nitrites (-.55 % or greater nitrate in forage), abortions may occur, especially in late gestation. Mycotoxins from the fungal agents are suspected to cause abortion. Bacterial endotoxin is responsible for sporadic abortions.Among the hormonal agents, estrogen, glucocorticoids, and prostaglandin are important. They cause abortion depending on the dose and the stage of gestation they are used. Stress may also cause abortion.

Important points to improve fertility

When reproductive efficiency reduces there is need to take help of veterinarian, artificial insemination (AI) technician, feed company representative and other resource people to troubleshoot the causes an efficiency, it is primarily essential to determine solutions to the problem.

For the maintenance of high breeding efficiency, it is primarily essential to maintain good level of general health,and thus prevent the onset of debilitating diseases. Animals do resist adverse influences which might interfere with the genital functions, but the resistance is largely depend on the inherent power of constitutional fitness to a great extent. Balanced feeding including greens and mineral mixture is essential for normal reproductive performance. Sufficient greens, mineral supplementation and correction of negative energy balance improve fertility. Additional feed allowance and mineral supplementation during last trimester of pregnancy and peak of lactation i.e. steaming up is advised to get a healthy calf and consistent yield. In selenium-deficient areas provide supplemental selenium as dry feed or injection to all dry cows. Provide about 160,000 units of vitamin-A (One milligram of carotene is equivalent to 400 units of vitamin-A from all sources (natural and supplemental).

Tips to avoid infertility (Senthilkumar,2014) under integrated farming system:

- Breeding should be done during the estrus period.
- Animals that do not show estrus or do not come to cycle should be checked and treated.
- Deworming once in 6 months should be done for worm infestations to maintain the health status of the animals. A small investment in periodic deworming can bring greater gains in dairying.
- Cattle should be fed with a well balanced diet with energy, protein, minerals and vitamin supplements from the system. This helps in increased conception rate, healthy pregnancy, safe parturition, low incidence of infections and a healthy calf.
- Care of young female calves with good nutrition helps them to attain puberty in time with an optimum body weight of 230-250 kg, suitable for breeding and thereby better conception.
- Feeding adequate quantity of green fodder during pregnancy will avoid blindness in newborn calves and retention of placenta (after birth).
- In natural service, breeding history of the bull is very important to avoid congenital defects and infections.
- Infections of the uterus can be largely avoided by having cows served and calved under hygienic conditions.
- After 60-90 days of insemination, the animals should be checked for confirmed pregnancy by qualified veterinarians.
- When conception occurs, the female enters a period of anestrous (not exhibiting regular estrus cycles) during pregnancy. The gestation (pregnancy) period for cow is about 285 days and for buffaloes, 300days.

- Unwarranted stress and transportation should be avoided during the last stages of pregnancy.
- The pregnant animal should be housed away from the general herd for better feeding management and parturition care.
- Pregnant animals should be drained of their milk two months before delivery and given adequate nutrition and exercise. This helps in improving the health of the mother, delivery of a healthy calf with average birth weight, low incidence of diseases and early return of sexual cycle.
- Breeding can be started within four months or 120 days after delivery to achieve the goal of one calf per year for economic and profitable dairy farming, according to them.

Summary

Among the common forms of functional infertility in cows are faulty oestrus manifestation including silent heat, inactive ovaries with anoestrus, cystic ovaries, abnormal oestrous cycle periodicity, and repeat breeding due to delay or failure of ovulation and fertilization or early embryonic death. These forms of infertility tend to affect individual animals, but they are becoming more important as attention is paid to the environmental and health constraints.

The best way to control many of these diseases is to prevent contact between herds. If this is not practicable, herd owners should buy only virgin heifers as replacement stock. All newly introduced stock should be quarantined for 3 to 4 weeks before joining the herd.

Farmers should not buy bulls that have been used for breeding in other herds unless they are proved to be completely free of important diseases. Semen for artificial insemination should be obtained from reputable centres.

All replacement heifers should be vaccinated with Brucella Strain 19 vaccine when 3 to 6 months old. Older animals should be given Strain 45/20. Once the incidence of brucellosis has been reduced, the herd should be regularly tested and infected animals culled.

The modified live virus vaccine for infectious bovine rhinotracheitis confers good immunity, and all heifers should be vaccinated.

Cattle can be vaccinated against both campylobacteriosis and leptospirosis annually. The vaccines are, however, expensive and may not be readily available, and even a vaccinated bull can act as a passive vector) after serving an infected cow.

Infections of the uterus can be largely avoided by having cows served and calved under hygienic conditions. Cows should be allowed plenty of room during calving, and the site should be clean. Bedding should be changed after each calving. When conducting obstetrical manipulations, use only disinfected instruments and disinfect both the operator and animal before and after any manipulations. If a birth is difficult, or otherwise abnormal, intra-uterine application of a broad-spectrum antibiotic will help prevent infection.

All infertile animals should be examined to determine the exact cause of their infertility. If a cow aborts, the aborted material should, if possible, be sent to a diagnostic laboratory to ascertain the cause.

Reference

Arthur G H. 1964. *Wright's veterinary obstetrics*. 3rd edition. Bailleire Tindall and Cassell, London, UK. 549 pp.

Arthur G H. 1982. *Veterinary reproduction and obstetrics*. 5th edition. Bailleire Tindall, London, UK. 616 pp.

Senthilkumar, T, 2014, CDAK Gateway, www.indig.in

Boyd H. 1977. Anoestrus in cattle. *Veterinary Record* **100**: 150-153.

Chauhan F S. Mgongo F O K and Kessey B M. 1984. Recent advances in normal therapy in bovine reproductive disorders: A review. *Veterinary Bulletin* **54**: 991-1009.

Kodagali S B. 1974. Report on study of infertility in cattle 1965-69. Veterinary Bulletin 45: 1285 (Animal Breeding Abstracts **43**: 3369).

Lagerlof N. 1963. Hereditary factors in infertility in cattle. In: *Infertility in livestock*. Animal Health Branch Monograph No. 5. FAO (Food and Agriculture Organization of the United Nations), Rome, Italy. pp. 63-77.

Lamming G E. 1977. Annual Report of the Milk Marketing Board of England and Wales, Thames Ditton, Surrey, UK.

Little D A. 1976. Assessment of several pasture species, particularly tropical legumes, for oestrogenic activity. *Australian Journal of Agricultural Research* **27**: 681-686.

Roberts S J. 1971. *Veterinary obstetrics and genital diseases*. 2nd edition. Published by the author. Distributed by Edwards Brothers, Ann Arbor, Michigan, USA. 776 pp.

Singh C S P. Singh S K and Singh B. 1981. Studies on the incidence of infertility in cows. *Indian Veterinary Journal* **58**: 909-912 (Veterinary Bulletin **51**: 7266).

Vandeplassche M. 1982. *Reproductive efficiency in castle: A guideline for projects in developing countries.* FAO Animal Production and Health Paper No. 25. FAO (Food and Agriculture Organization of the United Nations), Rome, Italy. 118 pp.

□□□

Chapter 11

Integrated Goat Production Strategies for Farming Systems

S.K. MAHANTA
ICAR - Indian Grassland and Fodder Research Institute, Jhansi-284003 (UP)

Introduction

Goat known as 'poor man's cow', has recently been recognized as the 'Future Animal' for rural prosperity under the changing agrogeoclimatic conditions and depleting resources. There is over 1005 million heads of goat in the world, out of which India possesses around 162 million goats (FAOSTAT, 2013) of 23 defined and nondescript breeds that are adapted efficiently in different agroclimatic conditions all over the country. Goat husbandry in India is essentially an endeavor of millions of small holders who rear animals on crop residues/grain by-products available from crop farming and using common property resources. The small holders produce milk, meat, fiber, skin etc for the community with virtually no capital and resource. Goat sector in India contributes around 8 per cent of the Gross Domestic Product (GDP) from livestock sector. In addition, the goat sector generates about 4% rural employment and around 20 million small and marginal farmers' and landless laborers' families depend on goats for their livelihood partially or completely. A substantial growth has been recorded in production of goat meat and milk during the last decade. The goat meat production has doubled (9.3% to 18.3%) and goat milk production has shown a growth rate of 31.53% during the last decade. The country stands first in goat milk production and is the second largest meat producer in the world, sharing 26.31% goat milk and 10.41% goat meat production (CIRG, 2011). Besides meat and milk, goats also produce good quality skin, valuable pashmina fiber and manure.

It is true that both demand and production of goat meat in India have shown steady increase during the last decade but country will need to double the number of goats to meet the projected requirement of goat meat for growing human population in the coming decades. However, the increase in livestock number would definitely put more stress over limited natural resources. Particularly management of feed and fodder resources is crucial for the future development of goat production in the country. There is very little scope for increasing the area under fodder production, keeping in view the priority for food grains, pulses and oil seeds. Hence, concerted efforts are required to develop production systems for enhancing per animal productivity per unit of land. Innovative strategies need to be formulated to ameliorate impact of climate changes on productivity of soils as well as animals. In this context integrated farming system is very important.

Integrated farming system (IFS) or integrated production system is an integrated approach to farming as compared to existing monoculture approaches. It refers to a production system that integrates livestock and crop production and may sometimes be known as Integrated Bio-system. At present, the farmers concentrate mainly on crop production which is subjected to a high degree of uncertainty in income and employment to the farmers. Integration of various agricultural and allied enterprises *viz.*, cropping, animal rearing, fishery, forestry etc. have great potentialities to boost up the farm income. These enterprises not only supplement the income of the farmers but also help in increasing the family labour employment. The IFS approach introduces a change in the farming techniques for maximum production in the cropping pattern and takes care of optimal utilization of resources. The farm wastes are also better recycled for productive purposes in the integrated system. A judicious mix of enterprises like dairy, goatry, fishery, sericulture etc. suited to the given agro-climatic conditions and socio-economic status of the farmers is expected to bring prosperity in the farming. Indeed, crops, animals and trees are the major components of any IFS. Crop may have subsystem like monocrop, mixed/intercrop, multi-tier crops of cereals, legumes (pulses), oilseeds, forage etc. Similarly livestock components include milch cow, goat, sheep, poultry, bees etc, while tree components may include timber, fuel, fodder and fruit trees. These components are decided or considered based on certain factors. Soil types, rainfall and its distribution and length of growing season are the major factors that decide the selection of suitable annual crops, trees and animal components. However, the needs and resource base of the farmers also controls the selection of IFS components in any farm. When adopted the following benefits are accrued from IFS (Anon, 2013).

- Higher crop/food production to meet the demand of famers
- Increased farm income through proper residue recycling and allied components
- Sustainable soil fertility and productivity through organic waste recycling
- Integration of allied activities results in the availability of nutritious food enriched with protein, carbohydrate, fat, minerals and vitamins
- Integrated farming helps in environmental protection through effective recycling of waste from animal activities like dairy, goatry, piggery, poultry and pigeon rearing
- Reduced production cost of components through input recycling from the byproducts of allied enterprises
- Regular stable income through the products like egg, milk, mushroom, vegetables, honey and silkworm cocoons from the linked activities in integrated farming
- Inclusion of biogas & agro forestry in integrated farming system solves the energy crisis
- Cultivation of fodder crops as intercropping and as border cropping results in the availability of adequate nutritious fodder for animal components
- Firewood and construction wood requirements may also be met from the agroforestry system without affecting the natural forest
- Avoidance of soil loss through erosion by agro-forestry and proper cultivation of each part of land by integrated farming
- Generation of regular employment for the farm family members of small and marginal farmers.

Prevalent goat production systems

In fact, a large number of production systems for small ruminants like sheep and goats are in practice and vary from country to country and region to region within a country; these can essentially be grouped under three systems viz. extensive, semi-intensive and intensive (Karim *et al.*, 2008; Mahanta *et al.*, 2009) The ultimate aim of adopting a particular production system is to bring improvement in productive and reproductive performances.

Extensive system

Small ruminants like goats are generally reared under extensive system of management (Fig 1). This system includes migratory, transhumance free range, pasture and range grazing management of goats. It is most common system throughout the Asian region because the small sizes of these animals have distinct advantages over other livestock species to contribute significantly to the sustainability of small farm situations that predominate in this region. They are usually owned by small farmers and landless labourers and grazed together and tend to be herded over long distances in search of feed and water. A low level of unpaid family labour represents the main input. The system is principally one of low resource use and a low level productivity emerges from poor nutritional availability. There is marked fluctuation in feed availability and its nutritive value among different regions and among years and seasons. Animals require feed all the year around. They produce all the time if it is always plentiful and of the right kind. If on the other hand , it is short or of the wrong kind only for a month or two there is less milk, less kids and less meat.

Fig. 1. Goats (along with sheep) under extensive system

Semi-intensive system

A kind of compromise between extensive and intensive is referred to as semi-intensive system of management. It is a combination of limited free range grazing and then feeding in stalls. Integration of goat rearing with arable cropping and tree cropping is also included wherein either the animals are tethered or cut and carry system of available fodder is employed. Children and women often take care of animals in this system. Animals belonging to several owners may be combined for grazing which is mostly done in morning and evening for 4-6 hours (Fig 2). The animals are supplemented with kitchen wastes, concentrate mixtures, crop residues, green and dry fodders and tree leaves etc, as per the availability. Thus animals utilize all available feed resources including natural grasses, shrubs, bushes, tree leaves, crop residues, stubbles, weeds cultivated fodders and concentrate etc. under this system. The level of nutrition is just optimum or low but surely better than that of extensive system. Production performance depends on the quality and quantity of grazing/browsing availability and supplementary feeding provided.

Fig. 2. Growing goats under semi-intensive system

Intensive System

The intensive system of animal management includes grazing on developed pastures and/or feeding completely in stalls (Fig 3) on cultivated fresh or conserved fodders, crop residues and concentrates. Although goats prefer to browse as compared to grazing, they are quite capable of making efficient use of sown pastures for meat and milk production similar to sheep (Table 1). Stoking rates of 16 to 60 animals per hectare are feasible depending on the type of grass, level of fertilization and presence and absence of legumes and fodder trees. This system requires high labor and capital investment and is suitable for only intensive meat production from sheep and goats. In addition to providing better milk, growth and carcass quality it removes pressure from the grazing lands. Under intensive feeding systems animals are fed based on the nutritional requirements of different physiological stages comprised of forages and concentrates.

Table 1. Performance of growing lambs and kids on sown pasture

Attributes	Lamb	Kid
Weight at birth (kg)	3.2	3.2
Weight at weaning (kg)	13.9	15.4
Weight at 6 months (kg)	20.6	18.0
Weight at 9 months (kg)	24.0	20.0
Weight at 12 months (kg)	27.0	24.0

Fig. 3. Goats under intensive system of rearing

Commercial goat production

The goat meat available in the Indian markets usually comes either from old and culled adults or from male kids slaughtered any time between 6 months to 1 year of age. The quantity and quality of this meat is poor due to poor market weights, lower dressing percentage and narrow bone: meat ratio as these kids are maintained on scrub vegetation like their dams and hence hardly attained body weights of 15 to 16 kg when they are normally marketed. The dressing percentage varies from 35 to 40 and bone: meat ratio from 1:3 to 1:4. The studies conducted in the country have revealed that a marked improvement both in the finishing weight and carcass quality through intensive feeding of the male kids. Based on these findings two packages of improved practices have been developed and recommended for adoption by the small ruminant farmers and the entrepreneurs for commercial goat production (Singh *et al.,* 2006).

Package for farmers

Cooperative Societies of small ruminant farmers may be formed in goats/sheep rearing areas. These societies in association with the village Panchayats should develop improved silvi-pastures on the available community grazing lands with financial assistance from the financial institutions and subsidies being provided by the central and state governments. The registered flocks may be allowed to graze on this pasture judiciously for 6 to 8 hrs daily. In addition to grazing, the pregnant animals during last 30 days of pregnancy and the lactating animals during first 60 days of lactation should be supplemented with 300g/h/d containing 18-20% CP and 65-70% TDN to ensure 2.5 to 3 kg birth weight and 14 to 15 kg weaning weights in male kids/lambs. To insure a weaning weight of 14 to 15 kg, these male animals are provided with ad lib suckling, creep rations and green/dry leguminous fodders during pre-weaning period and completely weaned at 60-90 days of age. These weaners may then be fed *ad lib* on complete feeds comprised of 50% concentrate and 50% roughage under stalls till they attain 25 to 30 kg finishing at around 6 to 7 months of age. Alternatively, the male animals may be allowed to graze on available pastures and supplemented with concentrate mixture @ 1.5 to 2% of the body weight till they reach the desired finishing weights. The finisher animals should then be sold by the farmers or their cooperative societies for slaughter.

Package for entrepreneurs

The male kids/lambs are procured from the farmers by the entrepreneurs at around 60-90 days of age and transferred from the villages to the meat production complexes established near the cities. This will help in reducing the grazing pressure on shrinking pasturelands, early rebreeding and easy management of the flocks. These complexes may be equipped with a feed compounding plant capable of incorporating higher production of cheaper low grade roughages and agro industrial by-products to manufacture economic complete feeds, feedlot animal houses for intensive feeding of the kids, modern slaughter houses, meat processing, product handling machineries. The kids/lambs procured from the villages may be intensively fed on complete 50:50 or 60:40 concentrate: roughage feeds under feedlot from 2 months to 6-7 month age when they attain finishing weight of 25-30 kg. These intensively fed kids may then be sold as live animals in the national or international markets or slaughtered for selling fresh meat or their meat may be processed and converted into different meat products for export.

Broiler goat rearing

Broiler goat production is also coming up in areas where green fodder is not available due to lack of lands either for grazing or fodder cultivation. It is one of the production technologies to improve the economy of rural farming community. Broiler goat rearing has been found to be highly remunerative compare to rearing other farm animals and it has been advocated as a better substitute of livelihood for the rural farmer. The kids produced from goats (whatever breed available in an area) can be used for broiler goat rearing (both male as well as female kids). The goat kids about 15 days to 1 month old i.e before starting to eat green leaves and are having higher birth weight and not used for further breeding are selected for broiler goat rearing. The selected kids are not be allowed to feed on green fodder/grazing in open spaces. Kids are reared intensively by providing concentrate feeds (goat feed) and also allowed for mother's milk twice or thrice in a day. Marketing is done at the attainment of 25-30 kg or at the age of 3-4 months whichever is earlier. Broiler goat meat is found to be soft with minimum goaty odour. A farm woman can manage about 10-20 kids at a time without any extra labour (Fig 4). It is highly profitable to the farmers who are already involved in goat rearing (Shanmughavel, 2014).

Fig. 4. Women farmers involved in managing broiler goats

Goat production strategies in IFS

It has now been established that goats can be reared in small yards just like poultry. Goats under stall fed conditions can ideally fit into the intensive integrated farming system (IIFS). These small animals are the most efficient converters of farm and crop residues into excellent organic manure. Several farmers have successfully run stall-fed goat farms, and they have found that such an integrated farming venture was more productive and profitable as well. Goats relish the stalks and residues of most of the nutritious cereals, and they do well particularly when mixed with green fodder such as grasses and nutritious tree leaves. Special goat-feeds can be formulated using farm-grown grains/millets and oilcakes. As the cost of the feed and also the labour gets distributed over other farming operations, the actual cost of raising the goats becomes minimal. The rich goat manure is ideal for fertilizing fishponds and all other crops. It is also a good base material for vermin-composting (Anon, 2002).

Indeed goat farming needs less capital when compared with dairying and the animals can be raised in small farms. The floor space requirement per adult animal is about one square metre. Stall-fed goat farming is an ideal occupation for the small, marginal and landless agricultural labourers. A properly fed and managed milch goat yields at least as much milk (on average two litres per day) as low yielding desi cows. She-goat (doe) delivers 2-4 kids at each parturition after a short gestation period of 150 days. A buck (male goat) usually provides service to 20-25 does in a single breeding season. The popular Indian breeds are Jamnapari, Surti, Tellicherry, Beetal, Malabari, Barbari, Sirohi, Gujarati etc. However, the milch-type animals are ideal for integrated farming system. A small shed with good cross ventilation is enough to keep a small herd. A deep-litter system with paddy husk and groundnut shell as bedding material is ideal for raising goats. The bedding material collects all the dung and urine efficiently and it is found to be enriched organic manure. An adult goat will add about a ton of rich manure to the farm every year.

Integrating goat farming with cropping systems like agro forestry models such as, agrisilviculture, hortipasture, boundary plantations, and silvipasture models with trees and grasses/legumes can be established in wastelands to supplement farm income. Being hardy animals, goats are fed on fodder trees like Neem (*Azadiracta indica*), Agathi (*Sesbania grandiflora*), Siris (*Albizia lebbek*), Subabul (Leucaena leucocephala), Babul (*Acacia nilotica*) etc and a wide variety of fodders and grasses/legumes such as Anjan grass (*Cenchrus ciliaris*), Guinea grass (*Panicum maximum*), Stylo (*Stylosanthus hamata*), Fodder cowpea (*Vigna uinguiculata*) etc.

Again there is a lot of variation in prevalent farming systems in the country depending upon the type of land available and climate. The challenge of an integrated farming system (IFS) is to upgrade technological and social disciplines on a continuous basis and integrate these disciplines to suit the region and the farm families in a manner that will ensure increased production with stability, ecological sustainability and equitability. During the recent past, following strategies or models with goats as an allied component have been tried for improvement in terms of productivity, income increase and employment generation, and nutrient recycling under different farming systems (Jayanthi *et al.*, 2009).

Lowland farming system

Crop + Goat + Fish integration was tried by allocating 0.32 ha, 0.04 ha and 0.04 ha for crop, fodder and fishpond, respectively (Fig 5). Inclusion of one unit of goat (20 females + 1 male) enhanced the net income by Rs 36,000 per year and productivity of Rs 98 per day. The goat unit provided around 11 tonnes of valuable manure apart from supplementing the feed requirement of fish (400 fingerlings). It also helped in generating employment opportunity to the family labour to the tune of 182.5 workdays per year, distributed uniformly over the year. About 10 kg of rice straw or crop wastes was also effectively utilized as substrate for producing 5 kg of edible mushroom per day. Cost of production per kg of mushroom in IFS was lower by 50% as compared with a commercial venture. Thus, at an annual production of 1,460 kg of edible mushroom, IFS offers an additional net income of Rs 11,680. It also helped in supplementing the crop activity with high-value organic manure (mushroom spent compost) of up to 3 tonnes. Employment opportunity was enhanced at the rate of 1/2 workday per day, or 182 workdays per year, to the family labour. Recycling no-cost/low-cost material of one enterprise as input to another enterprise helped to reduce the unit cost of production of output.

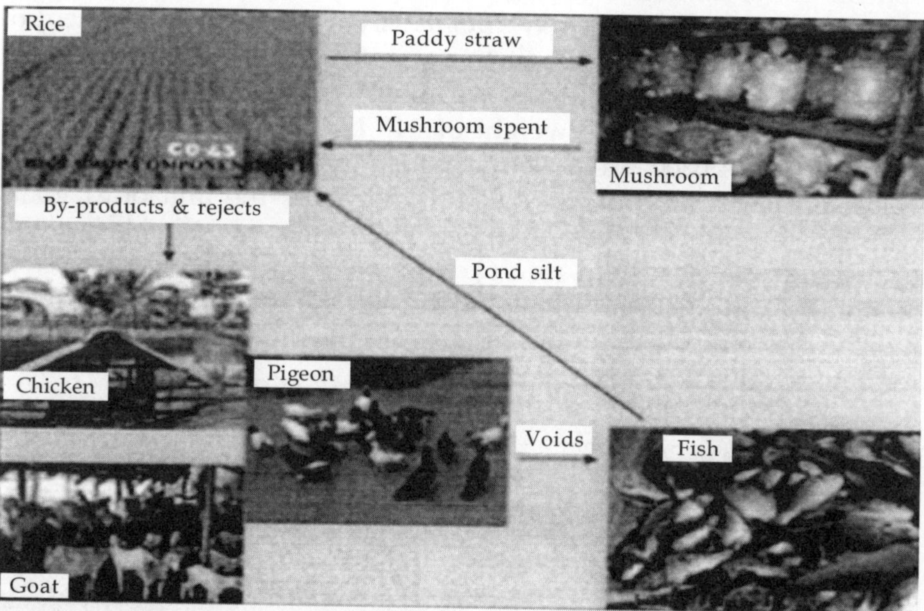

Fig. 5. Component integration and resource recycling in low land farming

Dry land farming system

The farmer often experiences complete failure of crops in rainfed areas because of nonavailability of moisture at the critical stage. This is why conventional rainfed agriculture is said to be an out and out gamble. Integrating different enterprises and utilizing the biomass built up have been identified as dependable paths to ensure regular income from rainfed farming. Rainfed soils are not only thirsty but also hungry. The linkage of other enterprises such as goat, buffalo and rabbit etc provides good amount of organic nutrients to the soil to enhance the yield substantially.

An IFS involving grain crop, fodder crop, fodder trees, perennial grasses and goat rearing in an area of 1 ha of rainfed land was investigated. It indicated that through short-duration field crops and perennial crops, the feed requirement for one productive unit of Tellichery goats, consisting of 20 does and one buck, can be satisfied for all the 365 days. After 5 years, the perennial fodder trees provided sufficient quantity of loppings to supplement along with millets, legumes and perennial grass linked in the system, which met the fodder requirement of 21adult goats. Indeed, Tellicherry goat is a good breed to build up body weight for every unit of feed secured through different sources under rainfed condition. It is a dual purpose animal and gives 80 to 100 ml of milk every day after satisfying the full requirement of dependent kids. Twenty productive females gave around 45 kids per annum. Each kid at the time of weaning weighed around 12 kg and thus gave a saleable live weight of 540 kg per year, producing an income of Rs 43,200. After deducting all the expenses involved, the net income was around Rs 35,000. In addition, the unit of 21 goats with kids at different ages produced 11.2 tonnes of valuable manure under deep litter system (Fig 6). This not only provided primary, secondary and micro nutrients for the crops but also absorbed more moisture, retained it in the soil and released it to the crop appropriately for better yields.

Fig. 6. Component integration and resource recycling in dry land farming

In a rainfed farming situation, integration of cropping, livestock (goat) and orchard crops was also found profitable (Fig 7). In an area of four acres, including livestock (1-5 Tellicherry goats) along with cropping 3 acres and orchard crops (one acre of ber or guava or any marketable drought tolerant fruit tree), the additional income obtained over cropping alone was on an average, Rs. 3300. However, depending upon the extent of holding, the size of the components may be altered. Such integration is not only for higher income in favourable years but also to safeguard the farmers from getting disaster income in poor rainfall years.

Fig. 7. Goats on deep litter rearing under IFS

Irrigated upland farming system

An IFS involving crops + livestock + vermicompost and biocompost was studied in western zone of Tamil Nadu (Fig 8). The crop activity consists of field crop (60 per cent), vegetable crop (10 per cent) and fodder crops (20 per cent) that are suitable for irrigated upland conditions. The cropping system in field crop was sunflower-maize + cowpea-green gram (0.60 acre); in vegetable crop, okra-chillies (0.10 acre); and in fodder crop, pearl millet Napier hybrid grass (CO3) + desmanthus (0.20 acre). Crop activities in farming system were taken in 0.90 acre. For comparison, one acre of land of the same farmer was taken where traditional practices were followed. The traditional practices were: sorghum/tomato-brinjal (aubergine)/ floriculture + 2 milch cows + vermicompost in Location I; maize-sorghum/tomato-Chillies + 3 milch cows in Location II; and cotton-fodder maize + 2 milch cows in Location III. Livestock components including vermicompost and biocompost were taken in 0.10 acre. Two cross breed milch cows + one calf were taken for the study in each location. Tellicherry goats comprising 10 does and 1 buck were maintained under deep litter system in each location with slated floor built above the ground at a height of 2-3 feet, designed for the convenience of manure collection without wastage. Twenty Guinea fowls (15 female + 5 male) were also reared in each location. The fowls were kept in a cage built above the vermicompost pit in all the three locations for collecting the manure and to maintain hygiene standards. Compost pits were made in each location for composting crop residues and farm wastes. The manure obtained from milch cows and field and fodder crop residues were used for making biocompost. Vermicompost was prepared utilizing goat and guinea fowl droppings and vegetable waste.

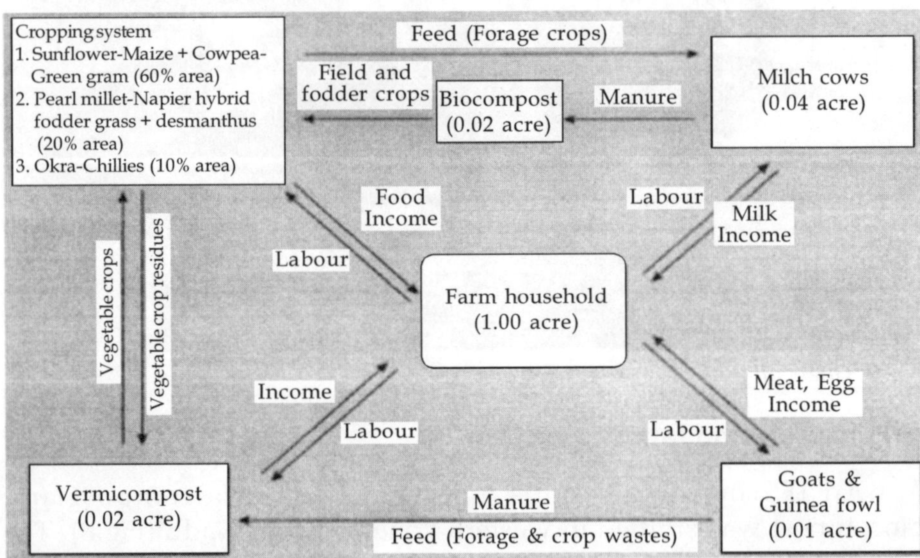

Fig. 8. Component integration and resource recycling in irrigated upland farming

Integration of cropping with components like milch cows, Guinea fowl and vermicompost resulted in higher productivity. The mean maize grain equivalent yield was about 9,417 kg/acre/year under traditional cropping system whereas under IFS, the maize grain equivalent yield was about 22,754 kg/acre/year. Inclusion of high-yielding varieties and allied components helped in increasing the productivity and as a result the maize grain equivalent yield increased.

Compared with traditional cropping system, IFS brought increased revenue, which was probably be due to resource recycling. The net return from inclusion of allied enterprises in IFS was about Rs 60,141, and the increase in income over traditional cropping system was about 43.6 %. Resource recycling by way of utilization of fodder cultivated in the field, which accounts for a major part of the cost of maintenance of cows and goats reduced the cost of production. The purchase of fertilizers for the crops was also reduced by way of recycling the manures from animal components, vermicompost and biocompost as organic fertilizers. The resource and residue recycling reduced the cost of production of an unit of economic produce viz., meat, milk and egg from goat, milch cows and Guinea fowl, than when produced with total dependence on external inputs (Jayanthi *et al.*, 2007).

IFS treatments generated more workdays of employment compared with the traditional system involving cropping and dairy. Cropping in traditional system generates 25 workdays per acre per year, while the various cropping systems under IFS generated 49 workdays of employment. A maximum of 183 workdays per acre per year was generated from animal components in IFS, whereas in traditional cropping system it is only about 80 workdays. Additionally, 3 workdays per acre per year was generated from vermicomposting and biocomposting. Employment generation in cropping is limited to the key operations of sowing, intercultural operations and harvest, and labour is not required during the rest of the year. Contrary to this, employment generation in a multi-enterprise farming system is spread uniformly throughout the year.

Another Integrated farming system (IFS) with canal irrigation facilities, comprising the components like cropping, vegetables, fishery, poultry and goat rearing was evaluated in Karnataka in terms of productivity, profitability, employment generation and water requirement of IFS over conventional rice-rice system (Table 2). Integrated farming system approach recorded 26.3 and 32.3% higher productivity and profitability, respectively over conventional rice-rice system. Among the components evaluated, the highest net returns was obtained from crop (63.8 %), followed by goat (30.9 %), fish (4.0 %) and poultry (1.3 %), respectively. Employment generation and water requirement was 275 man days/ha/year and 1247 mm, respectively under the integrated farming system (Channavasvanna *et al.*, 2009).

Table 2. Productivity (rice equivalent yield) and profitability of different components under IFS

Treatment	Area (ha)	Productivity (kg/ha/year)	Cost of cultivation (Rs)	Net Returns (Rs)	B:C ratio
Integrated farming system					
Rice-rice system	0.33	2175	8683	7387	1.84
Hybrid maize-sunflower	0.20	908	3697	3540	1.96
Vegetables	0.20	2136	4712	3673	2.00
Fodder + goat	0.21	1339	6289	7069	2.75
Fish	0.06	203	515	926	2.23
Poultry	0.005	327	2145	300	1.13
Total	1.00	7088 (19.42)*	18225	22887	1.97
Conventional cropping system					
Rice-rice	1.00	5611 (15.37)*	25503	17293	1.64

*Productivity in kg/ha/day

Conclusion

IFS approach has started gaining momentum with farmers of all types, and is expected to reach its peak of acceptance in near future. IFS will help in enhancing the productivity to satisfy the ever-increasing population of the country, and create confidence among the farmers through higher profitability. But the concept of sustainability is coming up as an important element in the development of integrated systems (Fig 5). The MEA (2005) defined it as a characteristic or state whereby the needs of the present and local population can be met without compromising the ability of future generation or population in other locations to meet their needs. The concept of agricultural sustainability initially focused only environmental aspects, but it has now been expanded to include broader socio-economic and political elements: Under *ecological consideration*, it focuses on environmental protection to enhance ecosystem resources and preservation of biodiversity, while in *socio-economic* aspects it concerns the value and management of the resources, their enhancement, socially acceptable technological improvements, farmers organizations and cooperatives, and improved livelihoods of poor farmers (Devendra, 2011). In this context, animals (either dairy or small ruminants like goats) as an allied component of IFS have substantial bearing on overall benefits accrued from a farming system.

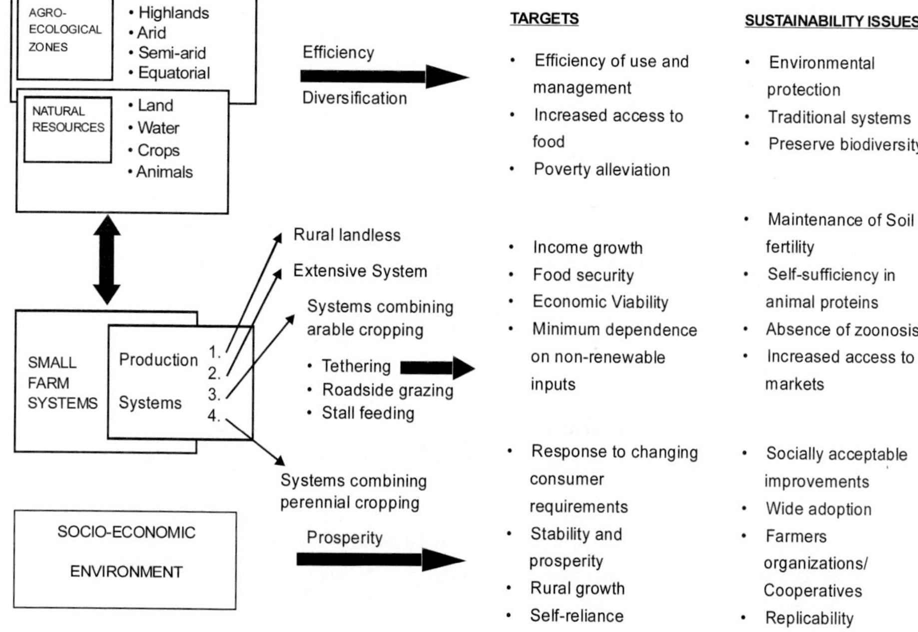

Fig. 9. Sustainability of a goat production system

References

Anonymous. 2002. Stall-fed goats in intensive integrated farming. http://www.thehindu.com/seta/2002/10/17/stories/2002101700140300.htm

Anonymous. 2013. Pallava integrated farm. http://indiaeng.com/Kaveripakkam/index-kaveripakkam.htm.

Channabasavanna, A.S., Biradar, D.P., Prabudev, K.N. and Hegde, M. 2009. Development of profitable integrated farming system model for small and medium farmers of Tungabhadra project area of Karnataka. *Karnataka J. Agric. Sci.*, **22**(1): 25-27.

CIRG. 2011. Vision 2030 document. Central Institute for Research on Goats, Makhdoom, Mathura.

Devendra, C. 2011. Integrated tree crops-ruminants systems in south East Asia: Advances in productivity enhancement and environmental sustainability. *Asian-Aust. J. Anim. Sci.* **24**(5): 587-602.

FAOSTAT. 2013. Livestock population data.faostat.fao.org/573.

Jayanthi, C., Vennila, C., Nalini, K. and Chandrasekaran, B. 2009. Sustainable integrated management of crop with allied enterprises: Ensuring livelihood security of small and marginal farmers. Technology Monitor, Jan-Feb 2009, pp. 21-27.

Jayanthi, C., Vennila, C., Nalini, K. and Vivek, G. 2007. Physical indicators on farm farming system for irrigated uplands of Western zone of Tamil Nadu. *Journal of Farming Systems Research and Development,* **13**(1): 17-25.

Karim, S.A., Tripathi, M.K., Chaturvedi, O.H. and Shinde, A.K. 2008. Small Ruminant Production in India: Strategies for enhancing. Satish Serial Publishing House, Delhi.

Mahanta, S.K., Pailan, G.H., Kumar Anil, Misra, A.K. and Verma, N.C. 2009. Final Report of the Project entitled 'Small ruminant production under grazing system'. IGFRI, Jhansi.

Millennium Ecosystem Assessment. 2005. Current State and trends (R. Hassan, R. Scholes and N. Ash, Eds.). **1**: 133-134.

Shanmughavel, S. 2014. Broiler goat rearing a profitable enterprise. http://www.kvkcalicut.gov.in/index.php/successstory/57-broilergoat.

Singh, N.P., Kumar, S., Goel, A.K. and Vaid, R.K. 2006. Proceedings of National Workshop-cum-Seminar on 'Commercial goat & sheep farming and marketing: Farmer-industry-researcher interface' at CIRG, Makhdoom during March 4-5, 2006.

□□□

Chapter 12

Poultry Production and Management under Integrated Farming System

KALYAN SUNDAR DAS[1], S BINDU MADHURI[2] AND N DAS[3]

[1] Zonal Project Directorate zone-II, Kolkata

[2] College of Veterinary Science, Rajendra Nagar, Hyderabad

[3] ICAR - Indian Grassland and Fodder Research Institute, Gwalior Road, Jhansi-284003

Summary

During last two decades, Indian poultry industry gets momentum of substantial growth due to the technological breakthroughs in breeding, feeding and health, and sizeable investments particularly from the private sector. This enterprise provides robust support to the national economy. Even then, there is demand and supply gap in the country probably due to huge human population and shrinking land resources. To bridge this gap, poultry production through integrated farming approach may prove a successful venture in the near future. This paper presents insights into current status of Indian poultry industry, conventional practices of poultry farming, scope of integration, approaches and way of integration of poultry farming with other agricultural and allied enterprises.

Key words: Integrated farming, Integration approaches, Poultry production.

Poultry is one of the fastest growing enterprises among the agricultural sector in India. In the year 2011-12, our country has emerged on the world poultry map as the 3rd largest egg (66.45 billions) and 5th largest poultry meat (3.6 million tons) producing country. This increase in poultry production has enhanced the per capita availability of 55 eggs and 2.4 kg poultry meat per annum (Economic Survey, 2012-13). The poultry industry contributes about Rs. 26,000 crore to the national income. The annual growth of total chicken population was recorded as 7.3% in the last decade. Organized sector accounts for nearly 70% of the total poultry output in the country. The current strength of layers and broilers in India is estimated to be 230 million and 2300 million, respectively. Poultry processing has been increased up-to 20% of total broiler production. Employment and rural income generation are also important contributions from this sector.

The available data indicated that more than 5 million people are directly or indirectly engaged in poultry sector, apart from numerous small poultry keepers in rural and tribal areas of the country. The domestic broiler meat demand is expected to grow at around *15-18%,* while table egg demand is expected to grow at 5-7% in medium to long term. The domestic poultry market size is estimated at more than Rs. 47,000 crore. Since 1970 global production, consumption and trade of poultry has grown faster than any other meat. Growth in India's poultry sector is mainly due to structural change in poultry sector like integrated farming. Other factors which enhanced growth in poultry sector are increase in income level of people, urbanization, easy availability of feed etc. The growth trend is likely to continue for the present decade due to steady market demand resulting from favorable socio economic factors like healthy GDP growth, rising purchasing power, changing food habits, and increasing urbanization. International trade in poultry meat and table eggs accounts for about 10% and 2.5% of their global annual output, respectively. India produces close to 27 million chickens a week, of which 95% is traded alive. The poultry sector in India is transforming from a backyard activity to large scale integrated poultry farming with wide spread adoption of modern technology (Poultry Fest, 2014). Due to investment friendly ambience, booming and emerging marketing opportunities, phenomenal economic growth rate, adoption of modern techniques in production etc., the Indian poultry production is considered to be the cheapest in the globe.

Further, if we look into the regional production pattern in the country, broiler industry is well dominated by the southern states comprising of nearly 60-70% total output. The layer industry once again is represented more in southern states especially, Andhra Pradesh, Tamil Nadu and Maharashtra producing nearly 70% of the country's egg production. The Department of Food Processing of Punjab reported that India's 75% of egg produce is consumed by the population of 25% living in urban and semi-urban areas. The concept of contract farming was introduced by the pioneer poultry group in the year 1980 and the concept of integration was introduced by Suguna Farms in 1990.

In spite of tremendous growth of this sector, the per capita consumption of these products is poor i.e. 37 eggs and 1 kg of poultry meat per capita per annum. Again, there is disparity in per capita consumption between rural and urban areas and also across the region. Per capita consumption of eggs is only 7.7 per annum in rural areas as compared to 17.8 per annum in urban areas. In seven states, per capita consumption is less than 3.5 per annum. Similarly, per capita consumption of poultry meat is 0.24 kg in rural areas and 1.08 kg in urban areas (Mehta *et al.*, 2003). Among many other reasons, it may be pertinent that our production system is not congenial enough for different categories of production levels i.e. for small and marginal farmers and large scale commercial production. Small units are now-a-days facing problems of high feed and transport costs, expensive vaccines, insufficient veterinary care services, non-availability of credit, improper markets for products, lack of proper government policy etc. As a result, the number of small poultry units (e.g. 200-500 chicks per cycle) is decreasing day by day and large units (e.g. 5000-50000 birds per week cycle) are common these days. Even, poultry farmers are shifting their farms from layer towards broiler production. Another reason to shift in broiler farming may be the quick economic return which can be realized only in six weeks.

Poultry management under traditional practices

Generally, poultry birds can be managed under extensive, intensive and semi-intensive systems. In extensive system, birds are exposed to pasture and sunshine. They are grazed on pasture land and are provided with shelter for sleeping at night and for protection against inclement weather. Either range system or fold unit or folding system of housing can be practiced. In intensive system, birds are reared under confinement with high quality balanced ration. It may be expensive at initial stage but in long run cost per unit is comparatively less. Birds can be housed either deep litter system or wire and slatted floor or straw yard system or battery cage system. Semi-intensive system has become obsolete now. It is characterized by the fixed building and attached runs or fenced pastures. Out of these systems, in our country, intensive management system is widely practiced for commercial production followed by backyard poultry farming for small scale production.

Integration and its approaches

Now-a-days, Indian agriculture is facing the challenge to bridge the gap of demand and supply of food for huge and growing human population with shrinking land resources. Because of declining per capita availability of land in India, there is hardly any scope for horizontal expansion of land for food production. The vertical expansion, by integrating appropriate farming components that require lesser space and time to ensure round the year income to the farmer, is the only solution. Therefore, the optimum use of available resources is vital for the sustainability of any production system. Again, as scientist our challenge should be to establish area specific stable production system for different regions considering available inputs from livestock, crop, fishery etc. for ecological sustainability, equitability and food security.

The International Organization of Biological Control (IOBC) describes '*Integrated Farming*' as a farming system where high quality food, feed, fibre and renewable energy are produced by using resources such as soil, water, air and nature as well as regulating factors to farm sustainably and with as little polluting inputs as possible. Rana and Chopra (2013) described the advantages/benefits of integrated farming system (IFS) which include pooling and sharing of resources/inputs, increase yield per unit area per unit time, efficient use of family labour, conservation, preservation and utilization of farm biomass including non-conventional feed/fodder resources, availability of diversified products, effective use of manure/animal waste, regulation of soil fertility and health, adoption of new technologies, minimization of environmental pollution, income and employment generation for many people and increase economic resources and others.

However, integration of livestock farming system is complex and generally based on the traditional socio-economic considerations and intelligent utilization of resources. An understanding of production factors viz. livestock, capital, feed, land and labour as well as processors viz. description, diagnosis, technology design, testing and extension that effect animal production is pre-requisite for livestock integration.

Integration of poultry farming in India

Backyard poultry farming was predominant until late 1980's. During the early part of 1990's, the concept of integration attributed a phenomenal role in the growth of poultry industry. Integration is nothing but contract farming. This concept was first introduced in the country by Suguna in Tamilnadu with the help of few willing farmers at that time. Earlier, the farmers had no control on different components of poultry industry like chick seller, feed seller, hatcher, medicine, transport, marketing etc. But, due to introduction of integration, all these above components come under one organization sharing the benefits of production among the farmers, consumers and the organization. It also eliminated the share of middlemen. The productivity has also been increased by adopting new technologies/package of practices particularly in respect to improvement in getting uniformity of flock, feed conversion ratio, use of balanced ration, housing, health and disease management, marketing etc. Even though the prices of basic raw materials have been increased, the production cost has come down due to this. By-products of poultry also fetches money for the farmers. Due to integration the cost of poultry production can be reduced leading to more increase in per capita consumption. Therefore, this integration acts as boon to the poultry farmers. Integration has paved way for employment generation. It prevented migration of rural youth from rural areas to cities and towns and helps to earn foreign exchange through exports. In India, there is very good scope of integration for poultry production. But, the level of integration is not uniform throughout our country. The National Coccidia Alert Network (NCAN) depicted that integration is high in southern region of India, slight in western region and negligible in northern and eastern region. Therefore, a suitable combination of integration should be found out for sustainable integrated poultry farming for a particular region.

Way of integration

The poultry farming can be integrated with fish production, other livestock farming, crop production, horticulture and others for achieving maximum economic benefits when it is practiced in the following ways.

A) Poultry-cum-fish farming

Intensive production of broiler meat and egg is now common in many parts of the world. In integrated poultry-cum-fish farming system the birds are typically fed complete diets and their excreta is used fresh or as dried poultry waste. The waste recycling is the key feature of the system, and integration of fish culture with poultry raising is one of the best ways of poultry waste management. Scientific reason behind the recycling of poultry waste is that about 80 per cent (dry weight basis) of feed stuff is digested and utilized by body of poultry birds, 20 per cent are excreted from the body as such through excreta. Again, due to their peaking habit while taking feed, about 10% feed is gone as waste. This wasted feed can be utilized directly as feed for fish in the pond. Poultry manure is a complete fertilizer, with the characteristics of both organic as well as inorganic fertilizers (Banerjee *et al.*, 1979) and fresh chicken manure contains 1.6% nitrogen, 1.5% phosphorous and 0.9% potassium (Woynarovich, 1979). Available report showed that one kg of fish can be produced by using about 17 kg of chicken manure (Fang *et al.*, 1986). The total protein content of dry chicken excrement can be as high as 30 per cent. Usually, good chicken feed stuffs have a protein content over 18 per cent. Table 1 represents the composition of poultry manure from different poultry-raising methods. Amanullah *et al.* (2007) reported the nutrient content of different types of poultry manure (Table 2).

Table 1. Composition (%) of poultry manure from different poultry-raising methods.

Constituent	Raising above the pond/in cage	Ground raising	
		Sawdust bedding	Dry grass bedding
Moisture	11.4	12.3	15.5
Crude protein	26.7	21.9	22.3
Crude fat	1.7	1.7	2.3
Non nitrogenous extracts	30.6	30.0	27.1
Crude cellulose	13.0	17.2	18.7
Minerals (Ca, P, etc.)	16.5	16.9	14.1

Source: The report of Food and Agricultural Organization on 'Integrated fish farming in Bangladesh'

Table 2. Nutrient content of different types of poultry manure

Particulars	Deep litter	Broiler house	Cage manure
C/N ratio	9.5-11.5	9.4-11.2	5.8-7.6
Total N (%)	1.70-2.20	2.40-3.60	3.63-5.30
Total P2O5 (%)	1.41-1.81	1.56-2.80	1.54-2.90
Total K2O (%)	0.93-1.30	1.40-2.31	2.5-2.90
Fe (ppm)	930-1380	970-1370	970-1450
Zn (ppm)	90-308	160-315	290-460
Cu (ppm)	24-42	27-47	80-172
Mn (ppm)	210-380	190-350	370-590
Ca (%)	0.90-1.10	0.86-1.11	0.80-1.02
Mg (%)	0.45-0.68	0.42-0.65	0.40-0.56

There are some additional benefits when poultry farming is integrated with fish farming on/near the pond/pond dykes which are highlighted as under :

- Minimum wastage of manure
- No transport cost
- Minimum labour cost
- High nutritive value of fresh manure
- No supplementary feed for fish
- Hygienic poultry production
- No extra space for poultry housing
- High production per unit area of land
- Maximum utilization of resources
- Maximum return

The integration of poultry with fish farming has now become an economically profitable venture under both rural farmers and commercial entrepreneurs. The cost of inputs can be reduced as costly feeds and fertilizers are not required to apply for such systems. Although organic fertilizers may be consumed directly or served as manure-derived detritus after heterotrophic microbial activity, the role of manure or manure-derived detritus as a source of food for fish is not universally agreed upon (Hoq *et al.*, 1999). Several studies showed that organic supplements contributed to fish yield by supplying inorganic P, N and C for algal growth, and by stimulating detritus production and heterotrophic utilization (McNabb *et al.*, 1990). The report of Agropedia (2012) envisaged that the production of poultry dropping in India is estimated to be 1300 thousand tons, which is about 390 metric tons of protein. Utilization of this huge resource as manure in aquaculture will definitely afford better conversion than agriculture. The application of poultry manure in the pond provides a nutrient base for dense bloom of phytoplankton, particularly nano plankton which helps in intense zooplankton development. The zooplankton has an additional food source in the form of bacteria which thrive on the organic fraction of the added poultry excreta. Thus, indicates the need for stocking phyto planktophagous and zoo planktophagous fishes in the pond. In addition to phytoplankton and zooplankton, there is a high production of detritus at the pond bottom, which provides the substrate for colonization of micro-organisms and other benthic fauna especially the chironomid larvae. The semi digested excreta of this fish forms the food of bottom feeders. For exploitation of the above food resources, poly culture can be done in fish cum poultry ponds. In the northern and north-western states of India, the ponds should be stocked in the month of March and harvested in the month of October-November as severe winter affects the growth of fishes. In the south, coastal and north-eastern states of India, ponds should be stocked in June-September months and harvested after rearing the fish for 12 months due to mild winter season.

Use of poultry manure for fish farming

Generally, two following methods are adopted to recycle poultry excreta for fish production.

a) The poultry droppings are collected from poultry farms, stored it in proper places and are applied in the ponds on regular basis. It is applied to the pond @ 50 Kg/ha/ day every morning after sunrise.

b) In this system, poultry house is constructed in such a way that the house partially covers the fish pond and droppings from poultry house directly fall on fish pond/tank which can be used by fishes. Care should be taken as excess droppings can cause decomposition and depletion of oxygen in the pond resulting to fish mortality. It has been estimated that one ton of deep litter fertilizer is produced by 30-40 birds in a year. Five hundred birds with around 450 kg as total live weight may produce wet manure of about 25 Kg/day which is sufficient for a hectare of water area under polyculture.

The intensive housing system usually deep litter system is preferred over the cage system due to higher manurial values. In deep litter system, birds are kept and the floor is covered with litter. Dry organic materials like chopped straw, dry leaves, hay, groundnut shells, broken maize stalk, saw dust, etc. are used to cover the floor up to a depth of about 6 inches. The birds are then kept over this litter and a space of about 0.3 - 0.4 square meters per bird is provided. The litter is stirred regularly for aeration. Lime can be used to keep the litter dry and hygienic. Fully built up litter is prepared by 10 months which can be used as fertilizer in the fish pond. The fully built up deep litter contain 3% nitrogen, 2% phosphate and 2% potash. It is available in the market also. The poultry birds under deep litter system should be fed regularly with balanced feed according to their age grower mash is provided to the birds during the age of 9-20 weeks @ 50-70 gm/bird/day, whereas, layer mash is provided to the birds above 20 weeks @ 80-120 gm/bird/day.

Pigs can be used along with fish and poultry in integrated culture under two-tier system. The upper panel may be used for keeping poultry and the lower for pigs. Poultry droppings are to be used as food source for the pigs which finally fertilise the fish pond.

B) Use of poultry manure in animal feed

Recent research findings depicted that poultry wastes/manure can make important contribution in animal feed preparation. The nutritive value of poultry manure is best preserved through the proper and careful management of litter although the quality of poultry wastes depends on production system, storage and treatment of waste, health status of birds and treatment provided, presence of drug residues, toxic minerals, climatic condition, nature of waste handling etc. Before using in animal feed, proper attention must be undertaken during processing of wastes to maintain animal health and welfare. A wide variation (15-35%) of replacement of feed ingredients in animal feed by poultry manure has already been established in several animal experiments without having any detrimental effects on production and reproduction. Therefore, if this waste is processed scientifically, it can reduce the cost of concentrate ration. The processing is generally done by deep stacking or ensiling or dehydrating or extrusion-pelleting. Out of these methods, deep stacking is most commonly practiced. However, different methods of processing are briefly described as under-

i) **Deep stacking method :**

"*Deep stacking*" refers to the process of stockpiling of litter. After stacking, the litter undergoes a combined composting-ensiling process. The action of bacteria generally heats the stack to a temperature of between 140°F and 160 ^{O}F. This temperature is sufficient enough to kill pathogenic organisms like *E. coli* and *Salmonella* which may be present in the raw litter. Overheating (>160 ^{O}F) may reduce the protein and carbohydrate content in the feed and this problem can be overcome by packing the stack, covering the stack with plastic, or both. Three weeks minimum heating period is must to use the litter as feed. Good product gives a smell of caramelized chocolate with fine texture. Grey colour product indicates underheating whereas black colour of overheating. The feeding value can remain unchanged as long as 5 years if properly maintained. For better production and storage either permanent structure with roof or temporary windrow should be constructed.

ii) **Ensiling :**

Poultry waste can be ensiled after mixing agricultural energy rich by-products like molasses, fruit and vegetable wastes etc. for better fermentation. Moisture content must be within the range of 40 to 75% and pH should be 3.8-5.2. The silage from waste can be prepared by 21 days. For small scale, silo bags and for large scale, silo pits can be used for ensiling.

iii) **Dehydration** :

Poultry waste/litter is thermally dehydrated to the moisture content not in excess of 15%. Dried poultry wastes must contain not less than 18% crude protein, not more than 15.0% crude fiber, 30.0% ash and 1.0% feathers. On the other hand, dried poultry litter should contain not less than 18.0% crude protein, not more than 25.0% crude fiber, 20.0% ash and 4.0% feathers.

iv) **Extrusion-pelleting :**

With the help of machines, poultry wastes are converted into extrusion-pellet to be used for animal feed.

Inclusion level in animal feed

The Food and Agricultural Organization under the topic "*Feeding animal wastes*" concluded the following points in different species when processed poultry waste/manure is used as animal feed.

i) **Sheep** :

- Incorporation of poultry litter into the ration of growing lambs can be as high as 70%, and for fattening stock up to 50%. Much however, depends upon the litter quality and counterpart forage and feed ingredients.
- Poultry waste fed at levels above 35% usually covers almost the total protein requirement of sheep, and contributes substantially to the energy of the total ration.

- About 70–74% organic matter and 80% crude protein contained in poultry waste are digestible.
- The only problem encountered in feeding processed poultry waste to sheep is the toxicity derived from the high copper level in poultry diets. It is therefore necessary to verify the copper content in the waste, and to include litter in the ration only to such an extent that copper will not exceed the tolerance level.

ii) **Pig** :

- The possibility of replacing a significant part of pig rations by poultry waste is of great economic importance to mixed farms (pig and poultry). An integrated layer/pig/fish cycle appears to be the most advanced and economically sound recycling system with zero pollution discharge. Research data are not available, but the system is successfully applied by farmers on small and large scales.

iii) **Turkey** :

- The recommended level of layer manure for turkey laying hens is about 10%, and for growing turkeys up to a maximum of 30%. A careful balance of energy and minerals (Ca:P) is of great importance, because the higher the level of poultry wastes in the turkey diet, the more critical becomes the energy supply and mineral imbalance.

C) Use of poultry manure as fertilizer

The use of poultry manure in the agricultural field as fertilizer for various field and horticultural crop production has now become a common practice. The poultry droppings are very rich in potassium, nitrogen and phosphorus. It can improve the nutritional quality of soil resulting to increase in productivity of crops, vegetables, grains etc. It is preferred to use manure after within 120 days of the harvesting of crops. It is not recommended to use fresh manure because it may contain certain microorganisms which are harmful for human health and may cause diseases. Poultry manure/wastes should be composted for at least for one month before use in the agricultural field as fertilizer to destroy harmful microorganisms. The advantages of poultry manure fertilizer over use of inorganic fertilizers are :

- ❐ More economical

- Maintains better soil condition
- Higher yield
- More environment friendly
- Less health hazards.

Predation on livestock ticks by chickens

Chickens are natural predators of ticks. Rhipicephalus appendiculatus were recovered in large numbers from the crops and gizzards of chickens which had scavenged for 30 min-1 h among tick-infested cattle. Other ticks recovered were Amblyomma variegatum. Boophilus decoloratus, Hyalomma marginatum rufipes, Rhipicephalus evertsi evertsi and Otobius megnini (Dreyer *et al.*, 1997; Hassam *et al.*, 1991). The numbers of ticks recovered ranged from 3 to 331, with an average of 81 per chicken. Cattle facilitated the predation of ticks by certain behavioural actions. Chickens also picked up both engorged and unengorged ticks seeded on vegetation, but unengorged ticks were preferred. Studies have confirmed that chickens are natural predators of livestock ticks and that chickens can be used as part of an integrated tick control plan in cattle-management systems in resource-poor farming communities.

Grazing of Poultry

"Consumption of grain decreases as consumption of grass increases. It all keeps the expense side of production down". But a chicken will not consume grass like a a cow. The consumption can be increased if we offered fresh and clean grasses and the chickens are allowed to move in the pasture. When chicken move in well developed pasture with legume preferably clover, they will eat more forage and more bugs and less grains. Pasturing can feed expenses up to an estimated 60 percent on layers and 30 percent on broilers. Also, the broilers attain marketable weight two weeks earlier than normal.

Cattle-Poultry Rotation

Cattle – Poultry rotational grazing is also beneficial. In this system cattle first graze the pasture; the cattle leave a trail of manure and long (4 to 6 inches) grasses stubble in their path. After grazing of cow, one to two inches of grasses are left in the pasture which is suitable for poultry. Chickens are put on that pasture after four days of cattle grazing. Both the layers and broilers love to pick through fresh manure for insects, including emerging fly maggots, and undigested food particles, both helpful sources of protein. The chickens sanitize the field by eating the parasites.

Conclusion

It can be concluded from the above discussion that both small and large scale poultry production through integrated farming system not only enhances productivity, profitability, and nutritional security for farmers but also improves soil condition through recycling of organic wastes. Intelligent resource management is the key factor for achieving maximum economic return from the integration. Side by side, animal welfare and human health aspect should not be neglected.

Further reading

Amanullah M M, Somasundaram E, Vaiyapuri K and Sathyamoorthi K. 2007. Poultry manure to crops: A review. *Agric. Rev.*, **28**: 216-222.

Banerjee R K, Roy P, Singit C S and Dutta B R. 1979. Poultry dropping- its manurial potentiality in aquaculture. *J. Inland Fish. Soc. India,* **2**(1): 94-108.

Dreyer K, Fourie LJ, Kok DJ. 1997. Predation of livestock ticks by chickens as a tick- control method in a resource-poor urban environment. *Onderstepoort J Vet Res.* **64**(4): 273-6.

Fang Y X, Guo X Z, Wang J K, Fang X Z and Liu Z Y. 1986. Effects of different animal manures on fish farming. In: Maclean J L, Dizon L B and Hosillos L V. (Eds.), The First Asian Fisheries Forum. Asian Fisheries Society, Manila, p. 117-120.

GoI (Government of India) (2013) Economic Survey 2012- 2013, Economic Division, Ministry of Finance, New Delhi.

Hassan SM[1], Dipeolu OO, Amoo AO, Odhiambo TR. 1991. Predation on livestock ticks by chickens. *Vet Parasitol.* **38**(2-3): 199-204.

Hoq M E, Das G B and Uddin M S. 1999. Integration of fish farming with poultry: effects of chicken manure in polyculture of carps and freshwater prawn. *Indian J. Fish.*, **46**(3): 237-243.

http://www.nationalchickencouncil.org/industry-issues/vertical-integration/

http://coccialertnetwork.org/indianpoultryindustry.html

McNabb, C D, Batterson T R, Premo B J, Knud-Hansen C F, Eidman H M, Lin C K, Jaiyen K, Hanson J E and Chuenpagdee R.1990. Managing fertilizers for fish yield in tropical ponds in Asia. In: Hirano R and Hanyu I. (Eds.),

Proceedings of the Second Asian Fisheries Forum, 17 - 22 April 1980, Tokyo, p. 169 -172.

Mehta R, Nambiar R G, Delgado C and Subramanyam S. 2003. Policy, technical and environmental determinants and implications of the scaling-up of broiler and egg production in India. Annex II of the IFPRI-FAO Final Report Livestock Industrialization, Trade and Social-Health-Environmental Impacts in Developing Countries, phase II, International food Policy Research Institute, Washington D C. Available at: http://www.fao.org/WAIRDOCS/LEAD/X6170E/X6170E00.HTM

http://www.indianmirror.com/indian-industries/poultry.html

http://www.dofppunjab.info/sector-specific-information/Poultry.php

http://coccialertnetwork.org/indianpoultryindustry.html

http://agropedia.iitk.ac.in/content/integrated-fish-cum-poultry-farming

http://www.fao.org/docrep/field/003/ac375e/AC375E04.htm

http://www.fao.org/docrep/004/X6518E/X6518E02.htm

Poultry Fest. 2014. Third Poultry Fest, 10-12 October, Luchnow, India.

Rana S S and Chopra P. 2013. Integrated Farming System. Department of Agronomy, College of Agriculture, CSK Himachal Pradesh Krishi Vishvavidyalaya, Palampur, 90 pages.

Woynarovich E. 1979. The feasibility of combining animal husbandry with fish farming with special reference to duck and pig production. In: Pillay T V R and Dill W A (Eds.), Advances in Aquaculture. Fishing News Books Ltd., Farnham, Surrey, England, p. 203-208.

□□□

Chapter 13

Integrated Fish Production and Management

M.M. DAS[1], G.H. PAILAN[2], S.B. MAITY[1] AND K.K. SINGH[1]

[1] ICAR-Indian Grassland and Fodder Research Institute, Jhansi-284003

[2] ICAR-Central Institute of Fisheries Education, Kolkata Centre, Kolkata-700091

Introduction

The farming practice where more than one component is involved and when the bye-product of one component is utilized for the benefit of other component, the system is known as integrated farming. In an integrated farming, when fish is one of the components, it is known as integrated fish farming. The principle of integrated fish farming involves farming of fish along with livestock or/and agricultural crops. The farming system practice offers great efficiency in resource utilization, as waste or by-product from one system is effectively recycled. It also enables effective utilization of available farming space for maximizing production. Its main benefits comes from the conversion of livestock waste into protein and more important is the indirect fertilization effects through livestock excreta entering the pond, thus enhancing microbial and plankton production which contributes to fish nutrition.

The fish farmers culturing fish in their ponds tend to leave their pond dikes and other farm spaces fallow. Generally, in the fish farm, actual water area is about 70% and the remaining 30% lies in the form of dikes and other farm spaces. This 30% of the area can be effectively utilized for food production including agriculture crops and animal components. As such, fish-pig, fish-duck, fish-poultry, fish-vegetable, fish-fruits etc. can be thought of as viable integration methods in the aquaculture-agriculture integration.

An integrated Fish farming enables the farmer to maximize the use of all available farm resources. This ensures the enhancement of overall food production of the farm from plant as well as animal components. The integrated system, which promotes crop diversification and nutrient recycling, can make even the most marginal lands more productive. In agro-aquaculture, especially in paddy-cum-fish culture, production of animal protein in the form of fish can be enhanced. Likewise in aqua-agriculture, the dikes and other fallow farm space of fish farm, can be utilized for the production of land crops including vegetables, fruits etc. and for the raising of animal components including pig, duck, poultry etc. In addition, the forest trees can be raised in the fallow farm space and also in the peripheral area of the fish farm. The multi-purpose tree like Subabul is very much beneficial in a fish farm as the leaves of this tree is an excellent food for the grass carp and also enriches the pond. The timber of this tree serves as a fuel wood for the farm family and also for fencing the farm. Thus, a higher economic return can be expected from a unit cultivable area and fish, fruit, vegetables, egg and meat enrich family diets. The integrated system reduces investment risks and provides year-round employment. Also, the soils are not continuously degraded, but made more fertile.

Types of Integrated Fish farming

1. Rice Fish Integration

Rice fish farming can contribute to household income, contribute to food security and nutrition and contribute to improved sustainability of rice production. Generally two production systems have been recommended for culturing fish in the rice fields. They are, simultaneous or concurrent method and alternate or rotational method.

The rice-fish system essentially comprises of a main rice field, dikes, side trenches and a pond refuge (Fig. 1). The side trenches and pond refuge serve as a water harvesting-cum-storage system. The field should hold water to a depth of 30 cm.The refuge is a pond within the rice field, which gives shelter to the fishes and also permits an extended period of fish culture even after the harvest of paddy. The semi-tall to tall rice varieties, which are pest and disease resistant, are preferred. With a stocking density of 6000 nos. of fingerlings per hectare, a farmer can expect a production of about 300 kg fish/ha in 4 months of culture in paddy fields without addition of much extra inputs. Application of fertilizers increases the nutrient concentration in water, which flourish the growth of phytoplankton, which are directly consumed by the fish or support good growth of zooplankton. Compost fertilizer for deep-water variety @30 tons per ha can be applied. In organic fertilizers for deep water variety paddy as nitrogen @120kg/ha, P_20_5 @ 60 kg/ha and K_20 @ 60 kg/ha can be applied.

To promote the growth of fish, rice bran and mustard oil cake in the ratio of 70:30 @2-3% of the body weight of fishes can be fed. After the harvest of paddy, the main rice plot can be used for growing crops, which require less water. Rice-fish farming can actually increase rice yields (up to 10% in some cases) while providing farmers with an important source of protein and extra income.

Selection criteria of fish species

- The fish should be capable of tolerating very shallow water level.
- Should withstand higher temperature (up to 40°C) and variable temperature fluctuations
- Those species, which can withstand fairly high turbidity of water and poor oxygen concentration.
- The species, which have faster growth rate and should have desirable characteristics to grow to marketable size in short duration at the time of harvesting the rice.

Stocking patters of fish

It may involve the stocking of young fry for the production of fingerlings or the growth of fingerlings to marketable size. Rice fish farming may either be the culture of only one species or a combination of two or more species of fish and crustaceans. Generally, stocking density of fish depends on its size, species and the fertility of the land. It is best to wait until the rice is well established before releasing fish seed. Fish can be stocked once two or three tillers have appeared for which the usual waiting is 1- 3 weeks after transplanting or 4-6 weeks after seeding.

Benefits of rice-fish culture

The rice-fish system has several advantages when compared to mono-cropping of either rice or fish.

- The excreta of fish and its left over food serve as manure for the growth of paddy.

- The rice-fish farming system is also a step towards Integrated Pest Management (IPM) as it is found that fish culture helps in biological control of insect pests on rice.
- Helps to increase the rice yield.
- An increase in income from the production of both rice and fish is possible
- Provides reliable source of protein for the farmers and their families
- Helps to reduce the use of fertilizers
- Conservation of water resources and plant nutrients
- Reduced operating costs relative to either system in isolation.

2. Integrated Livestock-Fish Farming

In a country like India, where a majority of the population lives in rural areas and of whom many are under-nourished, supply of protein rich food is essential. Therefore, development of integrated farming systems suitable to the village situations and based on the resources available with the farmer is important.

The main potential linkages between livestock and fish production concern use of nutrients, particularly reuse of livestock manures for fish production. The term nutrients mainly refers to elements such as nitrogen (N) and phosphorous (P) which function as fertilizers to stimulate natural food webs rather than conventional livestock nutrition usage such as feed ingredients. Integrated farming of fish with livestock including poultry birds is an old practice consisting of the culture of fish associated with the husbandry of domesticated animals such as pigs, cattle, goat, duck, poultry etc. Intensive farming of livestock would produce large quantities of manure and animal waste, which needs to be disposed in order to prevent serious environmental problems. These wastes can be effectively converted to manure for not only agricultural crops, but also for the production of fish. For 0.1 ha water area, one cow with a calf is sufficient. The cow should be brought about two months earlier to introducing fish into the pond.When organic wastes is used in fish ponds, about 40-50 kg are converted to 1 kg of fish. The end product is an improved production of animal protein, particularly needed in a developing country like India. The aim of integrated farming is the recycling of animal wastes (faeces, urine and spoiled feeds) to serve as fertilizers, and sometimes as food for fish grown in ponds. The basic principles involved in integrated farming are the utilization of the synergetic effects of inter-related farm activities, and the conservation, including the full utilization of farm wastes. Cattle are raised essentially for milk. Cow urine and cow dung are used for making compost or vermi-compost in permanent pits or large earthen vessels or on a plastic sheet laid on the ground. Compost, especially vermi-compost serves as an excellent fertilizer for the vegetable garden. Cow urine and cow dung are also used in the pond as manure. The maintenance of integrated crop livestock system is solely dependent on the availability of adequate nutrients to sustain livestock and plants and to maintain soil fertility. Animal manure alone cannot meet the requirements for crop, even if it does contain the kind of nutrients needed. This is because of its relatively low nutrient density and the limited quantity available to small-scale farmers. Alternative sources for the nutrients have to be arranged.

Fish husbandry

For profitable fish culture, the pond is to be managed scientifically. In order to obtain high fish production per hectare of water body, fast growing, compatible species of Indian and Exotic carps of different feeding habits are stocked & cultured in the same pond which is popularly known as composite fish culture or polyculture or mixed farming. The composite fish culture can be done in the following way:

2.1 Selection of Pond

The pond can be seasonal or perennial. The perennial ponds, which retain water throughout the year, are selected for culture of table size fish. Generally 0.1 ha or above water area of a pond are suitable for raising table size fish, preferably with nearby sources of water for replenishment as and when necessary.

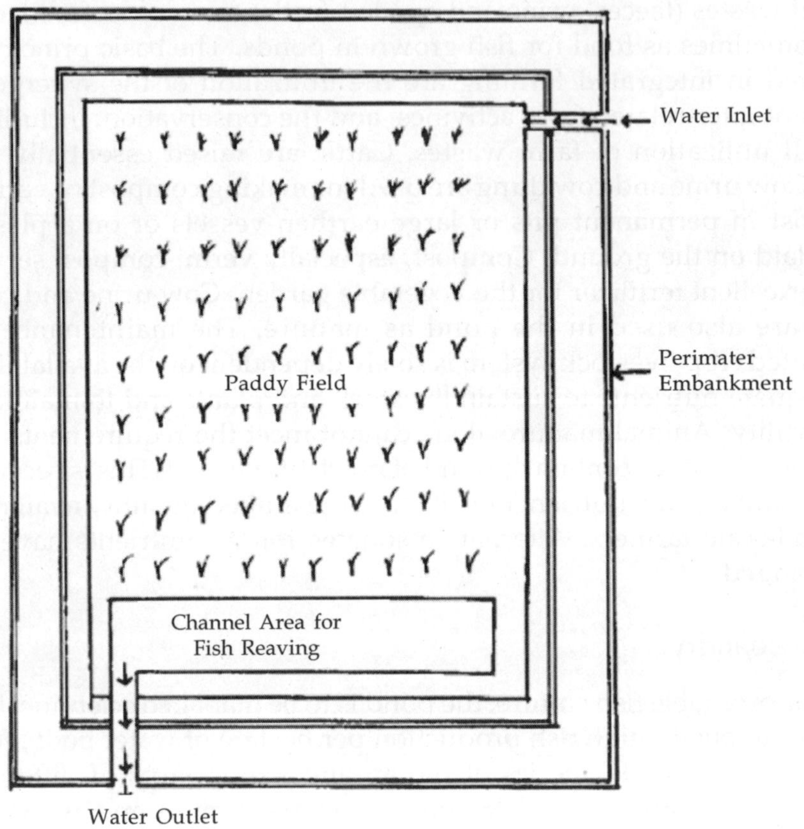

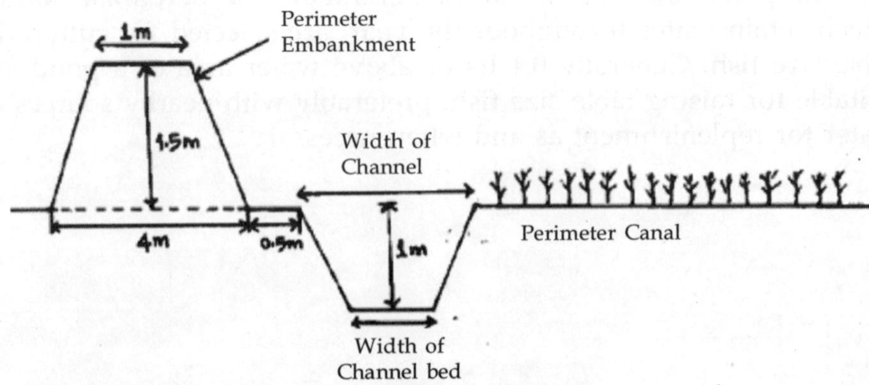

Fig. 1. Diagrammatic view of paddy-cum-fish culture plot

Paddy-cum-fish culture plot

2.2 Weed clearance

Aquatic weeds are detrimental to fish culture and cause various hazards for the normal growth of fish. The disadvantages are as follows:

a) Aquatic weeds reduce the living space available for fish and causes obstruction for their free movement.

b) Aquatic weeds prevent sunlight form entering the pond water and thus hamper natural food production for fish.

c) Aquatic weeds consume nutrient from the pond which otherwise could have been utilized for the production of fish food organisms.

d) Aquatic weeds provide shelter to fish enemies particularly insects.

e) Aquatic weeds cause obstruction for harvesting of fish.

f) Excessive growth of aquatic weeds, particularly during cloudy days causes serious problem by upsetting the oxygen balance, which may be detrimental to the fish stock.

These unwanted aquatic weeds could be removed manually, mechanically, chemically and biologically. Manual removal method is better. Some varieties of weeds such as Duck weeds, *Hydrilla*, *Najas*, *Ceratophyllum*, etc. can be effectively controlled by grass carp, if adequate number of these are introduced in the ponds.

2.3 Eradication of predatory fish

Presence of predatory fish in the culturable pond is not at all desirable. Similarly, the predatory fish directly prey on the culturable species of fingerlings. So, their eradication from the ponds before stocking is an important step in scientific management.

Dewatering ensures complete removal of unwanted fish. However, where dewatering is not possible, the ponds have to be poisoned using fish toxins of plant origin. Use of chemical pesticides (fish poison) is not desirable as their detoxification process is prolonged; they leave harmful residual effect in the pond and the fish thus killed are not safe for human consumption. Mahua oil cake is a popular fish poison of plant origin and is effective at a dose of 200-250 ppm (i.e. 2000-2500 kg/ha-m depth). This, at a later stage also serves as manure. Some time a thick layer of algal bloom of brown or green colour is seen over the water surface of pond. This can be removed from fish pond by using a piece of split bamboo followed by liming based on water pH. Chemicals like copper sulphate @0.1- 0.5 mg/lit. of water also helps in controlling this bloom. Noxious gases and the effect of other substances of pond bottom mud can be reduced by repeated netting or by moving a rope through the pond bottom mud.

2.4 Liming

Liming helps in maintaining the pH of fish pond water. This helps in increasing the natural productivity of the pond. Liming also helps in maintaining the cultured fish stock disease free. It is done based on the soil and water pH. After 15 days of mahua oil cake application, liming is followed to correct the acidity as well as to keep the pond hygienic. The dose of lime is adjusted according to the pH of soil and water (Table 1). Half of the quantity is applied before stocking the fish and the rest in 8 installments as and when necessary.

Table 1. The dose of lime for different soil types

pH of soil	Type of soil	Dose of lime (Kg/ha)
4.5-5.0	Highly acidic	2000
5.0-6.0	Moderately acidic	1200
6.0-6.5	Mildly acidic	1000
6.5-7.0	Near neutral	400
7.0-7.5	Mildly alkaline	200

2.5 Stocking

The species selected for culture should have the following characters- a) Fast growth rate b) Ability to withstand changing physico-chemical and biological conditions of the pond water c) Adaptability to withstand crowded conditions and resistance to diseases d) Good food conversion efficiency and good market value. The culture comprised of 3 indigenous species of fish viz. rohu, catla and mrigal and 3 exotic fish i.e silver carp, grass carp and common carp, keeping in view their different food habit and habitat. In some cases Tilapia species are included by balancing Silver carp and Grass carp combination. About 100-150 mm sized fingerlings are suitable for composite fish culture. To ensure that the water is no longer toxic, a simple test may be performed by keeping a few fingerlings in a 'hapa' fixed in the pond. If the fingerlings survive for 24 hours, the pond may be considered fit for stocking the young ones. A moderate density of 7500 fingerlings/ha is recommended, but a density of 6000 fingerlings/ha is also justified when the rearing period is limited to about 8-10 months. Due attention should be given on the biological production in the pond and the availability of aquatic/terrestrial weed for grass carps. In composite fish culture pond, fingerlings should be stocked in the following ratio

Catla : Rohu : Mrigal : Silver carp : Grass carp : Common carp = 2 : 2 : 1.5 : 2 : 1 : 1.5.

2.6 Fertilization and Supplementary feeding

Fertilization increases the natural food availability in the pond. Manuring alone can increase the production of the pond by 75%. Animal excreta is rich in nitrogen and phosphorous. In integrated fish farming system with livestock no supplementary feed is required for fish. The animal house washings, which encompasses of animal dung, urine and the leftover food particles are drained into the pond, which serves as manure as well as feed to the fishes.

However, grass carp, which feeds voraciously on macro-vegetation, fodder grasses and vegetable wastes, need to be provided. It should therefore be fed by supplying aquatic weeds, viz., Wolffia, Lemna, Azolla, Spirodella, etc. or fodder grasses such as Napier, Barseem, or even lawn grass or vegetable wastes such as cabbage or cauliflower leaves. It is better to provide feed to the grass carp in a feeding frame made up of four pieces of split bamboo poles tied at their corners and made into a rectangle.

2.7 Raking

Raking of pond bottom, the day after liming should be done for proper mixing of lime. It also helps to release the obnoxious gases formed in the bottom soil.

2.8 Growth and Health Care

The health of fish needs to be checked regularly by periodic netting at least once in two months. Healthy fish grow well and feed voraciously. If the feed provided is not being consumed, then it is indicated that the fish are suffering from some disease or the water quality is deteriorated. A check on the water quality and hygiene of the pond and the health of the fish is necessary to take remedial measures.

Catches of fish and prawn

Fish diseases are classified into i) infectious diseases, ii) invasive diseases and iii) Non-infectious diseases. The fish diseases are mainly caused by virus, bacteria, fungi or unicellular algae. Some of the common fish diseases are bacterial ulcerative disease, bacterial enteritis, bacterial gill rot/ fin rot etc. The loss of fish production from infectious diseases accounts for 60% of all diseased. As prophylactic treatment, the disease prone water areas can be treated with lime, CaO @ 50 kg/ ha followed by application of bleaching powder @ 0.5 ppm after one week during post monsoon season. At initial stage of the disease occurrence, lime (CaO) @ 100 kg/ha followed by application of bleaching powder @ 1 ppm after one week can be practised. For treating the disease, Oxytetracycline at 75mg/kg, Amoxycilin 80mg/kg, Furazolidone at 75mg/kg, Co-trimazine at 30mg/kg may be supplemented through feed.

3. Integrated Pig-cum-fish farming

Pig can fit into diverse systems of management. One of the best feed converter and prolific amongst the all domesticated livestock. A sow It can be reared on kitchen wastes, industrial or agricultural by-products and slaughter house offals. Pigs are voracious eater and can convert agro-waste in to edible and highly nutritious meat. Pigs can produce 16 piglets in a year. Dressing percentage is very high (65 to 70%) in comparison to sheep- goat (45 to 55%). Piggery can generate employment opportunity for rural population.

Housing Management

The pigs are very sensitive to adverse climatic condition hence, a well protected housing is necessary for pig farming. It should be well protected from direct sunlight, rain and wind. The walls and floor should be strong enough to withstand the rooting habits of pigs. To provide thermal comfort in hot summer days a wallowing tank may be provided in pig house. Pigs are kept under two housing systems viz. Indoor system and Open air system. A combination of these two systems is very advantageous and can be adopted in pig farming.

The pig sty facing the pond is constructed either on the pond embankment or nearer to the pond. The floor of the pen should be cemented with a drainage facility to the pond. A combination of open-air and indoor system is ideal for a pig sty with a thatched roof. The drain from the pig sty should be provided with a regulatory valve to control the release of wastes. For healthy growth of pigs, a living space of 1-1.5 m^2 per pig is ideal.

Integrated Pig-Fish Farming

3.1 Selection of pig

The exotic upgraded stock of pigs such as the large white Yorkshire, middle white Yorkshire, Berkshire Hampshire and Land race are almost suitable for raising in comparison to the local Indian variety since they are quick growers and prolific breeders. In pig-cum-fish farming system, 15- 20 days after bringing the piglets to the pig house fish pond are stocked with fish seed. The pig excreta during the period of 15- 20 days fertilize the pond water and produce the fish food organism like- phytoplankton and zooplankton which can be judged by observing the light greenish colour of water. They attain slaughter maturity size i.e. 60-70 kg within six months and give 6-12 piglets during every litter. During a period of one year, two crops of pigs (6 months each) can be raised.

3.2 Feeding

About 70% cost of the pig farming is feed cost. The pigs are fed three times a day with balanced pig mash as per their requirements. Green fodder are also provided to the pigs to reduce the feed coast. The green fodder can be incorporated to the extent of 15 to 30% in the feeding ration.

Composition of Pig Mesh Concentrate (PMC)

Ingredient	Parts (%)
Rice bran	30
Rice polish	15
Wheat bran	27
Maize broken	10
Ground nut oilcake	10
Fish meal	04
Mineral mixture	03
Common salt	01
Total	100
Vitamin supplementation	@ 20 gm/ 100 kg feed.

3.3 Health care

The resistance of pig is stronger than that of chicken. Proper care of health coupled with proper management practice, sanitary environment will keep pigs away from sick. Generally pigs suffer from diseases like swine fever, swine plague, swine pox and may be infested with parasites. Pigs are susceptible to heat stroke. Provision of wallowing or water sprinkling is very important during summer months. Hence, a healthy housing management is a pre-requisite for a disease-free environment for pigs and to avoid losses. Vaccinations for Swine fever and Food and Mouth disease must be done periodically. Regular deworming is advised for a pig farm. To save the cultured stock of pigs from anaemia (bloodlessness) they should be given iron injection like- Imferon.

3.4 Harvesting and Production

The most important nutrients like N, P, K are available in livestock manure along with other micronutrients. The average dung productions in pigs are 0.25 tones per pig per year. Pig manure is also a good source of nitrogen and phosphorus. In pig excreta N, P, K contents are 0.60%, 0.50%, 0.20% respectively. Pigs of 50 kg body weights on average produce 2.5 kg dung per day and 50 to 90 kg pig produce 5 kg per day and it contains 20 to 25% dry matter. On an average 30 to 40 pigs are suggested for one hectare water bodies.

Fish culture is generally practiced at least for 12 months in the integrated pig-fish system. In this period, generally, Catla attains a weight of 0.8-1.0 kg, Rohu 0.6-0.8kg, Mrigal 0.4-0.6 kg, Silver carp 1.0-1.2 kg, Grass carp 1.0-1.5 kg and Common carp 0.8-1.0 kg. In integrated pig-fish farming system, a fish production of at least 2.5–3.0 t per ha can be easily realized. Each pig attains slaughter maturity size of 60-70 kg in a period of 6 months. As such, pigs are sold out after 6 months of rearing and a fresh lot of weaned piglets are introduced for further rearing.

Table 2. Economics of Integrated pig-cum-fish farming (0.1ha area/12 months culture periods)

A. Expenditure		**Cost (Rs)**
i)	Rent of Water body	2,000.00
ii)	Pond preparation	2,300.00
iii)	Cost of fingerlings of 750 nos.	2,000.00
iv)	Animal house for rearing 3 no. of pig	7,000.00
v)	a. Fattening of 1st lot of piglets for 6 months	
Cost of weaned 3 nos. piglets @ Rs.1400.00		4,200.00
Cost of pig feed 750 kg @ Rs.15.00		11,250.00
Cost of medicine		400.00
Cost of green fodder		300.00
v)	b. Fattening of 2nd lot of weaned piglet of for 6 months	16,150.00
vi)	Labour charge	15,000.00
vii)	Lime 40 kg	400.00
viii)	Netting charges	500.00
	Total variable cost	62,200.00
B.	**Returns**	
I.	250kg fish @120.00/kg	30,000.00
II.	Pig meat (approximately 294kg @150.00/kg)	44,100.00
	Total return	74,100.00
C.	**Net Profit (B-A)**	**11,900.00**

4. Integrated Duck-fish farming

Among the different livestock based system, fish duck integration is one of the most popular farming practices among woman fishers.

Advantages of Fish Duck Integration

- Duck dropping act as feed and fertilizer for cultured fish in the pond.

- Duck collect 50% of their food naturally from the pond.
- Ducks keep the water body clean and increase dissolved oxygen content.
- Left over feed of duck used as supplementary feed for fish.

Duck-house

A low cost duck house, preferably made of bamboo or any cheap wood is constructed either above the pond or on the pond embankment. The duck house should be well ventilated. The floor of the house should be perforated so as to facilitate direct fall of droppings into the pond, if the duck house is constructed above the pond. Overcrowding in duck house should be avoided as this result in poor growth and egg production. About 0.3 to 0.5 m^2 of floor space per bird is required for night shelter. Again one male duck should be kept in every 5-6 female ducks. The ducks are reared for about 12-18 months after which they are sold.

4.1 Selection of duck

Some important varieties of ducks are Nageswari, Sylhet Meat, Indian Runner, Serachameli and Khaki Campbell.The ducks must be of good quality and the exotic Khaki Campbell would be more suitable for integration. One duck voids on an average, 125-150g droppings per day amounting to about 45-55 kg/duck/year and hence ducks stocked @200-300 ducklings/ha of fish pond recycle a quantity of 10,000 to 15,000 kg of droppings per hectare per year which besides fertilizing the fishponds and can be directly utilized as fish food. The droppings contain 81% moisture, 0.9% Nitrogen and 0.38% Phosphate.

Two to four months old ducks are suitable for stocking and necessary prophylactic measures need to be taken at the time of stocking.

Integrated Duck-Fish Farming

4.2 Feeding

Ducks find their natural food from ponds, but that is not sufficient for their proper growth. The natural feed is therefore supplemented with artificial feeds. A mixture of any balanced standard poultry feed and good quality rice bran in a ratio of 1:2 by weight can be fed to the duck. Generally 100 g feed is provided to each duck per day. Duck feed should be stored at cool and dry environment. If possible, manganese sulphate mixed with feed gives the better result at the rate of 10gms/100kg of feed. Duck weeds (Lemna, Wolfia, Azolla etc.) are also fed to the ducks. The left over feed given to the ducks and duck dropping fulfill more than 60% of food requirements of farm fishes.

4.3 Egg laying

The ducks start laying eggs after attaining the age of 24 weeks and continue to lay for two years. Local variety lays between 180 and 200 eggs/yr. Khaki Campbell about 300 eggs/yr. It is advisable to keep in some straw or hay in the corner of the duck house for egg laying. The eggs are collected at around 11.00 AM after the release of ducks into the pond.

4.4 Health care

A clean and hygiene environment is to be maintained in the duck pen to prevent them from any diseases. An experienced farmer can detect a sick bird among the flock by a careful look, listening to the sound of bird and by observing any reduction in daily feed consumption. A sick bird becomes restless, eyes lack brightness and water discharge comes out of eyes and nostrils. The ducks are to be vaccinated prior to stocking.

4.5 Harvesting and Production

Fish culture is generally practiced at least for 12 months and in this period, generally,Catla attains a weight of 0.8-1.0 kg, Rohu 0.6-0.8 kg, Mrigal0.4-0.6 kg, Silver carp 1.0-1.2 kg, Grass carp 1.0-1.5 kg and Common carp 0.8 - 1.0 kg. In integrated duck-fish farming system, a fish production of at least 2.0 – 2.5 t per ha can be easily realized. A total of 37,500 to 42,500 duck eggs and 500 to 750 kg duck meat (live) can be expected from duck rearing.

Table 3. Economics of Integrated duck-cum-fish farming (0.1ha area/12 months culture periods)

A.	**Expenditure**	**Cost (Rs)**
i)	Rent of Water body	2,000.00
ii)	Pond preparation	3,000.00
iii)	Cost of fingerlings of 750 nos.	2,000.00
iv)	Duck house for rearing 25 no. of duck	7,000.00
v)	Cost of duck 25 nos. @ Rs.150.00	3,750.00
vi)	Cost of duck feed 900 kg @ Rs.18.00	16,200.00
vii)	Cost of medicine	1,000.00
viii)	Labour charge (lump-sum)	15,000.00
vii)	Lime 40 kg	400.00
viii)	Netting charges	500.00
	Total variable cost	50,850.00
B.	**Returns**	
I.	230kg fish @120.00/kg	27,600.00
II.	Duck eggs, 4000 eggs @ Rs.6.00 per egg	24,000.00
II.	Duck meat (approximately 62 kg @ 65.00/kg)	4,650.00
	Total return	56,250.00
C.	**Net Profit (B-A)**	**5,400.00**

4.6 Integrated Poultry-fish farming

In this farming system, the fish crop is integrated by using only poultry droppings or dip litter. By adopting this technology, production of 3500 to 4000 kg fish, more than 20000 eggs and about 1250 kg (live weight) chicken meat can be obtained from a hectare of pond area in one year. Poultry can be reared either directly over the pond or on the pond embankment and the house can be constructed using cheaper and locally available materials. The poultry birds can be reared either in cages or through deep litter system. The deep litter system is preferred over the former due to higher manurial value of the built up deep litter. The poultry house must have adequate accommodation and well ventilated for supply of light and air. The house should be reasonably cool during summer and sufficiently warm during winter. The floor must be maintained dry throughout. The floor of the house should be perforated (4-6 cm^2 mesh size) and each bird requires a living space of 0.3 to 0.5 m^2.

4.7 Selection of poultry birds

The fowls of Rhode Island or other improved birds are suitable for farming. About 8 week old chicks after vaccination and other prophylactic measures are kept in poultry house. About 500-600 birds (layer) are required to produce manure for fertilizing one hectare pond area. The fully built deep litter removed from the poultry pens is stored in suitable places and is applied to the pond @ 50 kg/ha/per day every morning.

Integrated Poultry-Fish Farming

4.8 Feeding

The poultry birds under deep litter system is fed regularly with balanced poultry feed according to their age. The feed is provided to the birds in the feed hoppers to avoid wastage and for keeping proper hygienic condition in the house. Birds are to be fed with chicken mash @ 39g/bird/day during brooding period, 50 to 70 g/bird/day when the bird reaches the point of laying and 110 to 120 g/bird/day during laying.

4.9 Egg laying

Each house is provided with some nest boxes for egg laying. Egg production commences at the age of 22 weeks. The egg laying becomes irregular from the age of 18 months after which the birds are replaced by fresh stock.

4.10 Health care

Well managed deep litter kept in dry condition has a sterilizing action. The level of coccidiosis and worm infection is very low on a properly managed litter. The poultry house and equipment must be disinfected at least 30 days prior to bringing in the new flock. The birds are to be vaccinated against diseases like infectious bronchitis infected laryngo trancheitis, mark's diseses, ranikhet diseses, fowl pox, etc. at the appropriate age. Some of the bacterial diseases viz. salmonellosis, coryza, fowl cholera, etc. can be kept under control by maintenance of proper hygienic conditions. Broad spectrum antibiotics may be added to the water in case the infection is mild.

4.11 Harvesting and Production

Fish culture is generally practiced at least for 12 months in the integrated poultry-fish system. In this period, generally, Catla attains a weight of 0.8-1.0 kg, Rohu0.6-0.8 kg, Mrigal 0.4-0.6 kg, Silver carp 1.0-1.2 kg, Grass carp 1.0-1.5 kg and Common carp 0.8-1.0 kg. In integrated poultry-fish farming system, a fish production of atleast 2.3-2.8 t per ha can be easily achieved. Approximately 1,00,000 eggs and 1250 kg of live weight of meat can be expected from 500 to 600 birds.

Table 4. Economics of Integrated poultry-cum-fish farming (0.1ha area/12 months culture periods)

A.	**Expenditure**	**Cost (Rs)**
i)	Rent of Water body	2,000.00
ii)	Pond preparation	3,000.00
iii)	Cost of fingerlings of 750 nos.	2,000.00
iv)	Poultry house for rearing 50 no. of poultry birds	5,000.00
v)	Cost of chicks (6 weeks old) 50 nos. @ Rs.75.00	3,750.00
vi)	Cost of poultry feed 1800 kg @ Rs.20.00	36,000.00
vii)	Cost of medicine	1,000.00
viii)	Labour charge (lump-sum)	15,000.00
vii)	Lime 40 kg	400.00
viii)	Netting charges	500.00
	Total variable cost	68,650.00
B.	**Returns**	
I.	240kg fish @120.00/kg	28,800.00
II.	Eggs, 10,000 eggs @ Rs.4.00 per egg	40,000.00
II.	Meat (approximately 100 kg @ 75.00/kg)	7,500.00
	Total return	76,300.00
C.	**Net Profit (B-A)**	**7,650.00**

4.12 Benefits of Livestock-fish integration

In a nutshell, the advantages of Livestock-fish integration are -

- The livestock excreta act as excellent manure for fish ponds and increase the biological productivity of the pond and consequently increase fish production.
- Some fishes directly feed on livestock excreta, which contain digestible food for fish.
- No supplementary feed is required for fish culture, which normally accounts for about 60% of the total input cost in conventional fish culture.
- The pond dikes provide space for the erection of animal house and hence no additional land is required for erecting the animal house.
- The pond water can be used for cleaning the animal house and for bathing the animal.

- Pond silt may be utilized as fertilizer for fodder crops, which in turn is used to raise livestock and poultry or as fish feed.
- This system can improve the nutritional security and socio-economic status of the rural farming communities.

4.13 Economics of Livestock-fish integration

In integrated approach of fish farming by associating different components of livestock, especially piggery, poultry, duck would be more meaningful to utilize the available farm space and to boost up the farm income. A farmer who has 0.1 ha of pond and if he practices integrated livestock-fish farming, a net profit of approximately Rs.5000.00 to Rs.12000.00 can be easily realized from this system in a period of one year (Table 2-4). The farmers if they possess their own pond and utilize their own human resource, a net profit of more than Rs.20,000/- can be achieved by the farmer.

5. Integration of Fishery with Horticulture crops

Horticulture is another sector, which can accelerate the economic development. Even the rural farmers can maximize their farm income by growing wide variety of vegetables and fruit crops on the pond dikes and other areas of the fish farm. In this system, both the land crops as well as the aqua crops share the benefits. The success of the system depends on the selection of plants. They should be of dwarf type, less shady, evergreen, seasonal and highly remunerative. Dwarf variety fruit bearing plants like mango, banana, papaya, coconut and lime are suitable(Fig. 5.3), while pineapple, ginger, turmeric, chilli are grown as intercrops. Plantation of flower bearing plants like tuberose, rose, jasmine, gladiolus, marigold and chrysanthemum provide additional income to farmers. Studies have shown that vegetables like pumpkin, brinjal, radish, carrot, turnip, french bean, chilies, cabbage and cauliflower perform very well and these can be grown on the dikes and other areas of the farm. Growing horticulture crops on the dikes and other farm areas would be an important step in weed control and the fallow areas of the farm are also minimized.

A huge quantity of cabbage, cauliflower, turnip and radish leaves are generally thrown away during the time of harvest. These can be profitably utilized as supplementary feed for the grass carp. Green grass has an important role in the feeding of grass carp. Hybrid napier, once sown on the pond banks, can be cropped continuously for five years, needing little irrigation during dry seasons. The fodder grasses on the pond dike help in effective binding of the soil and prevent erosion. The ponds serve as a micro-watershed and the stored water can be used for irrigating the crops on the dikes and adjoining areas. The silt from the fish pond is an excellent fertilizer for the land crops.

5.1 Benefits of Horticulture Fish Farming

The benefits of the system include:

- The nutrient-rich pond silt can be used as fertilizer for growing crops, eliminating the cost of other manures to a considerable extend.
- Manured pond water can be used for irrigation of crops.
- Fruit and vegetable residues can be used as feed for fish.
- The plants on the embankment strengthen the dykes.

The farmer gets additional income from growing fruits and vegetables on the pond embankment and other farm space, which otherwise remain fallow.

Integration with Guava

Integration with Mango

Conclusions

- In addition to direct feeding of excreta, natural foods produced in integrated pond ecosystem results in the high yields of fish production without supplementary feeding and fertilizers from extraneous source.
- Composite culture of carps or prawn-based polyculture of carps integrated with animal farming is mostly viable and feasible at farmers level in Indian conditions.

References

B.K. Mahapatra, G. H. Pailan, S. Datta, P. Sardar, S. Munilkumar and S. Dasgupta (2014). Training Manual on Integrated Fish Farming (in Bengali).Published by Director, CIFE Mumbai, 25thMarch, 2014. pp.81.

http://www.vuatkerala.org/static/eng/advisory/fisheries/culture_fisheries/integrated_farming/introduction.htm

ý http://www.fao.org/3/a-y5098e.pdf

http://agritech.tnau.ac.in/fishery/fish_ifs.html

http://vikaspedia.in/agriculture/fisheries/fish-production/integrated-farming-systems-in-assam#section-2

Tamilnadu Agricultural University. *TNAU Agritech Portal*. Coimbatore: n.p., 2008. N. pag. Web. 25 July 2012.

www.krishisewa.com/cms/articles/livestock/439-fish-pig.html

ways.org/files/Integrated%20Fish%20Farming%20%20%202.doc

❑❑❑

Chapter 14

Integration and Management of Beekeeping in Farming Systems

N.K. SHAH

ICAR - Indian Grassland and Fodder Research Institute, Jhansi-284003 (UP)

Introduction

Integrated Farming System may be defined as a set of agricultural activities organized into a functional unit to profitably harness the solar energy while preserving land productivity, environmental quality and maintaining the desirable level of biological diversity and ecological stability. Integration of different agricultural allied enterprises with crop activity as base would provide ways to reuse and recycle produce/ waste material of one component as input in the other linked component and to reduce the cost of production of the economic produce of the component two and finally to enhance the net-income of the farm as a whole. Sustainable farm income from Integrated Farming system is to integrate different production systems like dairy, poultry, livestock, fishery, floriculture, horticulture, apiculture etc. with agricultural crops production as the base.

Indian agriculture has challenge of providing national as well as household food and nutritional security to its teeming millions. The human population of India has increased to 1210.2 million at a growth rate of 1.76% in 2011 and is estimated to increase further to 1530 million by 2030. On the other hand our national food grain production for the past 3-4 years is hovering around 234 million tonnes. There are projections that demand for food grains would increase from 234 million tonnes to 345 million tonnes in 2030. To meet multiple objectives of poverty reduction, food security, competitiveness and sustainability the answer lies in farming system approach. Farming system is a scientific integration of various interdependent and interacting farm enterprises for the efficient use of land, labour and other resources of a farm family which provide year round income to the farmers. Farming system envisages the integration of agroforestry, horticulture, floriculture, dairy farming, sheep and goat rearing, fishery, poultry, duckry/pigeon, apiculture, sericulture, biogas, mushroom cultivation and by-product utilization of crops with the main aim of increasing the income and standard of living of small and marginal farmers specially.

Integrated farming system holds promise for alleviating hunger, reducing poverty, preserving the environment and fighting climate changes. The core principles of agroecology include recycling nutrients and energy on the farm, rather than introducing external inputs; integrating crops and livestock; diversifying species and genetic resources in agroecosystems over time and space; and focusing on interactions and productivity across the agricultural ecosystems rather than focusing on individual species.

Bees are vital to agriculture and natural biodiversity. According to the Royal society, 76 percent of the world's most widely used food crops require pollination to be productive. A new Canadian initiative is looking to put bees to work to help conserve a fragile area. Trees are needed to protect watersheds. Trees and shrubs help to filter pollutants from water runoff and anchor the soil with their root binding, which helps to reduce soil erosion. Blending of apiculture (bee keeping) with integrated farming system supports for a bumper yield of agricultural and horticultural yields by facilitating pollination process of plants.

Sustainable Farm Income from Integrated Farming system

Beekeeping as an Integrated Activity

Beekeeping is an activity which meshes well with agricultural and rural development projects. There is a great developmental potential and possibilities of introducing and implementing apiculture ventures with farming systems. A majority of crops planted in agri or horticulture systems are sources of either pollen or nectar to bees. Thus the cropping systems can yield not only honey as a dividend but can impart indirect benefit by pollination of the crop. Thus beekeeping can provide additional income to the farmers.

Plants benefit from insect pollination by an increased seed set. This results in increased seed production and better quality of seeds/ fruits. Honey bees are beneficial as pollinators in those areas where natural insect pollinators are lacking or are insufficient to pollinate large areas devoted to a single crop.

Bee keeping is an agro based enterprise, which farmers can take up for additional income generation. Honey bees convert nectar of flowers into honey and store them in the combs of the hive. Collection of honey from the forests *i.e.* honey hunting has been in existence for a long time. The growing market potential for honey and its products has resulted in bee keeping emerging as a viable enterprise. Honey and wax are the two economically important products of bee keeping.

Nearly 70 percent of cultivated crops are cross-pollinated and depends on insects like bees for pollination. Dwindling population of such useful pollinating insects has now become a global problem. The importance of bees is often underlined with their role in pollination services and income generation by production of honey and there has been a common concern that population of indigenous bees are declining at an alarming rate. The benefit of honey bees as providers of pollination services for enhancing crop yields and maintaining biodiversity is thought to be much higher than their role as producers of honey and beeswax.

Bees are the most efficient pollinators on this earth. Various kinds of bees entangled in pollination of the plants are: The giant honey bee *(Apis dorsata F.),* the oriental hive bee *(A. cerana F.),* dwarf bee *(A. florae F.),* Bumble bees (*Xylocopa* spp.) and several species of stingless bees (*Trigona* spp. and *Melipona* spp.), are widely distributed in India. These species pollinate various plants and produce hive products such as honey, wax, pollen, etc. which are useful to mankind. Crop pollination is an essential ecosystem service which is efficiently provided by different pollinators. Amongst various pollinators available *Apis cerana* shows a disproportionately high mass-specific metabolic rate, their foragers make many more trips per day in the same habitat than do foragers of the other species. *Apis cerana* can therefore be considered as one of most efficient pollinators. In India crop pollination has been negatively affected due to reduction in density and population of efficient pollinators. With the help of beekeeping with indigenous honey bee, *Apis cerana,* its density and availability as a pollinator have increased in some areas of Himachal Pradesh, Punjab, West Bengal, Uttarakhand, Karnataka, Maharastra and Kerala. The current study investigated correlations between practice of beekeeping with the indigenous bee, its influence on pollinator composition and role in crop production increment.

Integrating beekeeping with other farming activities

Advantages of beekeeping as an income generation activity

- Bee keeping requires less time, money and infrastructure investments.
- Honey and beeswax can be produced from an area of little agricultural value.
- The Honey bee does not compete for resources with any other agricultural enterprise.
- Beekeeping has positive ecological consequences. Bees play an important role in the pollination of many flowering plants, thus increasing the yield of certain crops such as sunflower and various fruits.
- Honey is a delicious and highly nutritious food. By the traditional method of honey hunting many wild colonies of bees are destroyed. This can be prevented by raising bees in boxes and producing honey at home.
- Beekeeping can be initiated by individuals or groups.
- The market potential for honey and wax is high.

Beekeeping is an activity which meshes well with other agricultural and rural development projects. Regional development projects also offer possibilities of implementing beekeeping ventures. Certain crops planted in such projects can yield honey for the beekeeper as well as benefit from the pollination activities of the bees. Beekeeping can provide additional income to small farmers who plant these crops or have their bees nearby.

Plants benefit from insect pollination by an increased seed set. This results in increased seed production and better quality fruits. Honey bees are beneficial as pollinators in those areas where natural insect pollinators are lacking or are insufficient to pollinate large areas devoted to a single crop.

People involved in forestry projects are often interested in beekeeping. Beekeeping is an income-yielding undertaking based on a forest resource, yet it is not destructive to that resource. Someone who is earning an income from beekeeping quickly becomes an advocate for preserving the forest resource. Beekeeping also lessens the likelihood of brush fires being set by honey hunters when they burn bees out of a wild colony.

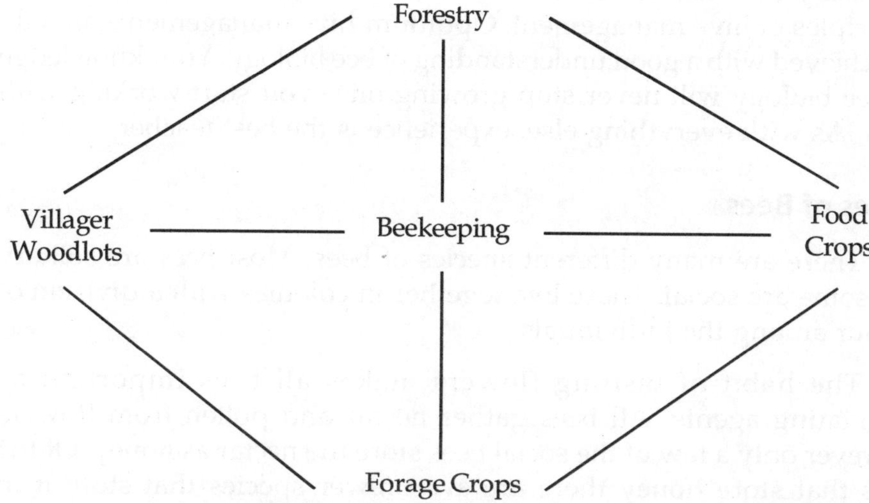

Tree species used in reforestation efforts which are also good bee forage can be instrumental in the establishment of a beekeeping industry. Beekeeping is a part of the utilization of a multi-purpose forest resource.

The following trees which are used for other purposes such as firewood, shelter belts, and shade also secrete sufficient nectar to produce yields of honey in some regions. As nectar secretion is dependent on many factors (climate, weather and soil), a tree may not be a good nectar producer when introduced into a new region. Check to see if a tree species is a good nectar producer under the conditions which it will be growing before advocating its use as a nectar source for bees.

Beekeeping as a Cooperative Activity

Beekeeping also works well in cooperatives. Many cooperatives have beekeeping projects as part of their activities. In some coops, beekeeping is the sole activity. These coops supply needed inputs, access to technical aid, and markets for honey and beeswax. In some cases, beekeeping crops have been very successful. Funding for beekeeping projects can be sought from the types of institutions mentioned above. Many missions and private donor agencies involved in small scale agricultural projects are often interested in beekeeping. This background information on the basics of honey bee biology is necessary in order to understand the nature of beekeeping and the principles of hive management. Optimum hive management can only be achieved with a good understanding of bee biology. Your knowledge of bee biology will never stop growing once you start working with bees. As with everything else, experience is the best teacher.

Types of Bees

There are many different species of bees. Most bees are solitary but some are social. These live together in colonies with a division of labour among the individuals.

The habit of visiting flowers makes all bees important as pollinating agents. All bees gather nectar and pollen from flowers however only a few of the social bees store the nectar as honey. Of the bees that store honey there are even fewer species that store it in sufficient quantity to make the effort of harvesting the honey worthwhile.

The true honey bee, *Apis* spp.

Although some "stingless" bees are robbed of their honey in tropical regions, bees of the genus *Apis*, the true honey bees, are the major producers of honey and other hive products. There are four species in the bee genus *Apis*— three which are native to Asia and one which is native to the Euro-African region. All of these are similar in appearance, though there are size and colour differences. All build vertical combs that are two cells thick.

In India four main source of honey are *Apis indica, Apis mellifera, Apis dorsata* and *Apis florae. Apis mellifera* the western hive bee has been introduced into most regions of the country for use in beekeeping. The giant or rock bee (*Apis dorsata*) and the little bee (*A. florea*) are found in Asia. Both of these bees build a single-comb, exposed nest. Nests are often seen hanging from branches of trees, roofs or ceilings. The adult bees hang in curtains around the nest to control nest conditions. Brood and honey stores are in the same comb - the brood in the lower section and the honey in the upper section.

Two species *viz.* Indian hive bee, *Apis cerana indica* and Europian bee, *Apis mellifera* normally build multi-comb nests in enclosed cavities.

These bees can be kept in hives and methods have been devised to allow for a more rational utilization of their potential. It is with these two species that a potential for beekeeping development exists.

The western hive bee (*Apis mellifera*) is native to western Asia, Europe, and Africa. There is tremendous variation in this bee across its range, and at least twenty different sub-species or "races" are recognized, broadly divided into European and African groups. Several races of this bee are considered especially desirable for beekeeping.

European races of the western hive bee have been introduced into most parts of the world, including the Americas, Australia and Asia. This bee has been studied intensely from both a strict biological and a beekeeping viewpoint. Under good conditions, desirable races build large colonies' and produce large surpluses of honey. Yields of 100 kg/year or better are possible under optimum conditions.

The eastern or Indian hive bee (*Apis cerana*, formerly *A. indica*) is native to Asia. Beekeeping developed with this bee in different regions of Asia, since it can be easily hived in man-made containers. Honey yields of up to 15-20 kg/year are obtained in some areas, but the average is much lower.

The issue of declining pollinator population

In recent years there is a world-wide decline in pollinator populations and diversity. The factors causing this decline could be the decline in the habitat, with the accompanying decrease in their food (nectar and pollen) supplies as a result of decline in pristine areas, land use changes, increase in monoculture-dominated agriculture, and negative impacts of modern agricultural interventions, e.g. use of chemical fertilizers and pesticides. Earlier, farmers used to grow a variety of crops, which bloomed during different months of the year and provided food and shelter for a number of natural insect pollinators and hence the pollination problem never existed. Monocropping also requires pesticide use to control various pests and diseases. Thus, it not only reduced the diversity of food sources of pollinator but also led to the killing of many pollinators due to pesticides. The insecticides have contributed to the extermination of both the diversity and abundance of pollinating insects.

Impact of decline in pollinator population and diversity

The decline in pollinator population and diversity presents a serious threat to agricultural production and conservation and maintenance of biodiversity in many parts of the world. One indicator of the decline in natural insect pollinators is decreasing crop yields and quality despite necessary agronomic inputs. Examples can be found in Himachal Pradesh in northwest India, northern Pakistan and parts of China where despite all agronomic inputs, production and quality of fruit crops, such as apples, almonds, cherries and pears, is declining. Extreme negative impact of declining pollinator populations can be seen in other areas, for example in northern Pakistan where both farmers and institutions have failed to understand the importance of managed pollination.

One implication of the decline in the pollinator populations as well as diversity is that it has created the need for managed pollination in order to maintain crop yields and quality. In fact, farmers engaged in cash crops' farming in those areas where pollinator populations have declined are forced to manage pollination of their crops through different ways.

Managed pollination as a solution to address the immediate problem of cash crop pollination

Promoting conservation and management of naturally occurring insect pollinators is very important for sustaining agricultural productivity in the long run. Efforts - both at research and extension level-must be continued to identify, assess and develop techniques/ methodology to rear and manage them for enhancing crop pollination. But as already explained the population of these bees are declining and there is a lack of scientific manpower and institutional infrastructure to rear and promote them in the countries of the Himalayan region. Moreover, the problem of pollination has already started in several areas. Therefore, promoting managed pollination is essential to address this immediate problem of inadequate pollination in several cash crops, for example, apples and pears. Our findings revealed two different cases of managed pollination of apple crops in the Himalayan region; one in Himachal Pradesh a small province in north-western Indian Himalayas where farmers are using manageable species of honeybees (*Apis cerana* and *Apis mellifera*) for pollination of their apple crop, and another in Maoxian valley located in the northwest of Sichuan Province of China where farmers employ "human bees" to pollinate their apples by hand (hand pollination). The details of these case studies are given in the following text.

Honeybees as the most efficient and manageable pollinators

As explained, many varieties of these cash crops are partially or completely self-incompatible and cannot produce fruit or seed without cross-pollination of their flowers. Moreover, it is not only self-incompatible varieties that benefit from cross-pollination, but self-fertile varieties also produce better quality fruit and seeds if they are cross pollinated. While other agronomic inputs, such as the use of manure, fertilizers, pesticides and irrigation are important, without cross-pollination desired crop yield and quality of harvest cannot be achieved.

Honeybees are the most widely known of all the bees because they provide honey, beeswax and other products and beekeeping is a prevailing tradition among mountain farming communities. They are the most efficient pollinators of cultivated crops because their body parts are especially modified to pick up pollen grains, they have hairy body, have potential for long working hours, show flower constancy and adaptability to different climates. Research has shown that pollination by honey bees increases fruit set, enhances fruit quality and reduces fruit drop. Among different species of honeybees, the hive-kept species (*Apis cerana* and *Apis mellifera*) are of special value because they can be managed for pollination and moved to fields/ orchards where and when necessary for pollination. Pollination using honeybees is the most cost-effective method for pollinating apple and other fruit crops. Use of beekeeping is, therefore, the most promising method of cash crop pollination in the Himalayan region.

Development of small-scale beekeeping with the eastern hive bee deserves attention in its native area. Although yields are considerably lower than with the western hive bee, this bee has the advantage of being well adapted to the area. It is more resistant to some of the disease and pest problems found in the area. Therefore, it is better able to survive under the minimal-management conditions which often characterize beekeeping at the small-farmer level.

In recent years, there has been an effort in Asia to replace the eastern honey bee with European races of the western honey bee. This has been successful only in temperate regions and only for large-scale, capital-intensive operations where the technology is available to control disease and parasite problems of the European races. For a small-scale development effort in this area, consider the likely beekeeping conditions in choosing a species.

The importation of bees is often fraught with problems. Importing bees for a small-scale project often makes the people psychologically dependent upon the outside source. They do not realize that there is a locally available bee resource, nor are they motivated to make use of it.

Table 1. Increase in crop yield due to bee pollination

Crop	Per cent increase in crop yield
Capsicum	227%
Tomato	160%
Cashew	157%
Pigeon pea	133%
Flat bean	128%
Chick pea	79.5%
Mustard	75%
Mango	68%
Banana	63%
Niger	60%
Papaya	60%
French bean	41%
Jowar	33%
Brinjal	31%
Ridge gourd	27%

Source: Rapid Impact Assessment Study, 2011, UTMT

References

Anonymous. 2014. Sustainable Farm Income from Integrated Farming. ICAR web site (www.icar.org.in) (*Source: CAZRI, Jodhpur*).

Chaturvedi, P. L. 1968. A tip for migratory beekeepers of Kumaon region. *Indian Bee Journal*, **31**: 48–51.

Dadant, C. P. 1924. First lessons in beekeeping. (*New Edition*, 2014)

Lal and Miller, 1990. Integrated farming systems for Sustainable Agriculture.

McGregor, S. E.1976. Insect pollination of cultivated crop plant. Agricultural Handbook, USDA-ARS, Washington D.C., New York, pp. 496.

Naim, M., & Bisht, D. S. 1978. Pollen gathering capacity of the Indian honey bee, *Apis cerana indica* F. *Indian Journal of Entomology*, **39**(2): 295–296.

Partap, U. & Partap T. 1997. Managed crop pollination. The Missing Dimension of Mountain Agricultural Productivity. International Centre for Integrated Mountain Development, Kathmandu. pp. 95–102.

Pastala , Vijaya and Krishnamoorthy, Sujana. 2012. Honey Bees - Farmers best friends. *LEISA India*, **14**(1): 1-35.

Rana S.S. and Chopra, Pankaj 2013. Integrated Farming System. HPKVV, Palampur, pp.90.

Shah, N. K. 1986. Hisalu, *Rubus ellipticus* Linn. a potential source of nectar to hive bees, *Apis cerana indica* Fabr. in northwestern Himalayas. *Indian Bee Journal*, **48**(1-4): 70

Shah, N. K., Melkania, N. P. 1982. Apiculture in the Himalayas: its practices, ecology and development. *Indian Bee Journal* **44**(3): 65-68

Singh, S. 1968. Beekeeping in India. Indian Council of Agricultural Research, New Delhi. 214 pp.

Suba Reddy, G. and Y.S. Ramakrishna.2005. Farming system modules in rainfed agriculture. Proc. of Symposium on '*Alternative Farming Systems*': Enhanced income and employment generation options for small and marginal farmers". (*Eds.* AK. Singh, B. Gangwar and S.K. Sharma). pp. 65-71.

Swamy, G. Madhava. 1985. Effect of diversified farming on income and employment. *Indian Journal of Agricultural Economics*, **40**(3): 333.

❑❑❑

Chapter 15

Quality Seed Production Issues and Strategies for Integrated Farming System (IFS)

D. VIJAY*, MALVIYA, A. MAITY, C.K. GUPTA AND V.K. WASNIK
ICAR - Indian Grassland and Fodder Research Institute, Jhansi-284003 (UP)

Introduction

Agriculture output is unsteady due to its dependence on complex factors viz., climate, market fluctuations, public policy etc. which are out of bounds of farming community. The emergence of Integrated Farming Systems (IFS) involving scientific agronomic practices and technologies has enabled development of alternative model to improve the profitability of farming activity. Integrated farming system (or integrated agriculture) is a commonly and broadly used word to explain a more integrated approach to farming as compared to monoculture approaches and refers to agricultural systems that integrate livestock and crop production or integrate fish and livestock etc. (Soni et al., 2014). Farming system requires new innovations and integrations of different available means for enhancing the sustainability of farming to small and marginal farmers. Innovation, at the level of an individual farming system, might be defined as the application of ideas that are new to the farming system, whether the new ideas are embodied in products, processes, services, or in work organisation, management or marketing systems (OECD 2005). Integrating seed production into farming systems is one such innovation.

Seed is a primary source for agriculture as well as animportant basis for human sustenance in the form of grain. Delivery of quality seeds of improved high yielding varieties ensures the improved crop production and enhances the ability of farming community to meet the upcoming environmental challenges. Food security is the fundamental objective of Integrated Farming System. Food security is dependent on seed security of farming communities. Integrated Farming system, a multi-disciplinary approach for solving the problems of small and marginal farmers, aims at enhancing their livelihood by integrating various farm enterprises and recycling of crop products and by products within farm itself. Seed production, a well-recognised farm enterprise fits well into the integrated farming system as it utilizes the crop products both for sustenance as input and as an income source. Farming system approach is one of the important solutions to face the challenges posed by present economic, political and technological environment as in this approach the different enterprises can be carefully undertaken and the location specific systems are developed based on available resources which will result into sustainable development (Dashora and Hari Singh, 2014). Thus,integratingseed production in the farming system approach helps in overall development of farming community. The assurance of quality of seeds helps in enhancing the productivity without dependence on additional resources and provides the back bone for the sustenance of the crop production activity in the integrated system.

Quality Seed production

Seed is the most common propagating material in maximum number of the crops and serves as the basic input of agriculture. As quoted in ancient Atharva Veda "Good seed in good soil realizes good yield". The quality of seed is a most important factor for realizing the yield potential of varieites.As per an estimate, all other factors being remaining same, the use of quality seed of high yielding varieties increased crop yield by 15-20% (Anonymous, 2007). Thus, production of quality seed paves the way for higher crop production. The quality assurance of the seed through certification is possible only in the released and notified varieties under section 5 of Seeds Act, 1966. The varietal release is a systematic procedure comprises of four major steps (i) varietal development (ii) testing (iii) identification and (iv) release and notification. Only the released and notified varieties comes under the purview of seeds act and can be certified by the certification Agency. The released and notified variety is maintained over years through maintenance breeding / nucleus seed production. Nucleus seed is the handful of seed developed by the original breeder while developing the variety. Later, it is maintained by ear to row method under stringent conditions to uphold the genetic purity. To produce sufficient quantity of quality seed manifold multiplication is done using generation system of seed production. The generation system consists of three seed classes viz., Breeder seed, Foundation seed and certified seed. Breeder seed is the progeny of the nucleus seed. It bears golden yellow tag. It is produced in the research station /institute by the original breeder or a designated breeder. The quality of the breeder seed is observed by monitoring team comprising of producing officer, crop breeder, representatives of AICRIP, NSP, State seed development corporation and seed certification agency. Foundation seed is produced from breeder seed under expert guidance by the National Seeds Corporation, State Farm Corporation of India, State Agricultural Universities, Co-operative societies of seed growers/ farmers, State Seed development corporations and private firms. It bears a white tag issued by the certification agency after monitoring the quality through field and seed inspection. In case of shortage, the Foundation stage II can be produced from Foundation seed stage I. Certified seed is the progeny of foundation seed and is produced by the foundation seed producing agencies like, NSC, SFCI, SSDC, SAUs, Co-operative societies and Private firms. It bears Azure Blue tag and its quality is assured by the Seed Certification Agency. While In case of crops with low seed multiplication ratio, certified seed stage II can be produced from certified seed stage I. But in any condition the Foundation and Certified classes should not be more than three generations. By following generation system of seed production seed multiplication can be done in sufficient quantities without any compromise in quality.

Quality concerns during seed production

The important quality parameters to be taken care of during seed production are- Genetic purity, Physical purity, Germination of harvested seed, Moisture content of harvested seed, Seed health and vigour. Among these at field level, genetic purity, physical purity and seed health will be taken care of during seed production.

a. ***Genetic purity:*** Maintenance of true to type of variety is called genetic purity and it helps in realizing the yield potential of the given variety. Genetic purity is to be maintained both in the field and after harvest.The genuineness of cultivars is to be maintained by maintenance breeding. All necessary precautions viz., isolation, growing suitable variety, roguing at regular intervals etc. are to be taken during seed production to avoid deterioration of varieties.

b. ***Physical purity:*** The harvested seed should be free from physical impurities and from obnoxious weedsso that the minimum physical purity as per the standards can be achieved. To maintain physical purity of seed lot mechanical mixtures are to be avoided during pre and postharvest operations along with roguing at regular intervals and isolation from other crops

c. *Seed health:* Theseedcrop should be free from diseased plants and even the seed should not contain pathogen infection. The source seed should be free from diseases and any diseased plants found in the field should be removed during rouging. Regular crop protection measures including prophylactic sprays are needed to avoid the diseases.

d. *Standards for seed quality:*To impose quality there is a need for standards, using which the quality can be assured. The Government of India has adopted certain standards for different quality parameters which needs to be invariably followed for assuring seed quality. These standards are called Indian Minimum Seed Certification Standards as the seed certification process completely depends on them (Tunwar and Singh, 1988). Certification is a process in which seed quality will be ensured by an independent organization called certification agency. Labels will be issued for the qualifying seed lot.During certification, the quality will be ensured both at field and seed level so that the ultimate product is of good quality. At field level for ensuring seed quality, field inspection will be carried out. During field inspection standards for isolation, designated diseases, objectionable weeds and off types will be ensured. Similarly during seed inspection in the seed testing laboratory, the quality standards for germination, physical purity and moisturewill be imposed for passing the testing lots. These standards vary from crop to crop. However the standards for genetic purity are similar for all crops and only differs among different classes of seed (Table 1).

Table 1. Genetic purity standards of different classes of seed

Seed Class	Standards for Minimum Genetic Purity (%)
Breeder seed	100.00
Foundation seed	99.00
Certified:	
i. Varieties, composites, synthetics &multilines	98.00
ii. Hybrids	95.00
iii. Hybrids of cotton, TPS, muskmelon, brinjal &tomato	90.00
iv. Hybrid castor	85.00

Principles of quality seed production

The principles of seed production are basically similar for all crops (Vijay, 2008). The principles can be categorized into two components viz. genetic principles and agronomic principles. The genetic principles are directed towards the maintenance of genetic purity of the seed produced whereas the agronomic principles includes crop management practices and other agronomic criteria specifically needed for seed production.

a. Genetic principles

The genetic purity (trueness to type) will get lost during multiplication resulting in deterioration of variety. The deteriorated variety is not able to provide the potential yield and quality as prescribed by the breeder during its release. Several factors are responsible for the varietal deterioration.

i) ***Developmental variations:*** The variations that creep into the varieties by growing them in other than the recommended climatic conditions, photoperiods, elevation, season etc. for several consecutive generations results in the deterioration of varieties.

ii) ***Mechanical mixtures***: These are most important source of varietal deterioration during seed production process. The mixture of unwanted materials including seeds of other varieties, weeds etc. during sowing, harvesting, threshing, winnowing, bagging and storage.

iii) ***Natural crossing***: This is also an important source of varietal deterioration both in self and cross pollinated crops. Natural crossing with other varieties, off types and weeds severely cast a doubt on genetic purity of seed produced.

iv) ***Mutations***: Even though mutations are not major source for deterioration of varieties but certainly they effect the genetic purity if not rouged out immediately after detection.

v) ***Minor genetic variations:*** The inherent variations present even in phenotypically uniform appearing varieties results in loss of genetic purity over a period of time. Due care during maintenance of varieties is must to avoid these variations to creep in.

vi) ***Selected influence of pests and diseases:***Susceptibility of varieties to newer pests and diseases sometimes causes varietal deterioration over a period of time.

vii) ***Techniques of plant breeder:*** Premature release of varieties results in segregation of certain characters. This is one of the reasons for faster deterioration of varieties.

Maintenance of genetic purity

Genetic purity is to be maintained for realizing the yield potential of released varieties. The above said factors results in natural deterioration of varieties over a period of time. Hence, necessary care should be taken for maintenance of genetic purity.

1. ***Generation system of seed multiplication:***Thegenetic purity of seed can be maintained by following the generation system of seed multiplication as each class is having defined standards for genetic purity as detailed earlier.
2. ***Isolation requirements***: The isolation distance helps in maintaining genetic purity by avoiding natural crossing and mechanical mixtures during various operations. The isolation distance depends on the type of pollination. For self-pollinated species it is as low as 3 m and is highest in cross pollinated crops up to 1600 m. Different types of isolation can be practiced based on the feasibility. The space isolation is the most common method and standards are there for each crop (Table 3). Time isolation, by planting at different times, is possible only in non tillering crops like maize. The last one, barrier isolation is created by erecting physical barriers between different varieties.
3. ***Roguing***: Removal of off-types or rogues or volunteer plants from the seed production plot is called rouging. Roguing is very important in maintaining genetic purity of a variety and should be followed without any excuse (Laverack and Turner, 1995). Rouging should be done at least in three crop phases i.e., during active tillering stage, flowering stage and maturity stage during field inspections.
4. ***Periodic testing for genetic purity:*** Regular testing of released varieties helps in checking the creeping variations due to genetic shifts, minor genetic variations and mutations.
5. ***Maintenance breeding:***Regular maintenance i.e. once in 2-3 years based on requirement strictly by following ear to row method avoids varietal deterioration. The changes occurred due to various factors can be detected and rectified during maintenance breeding with proper isolation distance and strict roguing throughout crop period.

6. ***Avoidance of genetic shifts:*** Genetic shifts are to be avoided by growing the varieties under suitable climatic conditions and regions for which they are released.

b. Agronomic principles

1. ***Selection of site***: The seed production site should have suitable climate for the selected crop and variety. It should be easily approachable with good transport facility. It is always better to take up seed production of same crop and variety in the whole area. This compact area approach results in production of good quality seed economically.
2. ***Selection of suitable variety***: Always suitable variety should be selected. This not only helps in avoidance of genetic shift but also helps in realizing full yield potential.
3. ***Source of seed***: The seed needs to be procured from authenticated source and of suitable class to ensure proper quality as per generation system of production.Source seed should be of higher class than the class of seed being taken up for seed production. The source seed should contain proper tag with quality details and should be from an authorized producer.
4. ***Land requirements***: The land selected should be non-problematic, well arable, fertile and properly levelled. The important precaution while selecting the land is the details of previous crop. Selection of land where same crop was not grown in the previous season helps the seed production plot free from volunteer plants.If previous crop is same, then invariably it should be of same variety and same or higher class of seed. The fertile soil helps in full expression of characters.
5. ***Land preparation***: Sufficient time should be spared for the preparation of land with proper ploughing/puddling and levelling. Initial irrigation followed by ploughing/harrowing for removal of weeds and volunteer plants should be adopted to maintain purity of the seed crop. Levelling not only helps in uniform growth and maturity of crop but facilitates easy identification of off-types.
6. ***Seed rate and spacing***: The wider spacing adopted in seed production facilitates favourable microclimate for the expression of characters and conducive for effective taking up of

management practices like roguing, weeding, fertilization etc. The wider spacing results in reduced seed rate in seed production compared to crop production.

7. *Gap filling and thinning* should not be practiced in seed production plots to avoid the variation among plants. The gap filled plants varies in their character expression in comparison to main crop. Thinning disturbs crop geometry thereby causing non uniform spacing.

8. ***Crop management***: All the crop management practices viz., irrigation schedule, pest and disease management, fertilizer application, weeding should be practiced in a timely manner as per the requirement of the crop and variety. This ensures proper development and enhanced productivity of seed crop.

9. ***Precautions during harvest and post-harvest operations***: During harvesting, threshing, winnowing and processing proper care should be taken to avoid mechanical mixtures as these stages are major source of mechanical mixtures.Harvesting by mechanical or manual means is a good source for admixtures particularly if more than one variety is harvested simultaneously. It is always advisable to take variety wise harvest and postharvest operations. Starting from carrying the harvested produce to threshing and winnowing utmost care should be taken to avoid contamination (Singhal, 2003). The threshing floor should be free from cracks and crevices. During mechanical processing maximum care should be taken for thorough cleaning of machinery during each operation.

10. ***Bagging and storage***:Baggingof processed seed is to be taken up in new gunny/ polyline bags to avoid admixtures. All bags should be labelled as per the type of seed for proper identification. The harvested seed needs to be stored under low temperature and humidity conditions for quality preservation. The moisture content of the seed should be as per the prescribed standard to avoid faster deterioration during storage. Regular checking of quality parameters during storage is required particularly under highly fluctuation conditions of temperature and relative humidity (RH).

Strategies for ensuring quality seed production

The seed quality is ensured through labelling as well as certification process. In India certification is voluntary but labelling is compulsory.

Thus all the seed material that are available for farmers in the market are labelled. In certification, an independent organization called seed certification agency duly formed by the state government will check the quality of seed lot both during field and seed level and issues the azure blue label for the qualified lots. In case of truthfully labelled seed, the producer himself guarantees the quality by mentioning the germination and physical purity on the opal green label.

a. Field level strategies

The pre harvest steps taken to ensure quality seed production comes under this category. This involves various measures taken at field level. The quality standards will be ensured by a process called field inspection, which will be carried out by the seed certification officer in case of foundation and certified seed production and by the concern seed production officer in case of truthfully labelled seed production.

Field inspection: The inspection of standing seed crops to verify whether they meet the standards or not is called field inspection. The field inspection is done mainly at three different stages of crop to identify and rogue out the off types, volunteer plants, pollen shedders, diseased plants and weeds so that the seed will be physically and genetically pure and true to type. The isolation distance will be monitored during field inspection to avoid varietal deterioration.

Stages of field inspection: The three main stages for filed inspection are vegetative, flowering and maturity stages. During vegetative / pre-flowering stage off types or rogues are to be identified based on morphological characters like plant type, height, leaf colour, leaf shape, stem colour, anthocyanin pigmentation, stipules/ auricle colour, stem and leaf hairiness, etc. During flowering stage the phenological characters viz., days to fifty percent flowering, no of days to flower (early/ late flowering) and floral characters like inflorescence type, its compactness, colour, length, shape, florets size, pigmentation if any, presence/ absence of awns, etc. should be observed. During Maturity/ Pre-harvesting stage the grain characters like length of grain, width of grain, grain type, colour, awns and the duration of maturity, plant height, type, etc. should be observed for identification of off types and rogues.

Field inspection procedure: After arrival to the seed production farm the seed inspector should verify the seed source, crop type, variety, cultivated area, class of seed, cropping history of the field, isolation etc. with the help of seed producer. Later he should monitor the field for off types, rogues, volunteer plants, pollen shedders, diseased plants, weeds etc. by moving through the field (Fig.1) and note their percentage to verify whether it meets the standards or not. The number of plants/ heads to be observed per field count depends upon the spacing. In case of wide spaced crops like maize, 100 number of plants/ heads are to be observed per count and in medium spaced crops like cowpea, 500 plants/ heads and in thickly sown crops like oat 1000 plants/ heads are to be observed per count. The number of field counts depends upon crop area as given below (Table 2).

Table 2. Minimum number of field counts

Cropped area	Minimum number of counts
up to 2 ha	5
2 to 4 ha	6
4 to 6 ha	7
6 to 8 ha	8
8 to 10 ha	9
Above 10 ha	10

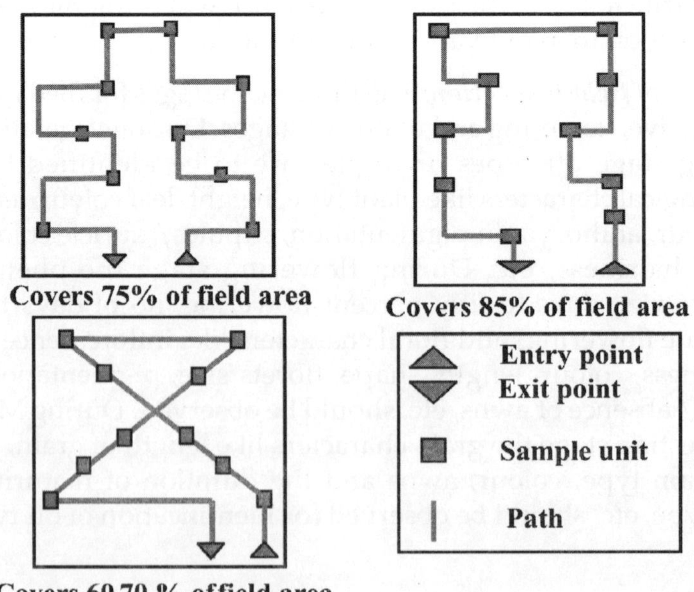

Fig. 1. Field movement designs and approximate percent field area coverage

If at any given field inspection the seed crop does not confirm the standards then the field will be rejected unless there is a scope for rectification and re-inspection. The number of inspections and standards for isolation and off types varies from crop to crop (Table 3).

Table 3. Field inspection standards some important crops

S. No	Name	Scientific Name	No. of Field Inspections	Stages for Field Inspection	Isolation Distance (m)		% Off types	
					F	C	F	C
I	**Cultivated crops**							
1	Sorghum & Sudan grass	*Sorghum bicolor*	3	Before flowering Flowering Maturity	200	100	0.1	0.2
2	Bajra	*Pennisetum glaucum*	3	Before flowering Flowering Maturity	400	200	0.05	0.1
3	Maize	*Zea mays*	2	Before flowering Flowering	400	200	1	1
4	Teosinte	*Euchlaena Mexicana*	3	Before flowering Flowering Maturity	200	100	0.1	0.5
5	Oats	*Avena sativa*	2	Flowering Maturity	3	3	0.05	0.2
6	Wheat	*Triticum spp.*	2	Flowering Maturity	3	3	0.05	0.2
7	Barley	*Hordeum vulgare*	2	Flowering Maturity	3	3	0.05	0.2
8	Paddy	*Oryza sativa*	2	Flowering Maturity	3	3	0.05	0.2
9	Cowpea	*Vigna ungiculata*	2	Before flowering Flowering and fruit stage	10	5	0.1	0.2

10	Pigeon pea	*Cajanus cajan*	2	Before flowering Flowering and fruit stage	250	100	0.1	0.2
11	Bengal gram	*Cicer arietinum*	2	Flowering Maturity	10	5	0.1	0.2
12	Black gram	*Vigna mungo*	2	Flowering Maturity	10	5	0.1	0.2
13	Guar/ Cluster bean	*Cyamopsis tetragonaloba*	2	Before flowering Flowering	10	5	0.1	0.2
14	Berseem	*Trifolium alexandrinum*	2	Flowering Maturity	400	100	0.2	1
15	Lucerne	*'Medicago sativa*	2	Flowering Maturity	400	100	0.2	0.1
16	Indian Bean	*Lablab purpureus*	2	Before flowering Flowering and fruit stage	10	5	0.1	0.2
17	Chickling vetch	*Lathyrus sativus*	2	Before flowering Flowering and fruit stage	10	5	0.1	0.2
18	Rice Bean	*Vigna umbellate*	2	Before flowering Flowering	50	20	0.1	0.2
19	Chinese cabbage	*Brassica pekinensis/ B. chinensis*	3	Before marketable head At marketable head Flowering	1600	1000	0.1	0.2
20	Sugar beet	*Beta vulgaris*	2	Before flowering Flowering	1600	1000	0.1	0.2
II	**Pasture species**							
21	Buffel grass	*Cenchrus ciliaris*	3	Before flowering	20	10	0.1	1

				Flowering Maturity				
22	Birdwood grass	*Cenchrus setigerus*	3	Before flowering Flowering Maturity	20	10	0.1	1
23	Daaraf Grass	*Chrysopogon fulvus* (*Andropogon montanus*)	3	Before flowering Flowering Maturity	20	10	0.1	1
24	Dinanath grass	*Pennisetum pedicellatum*	3	Before flowering Flowering Maturity	20	10	0.1	1
25	Guinea grass	*Panicum maximum*	3	Before flowering Flowering Maturity	20	10	0.1	1
26	Marvel grass	*Dichanthium annulatum*	3	Before flowering Flowering Maturity	20	10	0.2	1
27	Napier grass & Hybrid Napier	*Pennisetum purpurem & P. glaucum× P.purpureum*	2	Flowering Maturity	10	10	0.01	0.3
28	Pigeon/ Setaria grass	*Setaria anceps*	3	Before flowering Flowering Maturity	400	200	0.1	1
29	Indian Clover	*Melilotus*	2	Flowering Maturity	50	25	0.2	0.1
30	Stylo	*Stylosanthes. spp*	3	Before flowering Flowering Maturity	50	25	0.1	1

Source: Tunwar and Singh, 1988

b. Seed level strategies

Seed quality should be maintained and verified after harvesting the seed crop as the produce will be used for growing crop in next season. Quality assessment will be done by a procedure called seed testing. Seed testing will be done on a representative sample as it is not feasible to test the entire lot.

***Seed Sampling*:** The quantity of seed tested in the laboratory is very small compared to the size of seed lot which it is intended to represent. Hence, the sample collected should be a true representative of the seed lot and is of uniform in nature and of sufficient quantity to conduct the tests(Agrawal, 1997). The free flowing seed sample is collected by inserting a trier into the seed bag. In case of chaffy/ non-free flowing seeds hand sampling is the most satisfactory method. Sampling should be done from as many containers as possible to have a representative sample.

***Seed Testing*:** The science of seed testing, that is, the science of evaluating the planting value of seed has been developed to achieve the following objectives,

1. To determine the quality of seed i.e. their suitability for planting
2. To identify the seed quality problems and their probable cause
3. To determine if the seed meets the established quality standards
4. To establish quality and provide a basis for price and consumer discrimination among lots in the market.

Seed testing comprises the evaluation of quality attributes of the seeds and compares them with the given standards to assess their suitability for the development of next generation. If the assessed quality is equal to or above the standard value it may be considered as quality seed and is allowed for sowing purpose. If it fails to meet the standards it should be treated as non-seed. The quality of seed material is assessed by doing following tests in seed testing laboratory on a representative sample of seed lot.

❑ Physical Purity testing	Compulsory tests
❑ Seed germination testing	
❑ Seed moisture testing	
❑ Testing of genuineness of cultivar	Compulsory for hybrids
❑ Seed vigour testing	Optional test
❑ Seed health testing	Optional test

i) **Physical purity:** The physical purity is a significant quality parameter as it not only helps in maintaining the quality through reducing the weed seed, other crop seed, and un-necessary inert matter but also influences consumer by visual appearance. It directly impacts the yield by reducing competitive other crops and weeds from the main crop. Only the pure fraction obtained from this test will be used for further germination testing.

ii) **Seed Germination:** Germination is the most important quality test parameter evaluating the viability as well as the planting value of a seed lot. The viable seed when receives congenial conditions like sufficient water, oxygen, proper substratum and suitable temperature, imbibes water and starts respiration and other metabolic functions culminating in protrusion of radicle. This is considered as germination in botanical/ physiological sense. But from seed testing point of view the seed is considered to be germinated only when the resultant seedling has the capacity for continued development. Germination is tested in the laboratory under controlled conditions as field testing does not produce reproducible results.

iii) **Genetic purity:** The genetic purity helps in maintaining the trueness of variety. The genuineness of cultivar is tested either by other distinguishable varieties (ODV) test for self-pollinated crops or by grow out test for hybrids and cross pollinated crops.

iv) **Moisture content:** Seed moisture content is one of the important quality parameter influencing its viability particularly during storage. Seeds in general are hygroscopic and maintains equilibrium with surrounding relative humidity. In combination with high temperature, high moisture deteriorates seed quality rapidly and makes it unsuitable for propagation. Since seed storage till next sowing is inevitable, proper care is needed for maintaining standard seed moisture.

v) **Seed Health:** Seed being a highly complex biologically living substance, may contain diverse group of microorganisms and viruses including both parasites and saprophytes (Agrawal, 1997). Most of the dreaded diseases are seed borne. Seed borne pathogens drastically reduces yield, sometimes even up to 100 percent. Hence, it is necessary to detect the seed borne pathogens to evaluate its planting value and to advocate proper seed treatment to contain them.

vi) **Seed Vigour:** Seed vigour is relatively a new parameter in the seed testing laboratory. Seed vigour is the degree of liveliness and is the best indicative of field emergence potential. The decline in vigour precedes the loss in germination and two lots with same viability may differ in their field performance due to difference in vigour. There are many possible reasons for differences in vigour but the major explanation for vigour difference among different seed lots of same variety is seed ageing (Powell & Mathews, 1995).

Conclusion

Quality seed production is an integral part of effective and successful farming system. Seed production not only helps in utilizing the produce for the next crop but also remunerative if excess is shared with others. By incorporating seed production in the farming system the true objective of it i.e. food security can be achieved through seed security. The success of seed production depends mainly on following certain principles for quality maintenance. All the farmers adopting seed production should be invariably know the issues and strategies for quality maintenance both at field and seed level. The purity, both genetic and physical, should be maintained by following strategies like isolation and roguing during regular field inspections. Similarly postharvest seed testing helps in checking the quality parameters.

References

Agrawal R L. 1997. *Seed Technology*, pp.829. Oxford and IBH Publishing Co. Pvt. Ltd, New Delhi.

Anonymous. 2007. MSP Annual Report 2006-07 of Seed Production in Agricultural crops and fisheries. Directorate of Seed Research, Mau.

Dashora L N and Hari Singh. 2014. Integrated Farming System-Need of Today. *International Journal of Applied Life Sciences and Engineering* **1**(1): 28-37.

Laverack G K and Turner M R. 1995. Roguing seed crops for genetic purity, a review. *Plant varieties and seeds* **8**: 29-45.

OECD. 2005. Oslo manual: guidelines for collecting and interpreting innovation data, pp.163. OECD, Paris.

Powell A A and Mathews S. 1995. Detection of seed borne pathogens. *Techniques in Seed Science,* pp143-152. Agrawal P K and Dadlani M (Eds) South Asian Publishers, New Delhi.

Singhal N C. 2003. *Hybrid seed production in field crops (Principles and Practices)*, pp 308. Kalyani Publishers, New Delhi.

Soni R P, Katoch M and Ladohia R. 2014. Integrated farming systems: a review. *IOSR-Journal of Agriculture and Veterinary Science* **7**(10): 36-42.

Tunwar N S and Singh S V. 1988. *Indian Minimum Seed Certification Standards,* pp.388. The Central Seed Certification Board, Department of Agriculture, Ministry of Agriculture, Government of India, New Delhi.

Vijay D. 2008. Underlying principles for successful seed production. *Annals of Experimental Agriculture and Allied Sciences* **3**(1): 54-56.

❑❑❑

Chapter 16

Resource Conservation Technologies for Sustainable Agriculture

A.K. DIXIT AND T. KIRAN KUMAR

ICAR - Indian Grassland and Fodder Research Institute, Jhansi-284003

Introduction

In the past 40 years, per capita world food production has grown by 17%, with average per capita food consumption in 2003 of 2780 kcal/day (FAO, 2005), where a majority of the chronically hungry are small farmers who produce much of what they eat. Yet consumption in 33 poor countries is still less than 2200 kcal/day. As the new century begins, the world faces enormous challenges to meet the food, feed and fibre needs of a growing population with rising incomes. It is estimated that by 2025 the global population will be approximately 7.9 billion, up from 6 billion currently. Global cereal and meat demands will increase by 46% and 56% respectively (Rosegrant et al 2002). Food demand will both grow and shift in the coming decades, as population growth increases absolute demand for food; economic growth increases people's purchasing power; growing urbanization encourages people to adopt new diets and climate change threatens both land and water resources. Ever-growing demands and increasing pressures on land (Anonymous, 2002) and water resources mean that agriculture has to become even more productive, efficient and environmentally sound (Table 1).

Table 1: More people means less land per person

Year	World population (Billion)	Arable land and permanent crop (billion ha)	Farm land per person (ha)
1950	2.5	1.3	0.5
1975	4.0	1.4	0.4
2000	6.0	1.5	0.3
2020	7.5	1.5	0.2

Anonymous, 2002

Why Resource conservation technologies?

Land and water degradation is increasingly posing a threat to food security and the livelihoods of rural people who often live on degradation prone lands (Uphoff, 2002). Traditional farming practices normally involve intensive soil tillage and higher use external inputs. Farmers realised that in the longer run, under conventional farming system leads to lower productivity, multi nutrient deficiency, insect-pest building up, herbicidal resistance, reduces soil organic matter content and exposes soil to wind and water erosion.

Components of Resource conservation technologies

Resource conservation technologies (RCTs) are at the heart of sustainable agriculture, providing dynamic solutions to problems encountered in increasing food production without damaging the eco-system or depleting natural resources for future generations. RCT provides comparable yields to intensive conventional farming reduce water consumption, reduce negative impacts on the environmental quality. Hence we can say that it relies on the sustainable use of high-input technologies. Resource conservation technologies refer to those practices that enhance resource or input use efficiency such as new varieties that use nutrients more efficiently, zero or reduced tillage practices that save fuel and improve soil health, furrow irrigated raised bed system (FIRBS) or Broad bed furrow (BBF) and land levelling that save water, integration of trees and livestock in farming system that efficiently utilises farm inputs may be considered as RCTs.

Tillage management for sustainable agriculture

Conservation tillage is any tillage and planting system that covers more than 30% of the soil surface with crop residue after planting to reduce soil erosion by water. However, this is the minimum amount of residue generally considered to reduce soil erosion to acceptable levels. To reduce significantly reduce surface evaporation, 50% of crop residue cover is generally considered necessary. Under techniques developed to suit various crops and environments; zero tillage, minimum tillage, ridge tillage, mulch tillage, seed is sown directly into the previous crop's stubble, with little or no intermediate tillage. Later, crop rotation and maintenance of permanent or semi-permanent groundcover (live cover crop or crop residue mulch) are included as an integral part of this approach. Conservation tillage protects the upper soil layer from wind and water erosion and loss of ground moisture. It also improves soil biodiversity by providing a congenial environment for bacteria, insects and fungi. The abundance of soil organisms helps decomposition of mulch and its incorporation into soil as humus, contributing to stabilisation of soil structure and enhancement of soil fertility. Although benefits of conservation tillage are now well recognised, there are some areas of concern that generate resistance to changing over from conventional farming. One such area relates to weed management during the transition period, which may require a different management strategy. Similarly, farmers are concerned about potential for changes in weeds, diseases and insect pest problems.

Specific issued addressed by conservation technologies

i. **Soil structure deficiencies and compaction:** Restores soil using cover crops, residue management, crop rotation, and minimum/ zero tillage.

ii. **Nutrient deficiency:** Relies on integrated nutrient management through biological regeneration, plus targeted use of organic or inorganic fertilizers and includes integration of livestock production for nutrient recycling.

iii. **Water deficit:** Emphasizes management of soil organic matter (use of cover crops and mulch) for efficient capture of rainfall, soil moisture conservation and targeted irrigation.

iv. **Erosion control:** Minimizes erosion through reduced or no-till practices along with retention of cover crops/residue in arable agriculture and vegetation management in non-crop situations.

v. **Pest management:** Manages insect, weed, diseases and other pests using IPM approaches, which are economic, environmentally safe, and socially acceptable.

vi. **Environmental degradation:** Reduces soil erosion, water run-off from farms and emission of green house gases.

vii. **Loss of biodiversity and wildlife habitat:** Additional cover provided by crop mulch encourages micro-fauna and flora diversity and other wildlife species. Sustainable increases in productivity from existing areas avoid spread of agriculture to marginal areas.

Agroforestry for sustainable agriculture

The integrated use of suitable soil and water conservation practices, trees plantation and growing of crops/grasses gives self sustainability to the system. Substantial research work on rehabilitation of degraded lands has been carried out in India which includes adoption of various agroforestry systems such as silvi-pasture, agri-silviculture, hortipastures, agri-silvi-pasture and agri-silvi-hortipasture in culturable and non-culturable lands and rehabilitation of degraded and mine spoil lands in a specific habitat (Dhyani et al., 2005). In agroforestry, aspects related to tree and species selection, planting geometry, perenniality of the system, land use as per the topography, soil and water conservation, and soil fertility is important. Tree-crop interactions are a key consideration when designing an agroforestry system and a layout for planting. Basic principles included is how to increase the overall value of the system, maximize complementarities, decrease or eliminate competition; and minimize crop displacement, through appropriate tree management. It has been possible to increase land productivity from less than 1.5 t $ha^{-1}y^{-1}$ to about 10 t $ha^{-1}y^{-1}$ (10-year rotation) by developing silvipasture systems. The additional forage availability through these systems is likely to reduce grazing pressure and thus have important environmental implications. The following forage based agroforestry systems are suitable for achieving higher biomass per unit area and mitigate the ill effects of tillage based intensive agriculture;

i. **Silvipasture (Trees + grasses/legumes)**: Silvipastoral system which is mixed land use system for forage, livestock and wood production has paramount significance owing to forage supply from cultivated, marginal and degraded lands. The mixture of trees, grasses and legumes conserve soil and moisture. Establishment of successful silvi-pastures includes- selection of suitable species of trees, shrubs, grasses and legumes for different regions, adoption of measures for protecting the area from animals, suitable soil and water conservation measures and planting of tree and under storey components. Silvipasture system with trees Trees (*Acacia spp., Dalbergia sissoo, Leucaena lecocephata, Albizia lebbeck, Hardwickia binata, Grewia optiva, Melia spp.,* etc.) + Grasses/legumes (*Cenchrus, Chrysopogon, Panicum, Pennisetum, Dichanthium,* Stylo, *Macroptilium* etc) are becoming very popular with the farmers.

ii. **Hortipasture (Fruit trees + grasses/legumes):** Hortipasture system integrate livestock husbandry with fruit/plantation crops with main aim to enhance biomass productivity of degraded lands and in turn helps in increasing the livestock production. Hortipasture system supplies the protective food (fruit) for human being and fodder for animal and thus helps in bridging the wide gap between the supply and demand of fruit and fodder. Fruit trees (Anwala, guava, Ber, Citrus, Karonda and Mango etc.) along with Grasses/legumes (Anjan grass, Guinea grass, Dinanath grass, stylo, etc can be grown in wastelands under such hortipastoral systems.

iii. **Agrisilviculture (Trees + Crops):** In agri-silvicultural systems, MPTS (multi - purpose tree species) including fodder trees can be grown in association with field crops. Trees are normally pruned annually yielding fodder as well as fuel wood. In arid and semi-arid regions, hardy species like *Acacia tortilis, Albizia lebbeck A. nilotica, Dalbergia sissoo, Prosopis cineraria, Hardwickia binata, Azadirachta indica, Ailanthus excelsa, Zizyphus jujube, Grewia optiva,* etc. could be grown along with dry land crops. The timber-cum-fodder trees produced fuel, fodder and timber besides the crop production.

iv. **Alley cropping:** Alley cropping or hedgerow intercropping is an agroforestry practice in which perennial, preferably leguminous trees or shrubs are grown simultaneously with an arable crop. The trees, managed as hedgerows, are grown in wide rows and the crop is planted in the interspace or 'alley' between the tree rows. During the cropping phase the trees are pruned and the prunings used as green manure or mulch on the crop to improve the organic matter status of the soil and to provide nutrients, particularly nitrogen, to the crop.

Livestock integration in to farming system for sustainable agriculture

The crop–livestock combination offers farmers a more diverse source of food and income, reducing risk, diversifying labour use and recycling resources (FAO, 2001a & b). It is a common practice for small-scale farmers in tropical countries to practise both crop and livestock production from the same farm. There are obvious benefits of this practice such as nutrient cycling/transfer from one area to another, animal products (milk, meat, hides, skins, and manure) and animal power for the farm. However, the fact that livestock also feed on crop residue causes conflict with the conservation agriculture requirement that crop residues should be left or recycled back into the farm soon after harvesting. There is need, therefore, to include complementary livestock feeding strategies to avoid conflict between the two and ensure the success of both.

Benefits of resource conserving technologies

1. **Economic benefit:** Increases in land and labour productivity and lower operational costs, leading to increased farm incomes, have been commonly reported by farmers as one of the main reasons for adopting conservation technologies.

2. **Efficiency of natural resource use:** Trees and forages promote nutrient cycling and N fixation. Agroforestry system reduces runoff while increasing water infiltration and holding capacity in the soil.

3. **Risk management:** Long-term gains in productivity and more stable production can only be achieved by improving soil quality, which also reduces risks for farmers by stabilising yields, especially in dryland agriculture.

4. **Environmental impact:** Conservation tillage is improving soil quality, reducing erosion and water run-off from agricultural land, protecting water quality and reducing CO_2 emissions in to the atmosphere. These benefits help to meet demanding environmental protection regulations. Agroforestry can provide filtered shade which conserves water and reduces evapo-transpiration, keeps topsoil cool and helps maintain healthy soil biota activity. Tree wind breaks protect crops from wind damage and soils from wind erosion and drying. Wind breaks combined

with CA give more complete protection. Undisturbed tree, crop roots and mycorrhizal systems reduce nutrient leaching, bind soil and prevent erosion. Tree leaf litter enhances soil physical, chemical and biological conditions making soils more resilient to erosive forces.

5. **Soil quality:** Soil quality is largely determined by organic matter content and responds to soil management practices. Over time, conventional cultivation generally results in a reduction in the organic matter content of soils.

6. **Soil erosion:** Soil erosion is a major environmental threat worldwide. It is estimated that over the last 40 years, nearly one third of world's arable land has been lost to erosion. Quality soil continues to be lost at the rate of over 10 million hectares per annum. Intensification of conventional agriculture and adoption of cropping systems that leave the soil surface bare during the rainy season have largely contributed to this trend. Reduction or elimination of tillage along with retention of crop residue helps to reduce soil erosion.

7. **Reduced carbon dioxide (greenhouse gas) emission and improved air quality:** Maintenance of mulch under conservation tillage systems increases the ability of soil to sequestrate CO_2 and reduces emissions, protecting the atmosphere.

References

Anonymous, 2002. International Food Policy Research Institute,

Dhyani, S.K.; Sharda, V.N. and Samra, J.S. 2005. Agroforestry for sustainable management for soil, water and environment quality: looking back to think ahead. *Range Mgmt. & Agroforestry*, **26**(1): 71-83.

FAO, 2001a. Conservation Agriculture: Case studies in Latin America and Africa. FAO (FAO Soils Bulletin; 78).

FAO, 2001b. Mixed Crop–Livestock Farming. FAO, Rome.

FAO, 2005. *FAOSTAT Database*; Rome.

Rosegrant MW, Ximing Cai and SA Cline 2002. World Water and Food to 2025: Dealing with Scarcity International Food Policy Research Institute, Washington DC, USA

Uphoff N., 2002. *Agroecological Innovations*; Earthscan: London.

□□□

Chapter 17

Balanced Plant Nutrition for Quality Fodder and Milk Production

SB TRIPATHI AND AK DIXIT
ICAR - Indian Grassland and Fodder Research Institute, Jhansi-284003

Agriculture is the back bone of Indian economy and livestock sector is an integral part of agriculture. Although the milk producers in our country are more but the production rate per animals is low compared to world average and much below from the developed countries. The major constraints of low productivity of our milch animals are lack of adequate forage availability both in quantity and quality and genetically poor breed. It is well known that the fodder and feed constituents about 60% of total cost of milk production. Since fodder is a cheaper source of important feed constituents than traditional feeding concentrates *i.e.* grains and oil seed cakes mixtures. In this view, it is more important to put more fodder in feeding system of animals for increasing milk yield and maintenance of animal health. The scarcity of the quality forage and rapid decline grazing resources has made the livestock to suffer continuously from animal nutrition resulting in their production potential at sub-optimum level.

To achieve desired quantity and quality forages as well as animal origin products like milk, the use of cultivated lands in forage production under different production systems would be one of the options. Being forage plants especially the multi-cut and perennial ones are heavy feeder of plant nutrients and remove large amount of nutrients from the soils. The soil / agronomic management practices like addition of nutrients in soil through fertilizers and other sources in balanced form can be addressed to overcome nutrient deficiency / toxicity with increased quality fodder yield. The practices of feeding of enriched protein and mineral nutrient elements containing fodder to milch animals would not only produce economical more milk yield but well for general health of animal. In addition to this, green fodder based milk production would be environmentally free from pollution.

The research work carried out at IGFRI Jhansi and other places on "balanced plant nutrition for quality fodder and milk production" under various production systems are highlighted in the following ways.

Nutrient Status of Soil

Indian soils are mostly deficient in available major nutrients (NPK) except K. The available content of N and P varies from very low to low level (60-70% soils) while K status is ranged from low to medium (40-50% soils). Some soils are rich in K availability. Amongst, secondary nutrients, availability of S in soils are lower than Ca and Mg. In general, about 40-50 per cent soils are low in available S (below 10 ppm). The soils under intensive cropping system, particularly medium to course textural soils are most deficient in S. Calcareous and alkaline soils are also found in S deficiency. Mostly all soils except acid soils are rich in Ca. As regards Mg, availability, its deficiency is observed in acidic nature of sandy soils.

Micronutrients function largely in plant enzyme systems and are required in very small amounts. Four of the seven micronutrients are used as cations and three as anions. The amount available to plants is is closely related to soil conditions and plant species. Iron for example, is one of the most abundant elements in soils. Some soils are high in Fe, yet need its fertilization for profitable production. Copper availability also decreased with increased soil pH. Developed organic soils and leached sandy soils are most deficient. But its deficiencies are common in alkaline soils for all plants, because of the insolubility of the hydroxide and oxide forms of Fe and Mn. Cereals and grasses tend to have a manganese deficiency. Zn deficiency is common in calcareous soils where high pH reduces Zn availability. In acidic soils, Zn has been leached from the soils and shows its deficiency. Organic soils also low in Zn. Some cases, Zn deficiency is caused by high phosphorus fertilization.

In dry weather that limits decomposition of organic matter in the surface has caused boron deficiency. Fixation at high pH and leaching of boron from acid soils needs fertilization. The maximum availability of B is obtained near pH 7.0. Mo availability is observed to be low in soil having high pH *i.e.* calcareous and alkaline as well as low pH acidic soils (<5.5 pH). As regards chlorine, its requirement is very small, even though chlorine may be one of the most abundant anion in plants. It receives from the atmosphere and potassium fertilizer. The status of some important nutrients is given in Table 1.

Table 1. Soil nutrient Ranges.

Soil Nutrients	Nutrient Ranges		
	Low	Medium	High
Organic Carbon (%)	< 0.5	0.5-1.0	>1.0
Available nutrients (kg/ha.)			
N	<250	250-500	>500
P	< 10	10-20	>20
K	< 100	100-250	>250
S (ppm)	< 10	10-20	>20
Zn (ppm) Normal soil	< 0.5	0.5-1.0	>1.0
Zn (ppm) Calcarous soil	< 1.0	1.0-2.5	>2.5
Mn (ppm)	< 4.1	4.1-10.0	>10
Cu (ppm)	< 0.5	0.5-1.0	>1.0
Fe (ppm)	< 5	5-10	>10

Hence, soil deficient in available nutrients needed adequate fertilization of respective nutrients for the completion of crop growth and its development. The excess fertilization may show harmful effect e.g. excess of K on Mg, P on Zn and S and Cu on Mo availability. In general, the micronutrients applications may be based on soil test recommendations, otherwise their application may harmful by showing their toxicities and deficiencies of other nutrients on crop plants.

Soil, Fertilizers and Animal Health

It has been realized that plants are commonly deficient in mineral elements and exhibit deficiency (disease). It is not surprise that plant may contain inadequate amount of certain mineral elements for optimum animals' health. One of the most striking cases is the phosphorus deficiency seen in cattle grazing on low available P soils in tropics. In some cases plant contain toxic amount of elements that create animal diseases. In this direction there is need to take cautions in making a sweeping generalization about soil fertility and human health. There is great variation is plant composition due to genetic differences. In this situation the supplementation of the plant nutrients through fertilizers or any other sources may be done accordingly based on soil test, otherwise health of the soil will not sustain due to deficient or excess level of fertilizer nutrients.

Plant and Animal Requirements

Animal require about 20 different mineral elements, most of them the same as those required for plants. Boron however, is needed by plants but not by animals. The oldest and most common dietary supplement used by farmers is salt (sodium chloride) to supply sodium, which is not essential for the most plants. Animals, however, need chromium, iodine and selenium in addition to the elements needed by plants. These elements are in soils and are absorbed by the plants, even though they are not required, thus plants can be perfectly healthy and unable to satisfy the dietary need of animals.

In several nutritional surveys for forage quality evaluations shows inadequacy of mineral elements like Ca, P, Cu and Zn in the feed and forages of lactating cows, in general (Table 2). The requirement of other mineral elements, likes Mo, Se, Co etc. are location specific and depend upon the type of soil conditions and management practices, keeping in view, the role of some important plant nutrients are discussed below.

Nitrogen (N)

The role of N is known for the formation of proteins in forages, which are main source of food for human being as well as animals. In deficient N soils and need fertilization for producing good quality fodder in sufficient quantities. In the feeding of low quality fodder to milch animals, the supplement N of N in the form of 4% urea is required for promoting adequate level of protein in the body. Excess N in the form of NO_3 in the feed is not very toxic to animal; however, nitrates in the digestive tract of animal or in stored food converted into nitrites and are very toxic to animals by interfering transport of O_2 in blood stream. The nitrate accumulation usually is obtained with higher level of N fertilization particularly in leguminous fodders and cultivated grasses like hybrid bajra napier and guinea grass etc. Nitrate accumulation in plants is favored by drought and cloudy weather. The fodder which contains more than 2.5 percent NO_3 is not recommended to feed animals due to nitrate poisoning. The application of N through manures, crop residues and fertilizers in integrated manner is needed to crop plants.

Table 2. Critical limits of secondary and micro nutrients in plant and animal

Nutrients	Plant			Animal	
	Deficient	Sufficient	Toxic	Young stock	Lactating stock
Secondary Nutrients (%)					
Ca	< 0.1	0.2-1.0	-	0.50	0.70
Mg	< 0.1	0.1-0.4	-	0.18	0.20
S	< 0.1-0.2	0.1-0.4	-	0.16	0.32
Micronutrients (mg kg^{-1})					
Cu	< 2.5	5.30	> 200	8.0	10
Fe	< 50	100-500	>500	50	50
Mn	< 15-25	20-300	300-500	20	40
Zn	< 10-20	27-150	100-400	30	40
Mo	< 0.10	0.1-2.0	>100	< 2.0	< 2.0
B	< 5-30	10-200	50-200	-	-
Co	-	-	-	0.10	0.20
Ni	< 0.10	0.1-10	-	-	-
Cl	< 100	100-500	>500	0.50	0.20
I	-	-	-	0.25	0.60
Se	-	-	-	0.10	0.10

Phosphorus (P)

It is the most critical mineral element deficiency for grazing animals. The low soil P commonly limits plant growth. Weak bone of cattle grazed on grass is quite common due to phosphorus deficiency. Some of the complex relationship between P levels in soils and P deficiency is observed in many animals. As the dose of P fertilizer increased, the P contents in forage plants are markedly increased. Since cattle require about 0.3 percent P in their diet for optimum growth and it can be easily obtained by P fertilization in forage plants. The cattle which are totally on grass grazing need a phosphorus supplement to ensure adequate diet.

Potassium (K)

The role of K is very limited as constituent of plant compounds and function in regulatory mechanisms i.e. photosynthesis, carbohydrate transformation and protein synthesis. The deficiency symptom of K is found very rare in plants and animals, because majority of our soils have adequate amount of K. In low K soils, there is need of K fertilization for increasing K content in crop plants. The plants having sufficient K is assumed perfect diet of animals. In high K soil forages have adequate K content and no need of crop fertilization.

Calcium (Ca)

Plants and animals need relatively large amount of Ca. Rarely or only occasionally forage plants are not suffered Ca deficiency because some other soil factor becomes limiting before low soil Ca limit growth. Liming acid soils enable to grow more legume crops like berseem, alfalfa and peas etc. as they have more Ca content for feeding. It has been seen that cows deficient in Ca produce less milk than cows on adequate Ca in diets results more milk yields. As compared to graminaceous forages, leguminous forages supply sufficient Ca to grazing ruminants. Hence some care is required when animal fed of leguminous fodder and P rich supplement for maintaining desirable range of Ca: P ratio (1.5: 1 to 2:1).

Magnesium (Mg)

In animal production system, Mg behaves as Ca. The acid soils low in Ca usually low in Mg. The deficiency symptom on plants is fairly common in acidic and sandy soils. The higher levels of K inhibit Mg absorption. As with Ca, the accumulations of Mg in plants is strongly affected by species, where leguminous plants are highly rich in Mg than cereals and grasses. Grass tetany or grass staggers is common animal disease caused by low Mg in the grazed cattle and sheep. Pregnant and lactating animal are the most susceptible, Mg fertilizer can be applied to increase Mg and control tetany disease.

Sulphur (S)

It is an important secondary mineral element for the constituent of essential amino acids (methionine, cystine and cystein) and thiamine (vitamins). These amino acids are recognized to form proteins. Cattles on pasture requires 0.15% S in their diets. Low levels of dietary S results in poor utilization of non-protein nitrogen, which in turn reduces microbial growth and fermentation in rumen. The poor appetite, emaciation and dullness are sign of S deficiency. A diet high S and low in fibre induces S toxicity. This toxicity can interface with Se, Cu, Mo and thiamine metabolism of ruminants. A deficiency of S either limits protein formulation and lead to the accumulation of other ingredient in non protein form of low molecular weight. In excess level of N and limiting S, there is as accumulation of non protein organic forms of N, such as amides, amino acid and inorganic N (NO_3). On the other land the N supply is limited and S in abundant, S in non-legumes tends to accumulate in non protein forms of S. Therefore N/S ratio is an important parameter of quality determination. An indication of high N/S ratio assumes deficiency of S; a low ratio for adequate S or possible a deficiency of N. The dietary level of N: S ratio from 10:1 to 15.1 is adequate. Sulphur is also known to decrease toxicity of HCN and NO_3 concentration in certain forages i.e. HCN in sorghum and NO_3 in napier bajra hybrid and leguminous fodder (cowpea & berseem) under limited moisture. The excessive S supply in many grasses tends to accumulate sulphate in luxury amount which causes antagonistic effect with other nutrients like Zn, P and Mo etc.

Zinc (Zn)

Zinc is essential for enzyme systems and metabolic reactions as well as carbohydrate production; it is also involved of growth promoting hormones. Being an immobile in nature its deficiency symptom appears first on younger leaves. Plants and animals both develop Zn deficiency. In plants Zn deficiency frequently observed in sandy soils, where irrigation water are used in excess. The fertilization of Zn in forage crops may be potentially useful in improving plants as a source of animals dietary Zn. The soils which are heavily fertilized with phosphorus also produce Zn deficient forages. The application of Zn and P may be useful when they are supplement on soil test basis for avoiding antagonistic effect between Zn and P. Bio-fortified fodder proves better animal's dietary feeds.

Iron (Fe)

Iron is associated with enzymes like cytochrome oxidase, catalase and peroxidase. It also plays vital role as a catalyst to chlorophyll formation and act as an oxygen carrier in crop plant system. It is immobile nutrient and not easily transfers with the plant, so deficiency symptoms first show up on the younger leaves. But its deficiency in field crops is rare. The most of forages have adequate amount and no need of supplementation in feeding diet of animals.

Copper (Cu)

It plays important role in built up of red blood cells health, maintenance of reproduction and immunity. Also works with Mo and S for various enzymes involved in nucleotide and vitamin metabolisms. Cu deficiency in plants is most common on organic soils and much leached sandy soils. The Cu application in Cu deficient soils increases Cu content in plants for preventing its deficiency. Care must be taken in applying Cu in soils to prevent toxic levels in both plants and animals because it is not always deficient in soil. Ruminants are sensitive to Cu deficiency. The deficiency symptom of Cu develops more when levels of S, Mo and Fe are higher than 0.40 per cent, 150 ppm and 500 ppm, respectively. The dietary level of Cu exceed 100 ppm, it causes toxicities. The ratio of CU : Mo 2:1 to 4:1 allows for optimal performance of cattle. The dietary level of 8-10 ppm Cu in feed is adequate for meeting animals' requirement.

Manganese (Mn)

It is essential component of enzyme systems like caroxylases and hydrogenases. Mn takes part in activating several important metabolic reactions, chlorophyll synthesis and increasing availability of P and Ca. It is immobile nutrient and not transfers in the plant easily. Hence, the deficiency symptoms observe first on younger leaves. Gray speck disease in oat plant is most common symptom. Berseem leaves also affected with Mn deficiency. The deficient plants are stunted with delayed flowering and maturity and die at severe condition. The deficiency of the nutrient, generally, shows in forage growing areas under leached sandy soils. Some time toxicity symptoms appear in soil having excess Mn and low Ca content acidic soil pH having below 5.6. The application of lime reduces its toxicity.

Molybdenum (Mo)

Plants and animals require very small amount of Mo. Plant deficient in Mo needed its fertilization particularly in acidic and highly calcareous soils for adequate animal nutrition. The major nutritional problem is Mo toxicity called molybdenosis develop in grazing animal when forages have 10-20 ppm of Mo. The toxicity problem arises in the forage plants containing excess level of soil for Mo. Legumes accumulate more Mo than cereals / grasses. They absorb more Mo in wet soils than dry soils. Mo toxicity is also seen in cattle grazing pastures on organic soils.

The Cu: Mo ratio more than 2:1 in forage plant cause rapid accumulation of Cu in liver of animals.

Cobalt (Co)

Cobalt is required by bacteria that fix nitrogen in the nodules of leguminous fodder crops. This is the only known need of cobalt by plants. Micro-organisms in ruminants incorporate Co into vitamin B12. Cattle and sheep that are not supplied legumes in diet always need supplementation Co. The deficiency of cobalt in plant is obtained in fodder grown on poor drainage sandy soils, as it has higher downward movement rate than Fe.

Selenium (Se)

Selenium is a trace element responsible for immune system of the animal. It is not needed by plant but needed to animals in small amount. The forage grown in normal soils have sufficient Se. But the forages on acid soils as well alkaline soils formed from low Se parent materials are found to be deficient in Se (below 0.03 ppm) and caused white muscle disease. The toxicity symptom of Se is showed on the body of animals when forage plants having more than 10 ppm. Affected animals had sore feet and lost some of their hairs and in severe condition die. The certain groups of plant species are Se accumulators and extract more than 50 ppm Se from the soils. The Se accumulators are natively grown shrubs or weeds in semi arid and desert rangelands. But range grass contains Se less than 5 ppm. The toxicity level of Se in fodder berseem grown in alkali soil is also shown. The application of Se could help to prevent toxicity.

Research Work Done

Under limited forage production resources like land, water and nutrients as well as more availability of un-decomposed crop residues and animal wastes at farm and on farm, the integrated use of organic and fertilizer nutrients sources in balanced form will not only solve the problem of quality fodder production but also controlling environmental. The balanced dose of nutrients fertilization is not common to all the crops under different production systems, it differ depending upon plants genotypes, nutrient sources, reserve soil nutrient, introduction of legumes in cropping systems and production systems and problem soils, etc. The some important findings from the research work done on plant nutrition studies for quality fodder are summarized below:

Crop Genotypes

It has been well established that cereal fodder responds to N, legume fodder to P and cereal and legumes both to K and S application. An optimum dose of nutrient fertilization not only produce higher forage yield but maintain the level of other nutrients in the fodder also. The excess dose of nutrient showed adverse effect for examples P to S & Zn, K to Mg, S to P & Mo, Zn to P & S and Cu to Fe & Mo. The fertilizer dose of N 80-120 kg/ha to sorghum, bajra, maize and oat; 40-60 kg/ha to oil seed crops, 250-270 kg/ha to perennial grasses like hybrid bajra napier and setaria and 120 kg/ha to annual dinanath grass was worked out by Menhi Lal and Tripathi, (1989). They also reported optimum dose of PO5 80-100 kg/ha to berseem, 100-150 kg/ha to lucerne and 60-90 kg/ha to senji etc. The most of the forages including grasses required K dose of 30-60 kg/ha particularly in soil having available K below 100 ppm (Tripathi 2003). The dose of S was optimum to 40-60 kg/ha for leguminous and oil seed fodder crops and 20-40 kg/ha for cereal fodder crops including grasses. The dose of Zn 10-20 kg/ha in grass and 20-40 kg/ha in cultivated fodder is suggested particularly in Zn deficient soils (0.65 ppm). The necessity of Mo is also needed @ 10-15 kg MnO4/ha to berseem grown in acidic and calcareous soils (Tripathi, 2003 and Tripathi *et al.*, 1988).

Fertilizer nutrient application

It has been showed that the single fertilizer nutrient application not only declined forage yield but its quality also. The combined dose of 120 kg N/ha along with 80 kg PO/ha could be beneficial for higher fodder yield and nutrient uptake than N alone (Tripathi, 2003). According to Hazra and Tripathi (1986) the dose of 60 kg PO along with 40 kg N/ha and 40 kg KO/ha proved better than P alone for berseem production. Similarly, the dose of 40 kg N/ha and 40 kg S/ha in combined form was optimum than 60 kg N/ha in case of Chinese cabbage yield and quality (Tripathi, 2003). The yield of peal millets and cowpea increased with combined dose of NPK + Zn or NPK + S one NPK (Table 3) as reported by Hazra (1988). Menhilal and Tripathi (1987) observed HCN toxicity with added excess fertilizer N to sorghum and the lower dose of 60-90 kg N/ha along with S @ 30-60 kg/ha is considered to be useful for reducing HCN content and increasing fodder yield. The dose of 40 kg P and 20 kg S/ha in presence of recommended N&K is also suggested better to obtain maximum forage yield and quality (crude protein and digestive dry matter production) of forage sorghum. The application of 50 kg S/ha to cultivated perennial grasses (hybrid napier, guinea grass and setaria grass) was recommended for higher yield (Hazra, 1988). Rai, *et al* (2013) reported the balanced dose of 8 kg ZnSO and 15 kg FeSO when they were given along with NPK to fodder sorghum on marginal soil deficient in available Zn (0.65 ppm).

Table 3. Forage yield and S uptake by cowpea and pearl millet as influence by S application

Treatment	Forage yield (t/ha)		S uptake (kg/ha)	
	Pearl millet	Cowpea	Pearl millet	Cowpea
Control	0.8	0.7	2.6	3.6
100% N	1.6	0.9	5.1	3.8
100% NPK	2.0	1.1	6.1	4.9
100% NPK+S	1.9	1.4	7.4	7.3
100% NPK+Zn	2.0	1.2	6.4	5.4

In multi nutrient deficiencies (S, Zn, Mn, Cu), Tripathi *et al.* (2009) recommended dose of 40 kg S, 20 kg Zn, 10 kg Mn and 0.5 Kg Cu/ha when they were given along with NPK over NPK alone for higher quality fodder yield of sorghum (green fodder 385.181 ha, protein 10.3 % and in vitro digestibility 47.4 %, P-0.18 %, S-0.14 %, Zn-32 ppm, Mn-55 ppm and Cu-10 ppm over 100 % NPK. Similarly Rai *et al* (2013) reported beneficial dose of micro nutrients (10 kg Zn, 5kg Mn and 0.5 kg Cu/ha) along with NPK in presence of VAM and seed priming (ZnSO - 0.05 solution for 12 hrs) to oat crop for producing higher yield (517.6 q/ha green) and minerals (10 ppm Cu, 86 ppm Mn, 356 ppm Fe and 33 ppm Zn) in comparison to NPK alone (Table 4).

Table 4. Mineral content in fodder sorghum and cowpea as influence by nutrients application and management practices.

Treatments	Micro Nutrients (ppm)							
	Fodder Sorghum				Fodder Cowpea			
	Cu	Mn	Fe	Zn	Cu	Mn	Fe	Zn
NPK (control)	7	74	232	23	7	65	218	27
100% RDM	12	96	294	30	8	78	350	29
50% RDM+SP	10	101	334	29	9	80	186	30
50% RDM+ VAM + SP	9	102	360	28	11	86	356	33
CD 5%	2	26	55	5	1	8	70	5

RDM- Recommended Dose of Micro nutrients; SP- Seed Priming

Nutrient Resources

Efficiency of fertilizer varied with nature of soil and fertilizer. The research work showed that calcium ammonium nitrate is more effective in calcareous soils for oats and in acid soils for maize, whereas urea to all crops grown in normal soil (Tripathi, 2003). Similarly water soluble P sources viz; single phosphate in neutral to alkaline soils and citrate soluble P sources viz; rock phosphate in acidic soils is better than other sources. As regard S sources, elemental sulphur followed by gypsum in calcareous soil and gypsum followed by pyrites in non calcareous soils could prove better as reported by Tripathi (2003).

Reserve soil nutrient

The result of soil test crop response studies carried out by Tripathi, (2003) and Tripathi *et al.* (2007) noted optimum dose of N, P & K 120, 45 and 35 kg/ha in case of low fertility soils for sorghum and 80, 30 & 20 kg/ha in medium soils for sorghum + cowpea intercropping, respectively. In high P and K soils, the optimum dose of P was 20 kg/ ha, where as no need of K to both the cropping system (Table 5) as reported by Tripathi and Rai (2008).

The application of 40 kg S for oat pure and 60 kg S/ha for Oat + senji intercropped in low S soils (<10 ppm available S), whereas 20 kg/ha for oat pure and 40 kg S/ha for oat + senji in medium S soil (10-20 ppm available S) over NPK alone was optimum in increasing quality fodder yield (Tripathi and Tripathi, 2001). At higher level of S in soils (>40 ppm available S) then is no need of S application. The critical limit of available S was worked out to be 10 and 16 ppm for oat and oat + senji, respectively in deciding fertilizer S application for quality production. In another study on residual soil fertility, the application of 60 kg S made by elemental sulphur to summer fodder sorghum showed better residual response in producing kharif fodder cowpea than S applied to summer fodder cowpea and residual on kharif fodder sorghum (Tripathi *et al.*, 1995).

Integration of Nutrient Sources

The continuous use of single nutrient fertilizer application like urea has been found in lowering crop yield and quality. Tripathi (2003) reported balance dose of N @ 100 kg/ha half as urea (50%) and the rest half as FYM in presence of P @ 60 PO/ha to oat crop for higher quality yield (410 q/ha green and 132 kg/ha N uptake) compared to urea-N alone (390 q/ha green and 110 kg/ha N uptake). The application of N @ 240 kg/ha to hybrid bajra napier was given half through urea and rest half through FYM/ oil cakes along with 60 kg PO/ha and 60 kg KO/ha produced higher yield compared to 100% urea-N being maximum under mahuwa oil cake followed by neem cake (Tripathi and Hazra, 1999). The combined application of P sources (FYM and single super phosphate) in 1:1 ratio proved better balanced dose of 60 kg PO/ha in producing higher fodder (65.2 q/ha dry fodder) and P absorption (25.7 kg/ha) of sweet clover with net profit of Rs. 2.46 rupee investment as suggested by Tripathi (2003). Similarly grass legume mixed pasture (Table 6) had higher productivity when combined dose of 40 kg N half through urea (20 kg N/ha) and half through FYM slurry (20 kg N/ha) along with 80 kg KO/ha was applied together (Tripathi *et al* 2004). In deficient Zn soil, the application of S (20 kg), Zn (10 kg) & Mn (5 kg) along with 5t/ha FYM could be achieved better nutrient dose to fodder sorghum.

Table 5. Optimum dose of nutrients and fodder yield of sorghum and sorghum + cowpea intercropping (t/ha.) based on soil test.

Nutrient level in soil (kg/ha)	Optimum N dose (kg/ha)	Sorghum (t/ha)	Optimum N dose (kg/ha)	Sorghum + Cowpea (kg/ha)
Below 160			**Nitrogen**	
160 -225	120	34.2	120	44.3
	80	39.9	80	48.0
Below 12			**Phosphorus**	
12 -20	45	37.0	60	46.5
Above 20	30	39.8	40	50.5
	20	41.4	20	51.6
Below 800			**Potash**	
800 -1500	35	38.2	50	47.0
Above 1500	20	40.5	20	51.3
	No need	42.3	No need	51.6

The introduction of leguminous crop in nutrient supplement has greater role in deciding balance nutrient dose. The tree leaves (leucaena) @ 6.0 t/ha could replace 30 kg N/ha in dry land condition for cowpea, sorghum and setaria grass (Tripathi and Hazra, 1995). The higher quality fodders yield of oats obtained with 80 kg N/ha when it was sown after harvest of leguminous fodder crops i.e. sunhemp than oat grown in follow land at 120 kg N/ha as reported by Tripathi and Hazra, (1995). They also reported the dose of 90 kg N/ha over 120 kg N/ha to be optimum in presence of 60 kg PO under cowpea green manuring. The inclusion of legumes in hybrid bajra napier grass production could reduce dose of 240 kg N to 180 kg N/ ha/annum being maximum when hybrid bajra napier was inter cropped with berseem in rabi and sunhemp in summer as well as in kharif season than napier alone (Tripathi and Hazra, 1998). According to Menhilal and Tripathi (1987) the recommended dose of cereals could lower 45-50 kg N and 5-15 kg PO/ha with the introduction of legumes fodder crops (Table 7).

Table 6. Effect of combined dose of N and K on dry fodder yield and nutrient use efficiency of mixed Cenchrus + Stylosanthes pasture.

Treatment	Dry fodder (t/ha)	Nutrient use efficiency	
		K	N
Control	8.06	-	-
N 4O K 40	10.34	88.5	159
N 20 (4) + N 20 (FS)	9.68	-	185
N 20 (4) + N 20 (FS) + K 8O	11.40	116.0	260
CD at 5%	0.6	-	-

FS: Farmers Survey

Table 7. Balance nutrition through introduction of legume in intercropping system for forage yield and crude protein

Crop/ Crop Combination	Nutrients level		Green fodder	Protein yield
	N	P_2O_5	(G/ha.)	(kg/ha.)
Jowar	120	50	490	680
Jowar + Cowpea	70	55	384	1024
MP Chari	120	50	391	490
MP Chari + Cowpea	70	55	448	886
Maize	120	50	454	772
Maize + Cowpea	70	55	372	1049
Bazra	120	55	430	632
Bazra + Cowpea	70	55	430	971
Dinanath grass	90	30	463	658
Dinanath grass + Cowpea	55	45	435	942

Soil microorganism plays a vital role in augmenting plant growth and protein yield Rawat and Hazra (1997-98) noted an optimum dose of 45 kg/ha along with inoculation of Azotobactor and recommended P&K dose. They further reported a dose of 40 kg S/ha and 40 kg P/ha when applied together in presence of seed inoculation of

Rhizobium for berseem fodder yield. Yadav *et al* (2004-05) reported better dose of 120 kg N/ha applied 50% as urea N + 25 kg/FYM-N + 25 kg bio fertilizer–N along with 60 kg PO/ha and 40 kg KO/ha to fodder sorghum in a four years cropping compared to 100% urea-N. Tripathi *et al* (2009) also observed a beneficial dose of 180 kg N/ha to guinea grass and 60 kg PO/ha to berseem through FYM in grass legume mixed cropping.

Production Systems

It has been observed that responses of fodder crops to nutrient application are limited due to radiation curtailment under agro forestry production system and additional nutrient application to under story crops is needed to compensate such losses. The work carried out on chinese cabbage and safflower (Hazra and Tripathi, 1986a) and on oat and barley (Tripathi and Hazra, 1986) with N application under tree and non tree condition showed yield reduction curtailment under tree canopy and this could be compensated to some extent with 25% additional dose over recommended i.e. 80, 120, 60 and 40 kg/ha to oat, barley, chinese cabbage and safflower, respectively under open situation i.e. without trees. Similarly, the dose of 60 kg P/ha under *Hardwickia binata* tree is recommended to be optimum than 45 kg P/ha in open situation for higher yield of berseem (Hazra and Tripathi, 1986).

Under leucaena based pastoral system, the application of NPK @ 40, 20, 40 to *Cenchrus ciliaris* and 20, 40, 40 kg/ha to *Stylosanthesis hamata* in mixed grass + range legume pasture gave 2.65 time more yield having 124 kg N, 15 kg P and 71 kg K/ha in forage compared to nature *Cenchrus ciliaris* and *Stylosanthes hamata* mixed pasture (Tripathi *et al.*, 2006). In degraded land, the fertilizer dose of 40 kg N and 20 kg P/ha in presence of soil water conservation practices is suggested for high productivity of grass + legume pasture of *Cenchrus ciliaris* and *Stylosanthes hamata* (8-11 t/ha) compared to no such fertilizer application and soil water conservation practices like bunding & tranches and vegetative barriers, etc (Hazra, 1994).

Cropping Systems

Cereal and legume based cropping differs in their nutrient requirement. The research carried out by Tripathi and Tripathi (2003) indicated a balanced dose of 120 kg N, 60 kg PO and 40 kg KO/ ha for oats and sorghum mono cropping whereas 90 kg N, 60 kg PO and

40 kg KO/ha for oat + senji and sorghum + cowpea intercropping. Similarly the application of N, P and K @ 180, 60 and 60 kg/ha was found to be optimum under intercropping of hybrid bajra napier and cultivated legumes (berseem, sunhemp, pea and cowpea) than napier grown as pure under mono cropping system at fertilizer dose of 240 kg N, 60 kg PO and 60 kg KO/ha (Tripathi and Hazra, 1998). In intensive cropping system (Table-8), the application of S @ 40 kg/ha to fodder sorghum through ammonium sulphate was better than 80 kg S/ha through elemental sulphur in normal soil for producing quality fodder (Tripathi *et al.*, 1992).

Problem soils

The nutrient application varies with varied soil conditions i.e. sodicity, alkalinity and acidity. In sodicity and alkalinity condition the productivity of soils improved with the addition of gypsum along with recommended dose of NPK. Yadav (1987) suggested an optimum dose of 18 t/ha gypsum when it was applied with FYM to berseem. The application of pyrites @ 25 t/ha was equal to gypsum in berseem fodder crop as reported by (Tiwari *et al.*, 1984). Acid soils are generally low in Ca & P. According to Tripathi and Hazra (1996), the application of lime at 50-75 % of time requirement of soils (7.5 t/ha) as CaCO considered to be better balanced dose for increasing yield of berseem, maize, sorghum and dinanath grass by reducing the ill effect of excess concentration of Al & Mn (Table-9). In cultivated grass yield production, the dose of time @ 5.0 t/ ha is found effective in higher fodder yield of guinea, setaria and hybrid bajra napier grass over no lime. The phosphate application in acid soils needed more has normal soils. The dose of 160 kg PO through citrate soluble P fertilizers like triplesuperphosphate, bone meal and laccadive rock phosphate etc is found to be balanced dose for berseem and oats (Tripathi and Hazra, 1989). In highly P fixing soils (55-60%) the dose of P @ 120 kg PO/ha was more effective for improving fodder yield of lucerne compared to low P dose (30-90 kg PO5/ha). Similarly the dose of 160 kg PO/ha is worked out in bringingout higher production of berseem on acid soils (Tripathi and Hazra *et al.*, 1988).

In light textured calcareous soil, the deficiency of Mn on plant parts of oat crop caused poor growth and quality production. The dose of 10 kg Mn/ha to oats proved better by increasing 8.7-44.3 % than no such application (Sakal *et al.*, 1990). As regard S deficient calcareous soil, the higher quality yield of hybrid bajra napier, guinea, setaria produced maximum yield by 286, 624 and 152 kg/kg S due to 50 kg S/ha (Hazra, 1988). The Mo application is also beneficial in calcareous and acid soil. The dose of 1.0-1.5 kg MnO along with P (80-100 kg PO_5/ha) particularly to leguminous fodder crop is found useful in giving higher fodder yield and nutrient concentration (P & Mo), as reported by Tripathi *et al* (1989). Besides correcting alkalinity and sodicity of soil, the use of pyrites as a source of S @ 120-160 kg/ha could show positive effect in higher yield of cowpea and sorghum fodder crop as obtained with elemental sulphur (Tiwari *et al.*, 1984).

Table 8. Effect of levels and sources of S on fodder yield and quality constituents of fodder sorghum.

Tri Frequently	Protein (%)	Sugar (%)	Methionine (%)	N:S ratio	S:P ratio	Ca:P ratio
S- Level (kg/ha)						
20	8.96	6.5	0.052	12.25	1.92	4.95
40	9.81	7.1	0.064	11.18	1.99	4.51
60	10.1	7.4	0.071	10.89	2.11	4.50
80	10.2	7.5	0.076	10.86	2.19	4.54
CD at 5%	0.50	0.60	0.010	0.75	0.25	NS
S- Source						
Ammonium sulphate	9.93	7.3	0.073	11.57	2.05	4.72
Pyrites	9.55	7.0	0.062	11.06	2.09	4.64
Elemental sulphur	9.51	6.8	0.059	11.71	2.10	4.85
CD at 5%	NS	NS	NS	NS	NS	NS

Table 9. Forage yield and its quality of fodder crops grown in acidic soils as influence by liming

Fodder Crops	Lime level (t/ha.)	Nutrient contents			
		Ca	Ma	Al	Mn
Maize	0	0.45	0.33	190	66
	7.5	0.56	0.41	108	45
Berseem	0	1.17	0.35	186	133
	7.5	1.45	0.40	150	99
Lucerne	0	1.14	0.33	167	156
	7.5	1.30	0.35	145	128
Sorghum	0	0.38	-	254	64
	7.5	0.47	-	120	31
Teosinte	0	0.40	-	305	123
	7.5	0.44	-	129	47

References

Hazra, C.R. (1988) Sulphur fertilization of forages for yield and quality. Proc. TSI-FAI symposium on sulphur in Indian Agriculture, New Delhi, 5(2)/ 4/5

Hazra, C.R. (1994). Soil and water conservation for natural resources in agroforestry. *Range Management & Agroforestry* **15**(2): 230-239.

Hazra, C.R. and Tripathi, S.B. (1986). Soil properties micro- meteorological parameters, forage yield and phosphate uptake of berseem as influenced by P application under agroforestry system of production. *J. Agron. of Crop Sci,* **186**: 145-152.

Hazra, C.R. and Tripathi, S.B. (1986a). Influence of nitrogen on some soil properties and forage production of safflower and chinese cabbage with and without tree association. *J. Indian Soc. Soil Sci.,* **34**: 275-280.

Menhi lal and Tripathi, S.N. (1987). Role of fertilizer in production of forage and fodder crops-gaps and future need. *Fertilizer News,* **32**(12): 77-83.

Pahwa, M.R. (1995) Effect of different level of S & P on yield and content of crude protein in inoculated berseem at different cuts. *Agric. Sci. Digest* **5**(1): 26-28.

Rai, A.K., Singh, K.K., Das M.M., Mahanta, S.K., Sunil Kumar, Dixit, A.K. and Tripathi, S.B. (2013). Micronutrients management in forages. Research Bulletin No3, IGFRI, Jhansi, **1**: 13.

Rawat, C.R. and Hazra C.R. (1997-98). Effect of nitrogen and Azotobactor on forage yield of oats. *Forage Res.* **23**(3&4): 241-243.

Sakal, P., Singh, A.P. and Bhopal (1990). Evaluation of critical limit of Mn for oats in calcareous soils using different sources. *J. Indian Soc. Soil Sci.* **38**: 254.

Tiwari, K.N., Dwivedi, B.S., Upadhayay, G.P. and Pathak, A.N. (1984). Sedimentary iron pyrites as amendment for acidic soil and carrier of sulphur and iron- a reviewed. *Fert. News* **29**(10): 31-41.

Tripathi, S..B and Tripathi, S.N. (2007). Sulphur application in relation to productivity and quality of herbage in napier bajra hybrid and seasonal legumes intercropping system. *J. Range Management & Agroforestry* **28**(1): 116-120.

Tripathi, S.B. (1986). Response of fodder oat grown as pure and intercropped with senji to phosphate application in medium black soil. *Forage Res.* **22**(4): 60-64.

Tripathi, S.B. (2003). Nutrient management in forage based cropping systems. *Fert. News,* **48**(10):41-48.

Tripathi, S.B. and Hazra C.R. (1986). Forage yield of oat and barley as influenced by nitrogen application under agroforestry production system. *J. Indian Soc Soil Sci.* **34**: 774-779.

Tripathi, S.B. and Hazra C.R. (1988). Forage production on problem soils. *Indian Farming* **45**(1): 9-13.

Tripathi, S.B. and Hazra, C.R. (1994). Response of oat to nitrogen application in relation to proceeding legume crops. *Forage Res.* **20**(2&3): 177-181.

Tripathi, S.B. and Hazra, C.R. (1998). Nitrogen requirement for napier bajra hybrid legume intercropping and soil fertility. *Forage Res.* **23**(3&4): 227-230.

Tripathi, S.B. and Hazra, C.R. (1999). Effect of N sources on forage yield of hybrid bajra napier grass and nutrient uptake. *J. Indian Soc. Soil Sci* **47**(2): 375-376.

Tripathi, S.B. and Rai, S.K. (2008). Soil test based fertility K recommendation to fodder sorghum alone and intercropped with cowpea. *J. Range Management & Agroforestry* **29**(1): 48-52.

Tripathi, S.B., Gupta, J.N., Suresh, G. and Sharma, D.K. (2006). Soil fertility improvement in relation to fertilization. *J. Range Management & Agroforestry* **26**(2): 101-107.

Tripathi, S.B., Hazra, C.R. and Tyagi, R.K. (1988). Soil management practices for higher production. *Indian Farming* **37**(11): 160.

Tripathi, S.B., Manmiker, N.D. and Hazra, C.R. (1989). Effect of molybdenum and phosphate on berseem and residual effect on fodder maize in acid soils. *J. Indian Soc. Soil Sci* **37**(1): 200-201.

Tripathi, S.B., Pahwa, M.R., Attar Singh and Patra, A.K. (2004). Effect of nitrogen and potassium fertilization on herbage yield of *Cenchrus ciliaris* + *Stylosynthesis hamata* mixed pasture and soil fertility. *J. Range Management & Agroforestry* **25**(2): 92-97.

Tripathi, S.B., Rai, A.K., Dixit, A.K. and Singh, K.A. (2009). Improving yield and quality of forage through secondary and micronutrients. *Indian J. Fertilizers,* **5**(4): 81-96.

Tripathi, S.B., Singh, R.S. and Tripathi, R.K. (1992). Effect of levels and sources of S on forage yield nutrients content and their uptake by sorghum. *Forage Res.* **18**(1): 15-25.

Tripathi, S.B., Singh, R.S. and Tripathi, R.K. (1995). Effect of S fertilization on forage yield , S uptake and soil fertility under different crop sequences. *J. Indian Soc. Soil Sci* **43**(4): 641-645.

Tripathi, S.B., Tripathi, S.N. and Singh, K.K. (2007). Influence of S and micronutrients application on yield and quality of fodder sorghum with and without FYM. *J. Range Management & Agroforestry* **28**(2): 162-164.

Tripathi, SB and Tripathi, SN (2001). Sulphur in soils and its effect on yield and quality of fodder oats in sole and intercropping system. *J. Range Management & Agroforestry,* **22**(1): 20-27.

Yadav, JSP (1987). Management of saline and alkaline soils for fodder and forage production. Proc. National Rangeland Symposium at IGFRI Jhansi.

Yadav, R.B., Pahwa, M.R. and Das, S.K. (2004-05). Annual Report IGFRI Jhansi.

□□□

Chapter **18**

Weed Management Practices for Integrated Farming Systems

G. PRABHU, R. SRINIVASAN, MANOJ CHAUDHARY, KIRAN KUMAR, C. SARATHAMBAL AND G.K. SUJAYANAND

ICAR - Indian Grassland and Fodder Research Institute, Jhansi-284003 (UP)

Weeds have been existing on the arable and non-arable lands ever since the man started to plough and rear the animals around 10000 B.C. (Macneish, 1964). It is being underestimated crop pests in agriculture, although they cause higher reduction/loss in the yield of crops than other pests and diseases. The annual loss of agricultural produce due to weeds for 37 percent, insects for 29 percent, diseases for 22 percent and other pests for 12 percent annually in India reported by Yaduraju (2006). Besides, they also remove 30-40 kg N, 10-15 kg P_2O_5 and 20-40 kg K_2O per hectare (Alagh, 1988).

Critical period of crop-weed competition

The shortest time span in the ontogeny of acrop, when weed causes maximum reduction in yield, orin other words, when weed control may fetch near maximal or maximum acceptable yield Critical period of weed competition (period threshold) is defined as critical period of crop – weed competition (Das *et al.*, 2012). The research on weed management has been carried out in India has find out the critical period of crop-weed competition (Table 1) in a large number of crops.

Table 1. The critical period of weed competition in crops

Crops	Critical period	Crops	Critical period
Rice (upland) direct-seeded	15-45 DAS	Chickpea, lentil	30-60 DAS
Rice (lowland transplanted)	30-60 DAT	Green gram	15-45 DAS
Rice (lowland direct-seeded)	40-60 DAS	Black gram	15-30 DAS

Dwarf wheat, barley, pea	30-45 DAS	Cowpea	First 30 days
Tall wheat	35-50 DAS	Cotton (rainfed)	20-60 DAS
Maize	30-60 DAS	Jute, Mesta	15-45 DAS
Sorghum	15-45 DAS	Sugarcane	30-60 DAS
Pearl millet	15-30 DAS	Potato	First 4 weeks
Soybean	First 60 days	Rapeseed	10-40 DAS
Groundnut	First 45 days	Seasonal forage crops	4-6 weeks

Yield loss

The losses caused by weeds in various cropsand cropping systems were quantified, indicating the needfor weed management for realising optimal crop yields (Table 2). Rao and Nagmani, (2010) reported that the yield loss estimation from different studies shows that the loss from 2.5 to 77.4 % in different crops in India.

Weed flora

Weeds composition and competition for nutrients, moisture and space is dynamic and is dependent on soil, climate, cropping andmanagement factors (Rao and Nagamani, 2010). Several studies were conductedon weed flora in India/World which include : rice (Ravisankar *et al.*, 2008) in Tamilnadu; wheat in Ethiopia (Amare *et al.*, 2007); rice-wheat system in Uttarpradesh (Kumar *et al.*, 2014); maize-vegetable system in Karnataka (Hugar and Palled, 2008); green gramin Gujarat (Chhodavadia *et al.*, 2013); sunflowerin Tamilnadu (Baskaran and Kavimani, 2014); groundnut in Rajasthan (Kumar *et al.*, 2013); and brinjal in Chhattisgarh (Kunti *et al.*, 2012). Weedy rice is emerging as amajor problem in direct-seeded rice (Rao and Nagamani, 2007). The list of important weed flora in different crops is been mentioned in the (Table 3).

Table 2. Estimated yield losses caused by weeds in different crops and cropping systems in India

Crop	Weeds	Yield loss (%)	References
Rice	Various	12-98	Rao, 2011
Rice-wheat cropping system	In farmers field	13.1-22.4	Singh *et al.* (2005)
Wheat	Broad leaves	17.00	Walia and Brar (2001)
	Wild oats	36.6	
	Both	45.1	

Maize	Grassy	77.4	Pandey *et al*. (2002)
	Non-grassy	44.2	
	Sedges	38.4	
Lentil	Season long weed	46.6	Pandey *et al*. (1998)
Barley	competition	28.1	
Wheat		40.6	
Chickpea		28-100	Moorthy *et al*. (2003)
Green gram	Cuscutta 1 -10 / m^{-2}	27.7-88.3	
Soybean-chickpea cropping system	*Euphorbia geniculate* at 10 to 120 plants / m^{-2}	12-30 (Soybean) 18-53 (chickpea)	Mishra and Singh (2003)
Lucerne		11.7	Menhi Lal *et al*. (1994)
Oat		8.3	
Sorghum		54	

Table 3. Important weeds of different crops and cropping system

Crop	Grassy weeds	Broad leaves	Sedges
Rice	*Echinochloa colona* L.) Link.*E. crusgalli* (L.) Beauv. and *Leptochloachinensis* L.) Nees.	*Eclipta alba* (L.) Hassk., *Ammaniabaccifera*L. and *Ludwigiaparviflora* Roxb.	*Cyperus difformis* (L.) and *C. irria* (L.)
Wheat	*Avenafatua* L., *Phalarisparadoxa* Retz.	*Chenopodium album* L. *Polygonumnepalense* Meisn. *Spergulaarvensis* L.	*Oxalis latifolia* L.
Rice-wheat cropping system	*Phararis minor* Retz. *Avena ludoviciuna* L. *Polypogonmonspeliensis* (L.) *Cynodon dactylon* (L.) Pers. *Paspalumconjugatum* Bergius.	*Chenopodium album* L. *Anagallis arvensis* L. *Fumariaparviflora* L. *Lathyrusaphaca* L. *Medicago denticulate* Willd. *Melilotus indica* (L.) All. *Rumexacetosella* L. *Vicia sativa* L. *Coronopus didymus* L. *Eclipta alba* L. *Ageratum conyzoides* L	*Cyperus rotundus* L.
Maize-vegetable system	*Elusine indica* (L.), *Cynadondactylon* (L.), *Setariaglauca* (L.)	*Ageratum conyzoides* L., *Amaranthusviridis* (L.), *Leucasaspera* (L.), *Phyllanthusniruri* (L.)	*Cyperus rotundus* L.
brinjal	*Cynodon dactylon* L.	*Alternentheratriandra* L.	*Cyperus iria* L.

		Sonchus arvensis (L.) Hill., *Chenopodiumalbum* L., *Physalis minima* L., *Euphorbia hirta* L. and *Parthenium hysterophorus* L.	
Sunflower	*Cynodon dactylon* L., *Chloris barbata* L. *Dinebraretroflexa* (Vahl) Panzer., *Panicum repens* L.	*Trianthemaportulacastrum* L., *Digera arvensis* Forssk., *Parthenium hysterophorus* L,	*Cyperus rotundus* L.

1. Weed management in sole cropping systems under IFS

Rice

Rica is an important food crops of more than half of the world population. Weed growth in rice is a major menace under low and upland rice ecosystem because of its anoxic soil environment and management practices. Yield losses incurred when weeds are not controlled can be very high (Chauhan and Johnson, 2011), so deriving workable strategies to suppressweed growth is fast becoming a major research priority (Ladha *et al.*, 2007). Each management strategies had a positive effect of accumulation of weed biomass (Fig 1). Four times in a season hand weeding is an effective method of weed management under direct seeded rice in 2008 and 2009 reported by (Saqib *et al.*, 2015).

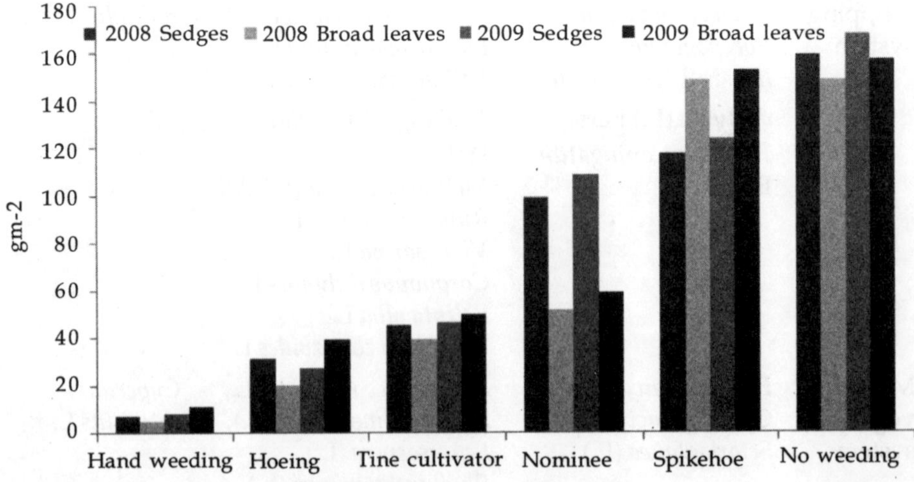

Fig. 1. The effect of different weed management strategies on the accumulation by 45 days of sedges and broad leaved weeds.

In 2008 and 2009 the lowest population of sedges (7.29 and 8.35 g m^{-2} respectively) and broad leaved (5.1 and 10.97 g m^{-2} respectively) weeds were observed in four hand weeding (15, 25, 25 and 45 days after planting). Similarly, manual weeding at 20 and 40 days after planting gave significantly highest grain yield of 7853. 50 kg ha^{-1} (Antralina *et al.,* 2015). This was due to better vegetative stage and number of filled grains, which was greatly achieved by manual weeding at 20 and 40 days after planting without disturbing vegetative and reproductive stage (Table 4). Subramanian *et al.,* (2006) reported that the highest weed-control efficiency (WCE) was achieved with integrated application of glyphosate + pretilachlor with Safener along with two manual weedings (25 and 45 DAP), which was comparable with weed-free check. This could be attributedto greater reduction of dry weight of weeds by the combined effect of herbicides and hand-weeding. Performance of early post emergence application of butanilon weed management ranked next to the pre-emergenceapplication of pretilachlor with Safener (Fig 2). Integration of suitable herbicides with two hand weeding at 25 and 45 days after planting could be suitable option to get quality yield and effective control of all kinds of weeds in low and upland ecosystem.

Table 4. Effect of Difference Weed Control Methods on Rice Yields Weed Control Methods Grain Yield (kg ha^{-1})

Treatment	Yield (kg ha^{-1})
Manual weeding	7853.50a
Un weed control	3175.00b
Penoxulam + Cyhalofop-butyl	3271.50b
2,4-D + methyl metsulfuron	6593.50ab

Note: The average numbers followed by the same letter in the same column are not significantly different according to Duncan's Multiple Range Test at the 5% level.

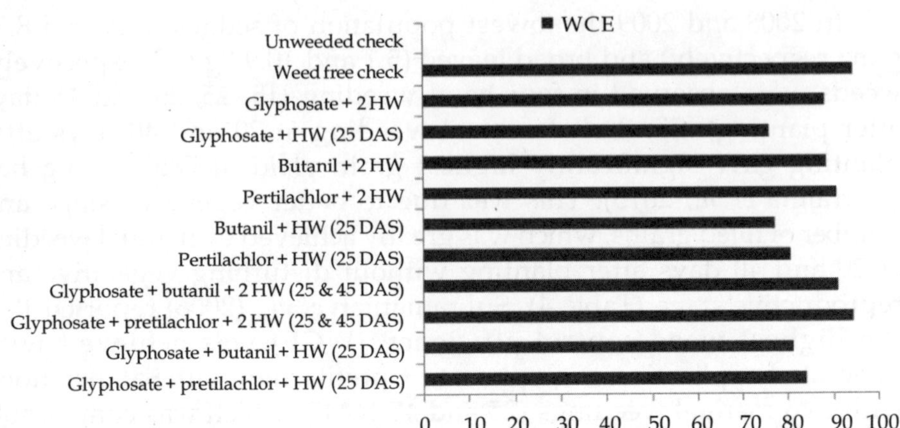

Fig. 2. Effect of integrated weed management strategies on weed control efficiency in wet seeded rice.

Wheat

Wheat grown area in India is about 27.99 million ha with aproduction of 75.81 million tons. After the independence (1950–51) the production of wheat in India was only 6.46 million tons and productivity was mere 663 kg/ha, insufficient to feed the Indian population (almost 1214.3 million).The main factors which influence crop production are radiation, soil moisture, nutrienta vailability and length of growing season. Yield reducing factors also encompass disease, insects and weed infestation (Varshney *et al.*, 2012). Line sowing of wheat crop with high seed rate (150 kg ha^{-1}) and an application Buctril super.

Herbicide (Fig. 3) could be a best management technique to control the weeds in wheat crop (Marwat *et al.*, 2011). Line sowing was provided better environment to germinate and earlier good stand whereas, buctril super is effective against broad leaves weeds, the field full of broad leaves weeds which could have been improved its efficiency of the herbicide.However, Fahad *et al.* (2013) reported that the maximum biological yield of (13200 kg ha^{-1}) was reported (Table 5) in Isoproturon 50 WP and minimum value of (9810 kg ha^{-1}) was reported for weedy check and the maximum value of grain yield of 3210 kg ha^{-1}was recorded in Isoproturon 50 WP treated plot followed by buctril super 60 EC (2850 kg ha^{-1}). Similarly, Amare *et al.* (2014) reported that the highest grain yield (2102.4 kg ha^{-1}) was recorded in weed free check followed by Isoproturon 1.50 kg ha^{-1} (2027.1 4 kg ha^{-1}) which was due to their weed free condition and reduction of weed dry weight in the above treatments (Table 6).

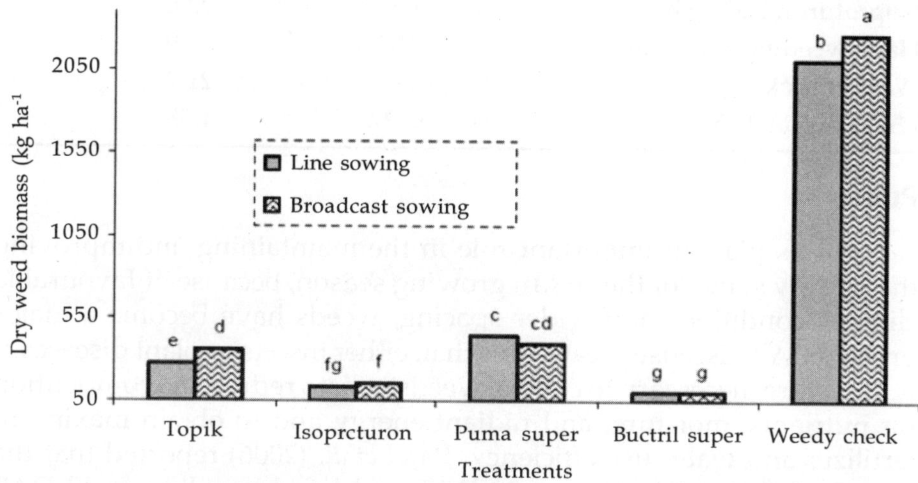

Fig. 3. Interaction of sowing methods and herbicides (Marwat*et al.*, 2011)

Table 5. Bioological yield (kg ha^{-1}) and grain yield (kg ha^{-1}) as affected by different herbicide treatments in wheat crop in pakistan.

Herbicides	Biological yield (kg ha^{-1})	Grain yield (kg ha^{-1})
Agritop 500 G/L	11480	2750b
Aim 40 DF	10230	2330c
Buctril super 60 EC	11550	2850b
Puma super 75 EW	10690	2351c
Topik 15 WP	11040	2498c
Isoproturon 50 WP	13200	3210a*
Weedy check	9810	1820d
LSD value at 5 % α level	NS	267

*Means followed by different letters in the respective columns are significantly different at 0.05 α level according to LSD test.

Table 6. Effect of weed management practices on grain yield (kg ha^{-1}) and Harvest index (%) of wheat

Weed management practices	Grain yield	Harvest index
Clodinafop-propargyl 0.065 kg ha^{-1}	1001.6	23.3
Clodinafop-propargyl 0.080 kg ha^{-1}	1169.4	24.9
Clodinafop-propargyl 0.105 kg ha^{-1}	1298.5	25.8
Isoproturon 1.00 kg ha^{-1}	1442.1	25.9
Isoproturon 1.25 kg ha^{-1}	1719.9	26.3

Isoproturon 1.50 kg ha^{-1}	2027.1	27.6
Hand weeding at tillering	2102.4	27.9
Weedy check	819.2	21.7
LSD value at 0.05	52.27	1.38

Pulses

Pulses play an important role in the maintaining andimproving the fertility status of the soil.In growing season, because of favourable climatic conditions and wider spacing, weeds have become a major problem. Weedscause great losses than either insects or plant diseases.It is, therefore, necessary to control weeds so asto reduce the competition for nutrients, moisture, and radiant energy and to obtain maximum fertilizer and water use efficiency. Patel *et al.* (2006) reported that the treatment intercultivation at 20 DAS and hand weeding at 40 DAS recorded significantly lower weed biomass and density in chickpea than rest of the treatments (Table 7) whereas, oxadiargyl @0.075 kg a.i ha^{-1} which was at par, this might be due to its longer persistence could have provided weed control to the long time. Maximum weed control efficiency was achieved under IC + HW at 20 and 40 DAS (76.5%) followed by oxadiargyl @0.075 kg a.i ha^{-1}(71.5%). However, in another experiment Pedde *et al.* (2013) reported that two handweedings resulted in the highest plant height, plant spread, branches, root nodules and dry matter. Among the herbicidal treatments, the pre-emergence application of pendimethalin 0.75 kg/ha was effectivein recording higher growth parameters followed byquizalofop-p-ethyl 40 g/ha as post-emergence. The highest grain and straw yields were recorded with one hoeing+ two hand weedings, followed by 2 hand weedings. In green gram (Chhodavadia *et al.,* 2013) two hand weeding with two Interculturing at 20 DAS and 40 DAS proved its superiority over rest of the weed management in summer green gram. Among herbicidal treatment, application of quizalo fop-ethyl at 20 DAS and fenoxaprop-p-ethyl at 20 DAS was found to be relatively more effective in controlling weeds than their sole application.

Oilseeds

Groundnut is an important oil seed crop of southern and western India, keeping groundnut free of weeds upto critical period of weed management, pre-emergence application of pendimethalin at the rate of 1 kg a.i. ha^{-1} + post emergence application of imazethapyr at the rate of 0.50 g a.i ha^{-1} at 20 days after sowing + one hand weeding at 45 days after sowing produced higher pod yield of 4147 kg ha^{-1}, highest weed control efficiency of (88.25 %) and lowest weed index of (0.19) by kumar *et al.* (2013).Dhakar*et al.*(2000) reported that the highest yield of (1717 kg ha^{-1}), net return of (Rs. 11017 ha^{-1}) and cost benefit ratio of (1.23) resulted from weeding at 20 days after sowing and hoeing at 20 and 40 days after sowing. Senthilkumar *et al.* (2009) reported that maintain plant population of 5.0 lakhs per hectare reduced weed density, weed dry weight and then resulted in higher weed control efficiency. For sunflower application of pendimethalin at the rate of 1 kg ha^{-1} followed by one hand weeding at 40 days after sowing resulted in lowest weed density of 377 m^{-2} in 2012 and 76.2 m^{-2} in 2013, 377 g m^{-2} and 391.3 g m^{-2}of weed biomass in 2012 and 2013 respectively and highest seed yield of 1.64 and 1.56 tonnes ha^{-1} in 2012 and 13 respectively by Baskaran and Kavimani (2014).

Vegetables

Weed management in brinjal, pre-emergence (3-5 DAT) application of pendimethalin at the rate of 1.5 L ha^{-1} + hand weeding at 30 DAT was recorded highest fruit yield of 21.43 t ha^{-1} and weed control efficiency at 30 DAT (84.33 %), 60 DAT (73.39 %) and 90 DAT (47.61 %) reported by Shubra and Nath, 2015. Similarly, Pendimethalin (extra) (0.64 kg/ha) pre-transplanting+ one hand weeding at 40 DAT + pendimethalin (extra) (0.64 kg/ha) at 45 DAT recorded highest fruit yield of 25.20 t ha^{-1} and weed control efficiency of (79 %) by Kunti *et al.* (2012). This was due to lower accumulation dry matter of weeds at all the observational stage. In green chilli Singh *et al.* (2011) reported that weed free check recorded significantly more number of fruits plant-1 (40.10),fruit weight plant-1 (231.20 g) and green chilli yield (4,393 kg ha^{-1})respectively at harvest than rest of the treatments. Butachlor1.0 kg ha-1 was next to weed free check in terms of fruits numberplant-1 (37.35), fruit weight plant-1 (206.35 g) and green chilliyield (3,948 kg ha^{-1}). The weed index (10.12 %) was the lowest inbutachlor indicating the lowest weed- crop competition.

Table 7. Effect of weed management practices of weed density, weed biomass and weed control efficiency

Treatment	Weed density (NO./m^{-2} at harvest)			Weed biomass Kg ha^{-1} at harvest			WCE onpooled basis
	2003-04	2004-05	Pooled	2003-04	2004-05	Pooled	
IC+HW at 20 and 40 DAS	2.60 (5.81)	3.08(8.56)	2.84(7.19)	427	884	656	76.5
oxadiargyl @0.075 kg a.i ha^{-1}	2.76(6.69)	3.60(12.00)	3.18 (9.35)	539	1060	799	71.5
Pendimethalin @0.75 kg a.i ha^{-1}	3.08(8.56)	4.01(15.13)	3.54(11.85)	745	1335	1040	62.9
Fluchloralin @ 1 kg a.i ha^{-1}	3.98(14.94)	4.91(23.13)	4.45(19.04)	1153	1568	1360	51.5
Weedy check	9.31(85.69)	9.99(98.88)	9.65(92.29)	2142	3462	2802	-----
C.D. (p=0.05)	0.20	0.16	0.37	215.30	297.16	735.57	-----
--	0.07	0.06	0.09	76.12	105.06	187.36	-----

Table 8. Weed density, weed dry matter and weed control efficiency of wheat crop in rice-wheat cropping system influenced by weed management

Herbicide	Weed density (No. m^{-2}) at 60 DAS		Weed dry matter g m^{-2} At 120 DAS		Weed control efficiency (%)	
	2004-05	2005-06	2004-05	2005-06	2004-05	2005-06
Metribuzin @ 250 g ha^{-1}	6.71 (45.28)	6.52 (42.62)	6.64 (43.88)	6.59 (43.15)	92.30	92.25
Sulfosulfuron @ 33.3 g ha^{-1}	6.13 (37.52)	5.86 (34.37)	6.39 (40.62)	6.30 (39.50)	92.88	92.91
Weedy check	35.82 (1301.67)	35..21 (1260.17)	23.82 (568.28)	23.56 (556.00)	0.00	0.00
S.Em±	0.37	0.29	0.23	0.20	1.33	1.29
LSD 0.05	1.07	0.87	0.67	0.59	3.90	3.74

Forage crops

Weed competition is more prominent at early stages of forage crops. Crop weed competition up to 28- 40 days after sowing in annual forages. The loss in fodder yield due to weed competition has been reported to the extent of11.7% in lucerne and 8.3% in oat. In crops like sorghum, magnitude of yield loss was as high as 54 %, while uptake of nitrogen, phosphorous and potassium by weeds was to the extent of 48.8, 22.0 and 55.0 kg/ha, respectively (Menhi Lal *et al.*, 1994). For effective control of weeds, integration of weed management practices could be an important idea. Whereas, works on weed management measures is meagre in forage crops. In berseem, imazethapyr at the rate of 0.10 kg a.i. ha^{-1}was significantly superior in controlling *Chicorium intybus* L.along with the other weeds, weed control efficiency of (74.2 %), higher forage yield (88.2 tonnes ha^{-1}), crude protein (2.6 t ha^{-1}) and benefit cost ration (1.86) by Sunil kumar and Dhar (2008). The highest weed control efficiency (88.5%) was observed with the application of imazethapyr 0.125 kg/ha + hand weeding at 6 weeks crop stage of Berseem (Dhar *et al.*, 2004). Ishaya *et al.* (2008) reported that the application of metolachlor + prometryn at 0.75 + 0.48 kg a.i. ha^{-1} (2.13) and metolachlor+ prometryn at 1.00 + 0.64 kg a.i. ha^{-1} (2.13) resulted in significantly lower crop injury score than all other herbicide treatments in cowpea.

Weed management in Inter/multiple cropping systems under IFS. When two or more crops are grown in rotation is called multiple cropping system, the herbicides recommendation also requires due consideration like intercropping system. Higher persistence may leave toxic residues of herbicides in soil even beyond the existing season, which if not selective, may affect next crop in sequence.

Kumar *et al.*(2014) reported that the application of metribuzin and sulfosulfuron was significantly reduced the weed density and weed dry matter (Table 8). Metribuzin controlled grasses andbroad-leaf weeds including deep rooted weeds,whereas the sulfosulfuron was slightly moreeffective in controlling the grassy weed. Rao and Bhardwaj (1986) reported that metribuzin wasreadily taken up by the plant roots through diffusion and translocated upward apoplectically. Whereas, grain yield (6907 kg ha-1) was also the highest with Affinity 50 WDG followed by 6348 kg ha-1 in 2, 4-D+ Puma super, while control showed the lowest grain (Usman *et al.*, 2012).

Hand hoeing gave the highest weed control efficiency of 98.62 and 98.13% of narrow and broad leaved weeds, respectively. Results showed that the application of Stomp with a dose 3.5 l ha-1 at the time of seed bed preparation controlled narrow leaved weeds efficiently (97.45 %) but its effects were weaker on broad leaved weeds (81.60 %). On the other hand when Stomp @ 3.5 l ha-1 was applied just after sowing, it controlled broad leaved weeds efficiently (95.56 %) but weak for the control of narrow leaved weeds (86.97%). Spray of Topik @ 0.250 kg ha^{-1} also gavethe good control of grassy weeds (96.18%) but failed to control the broad leaved weeds under rice-chickpea based cropping system (Chaudhary *et al.*, 2011). Hassan *et al.* (2003) reported that the interaction of herbicides with the herbicides exhibited that highest numerical yield of 2095 kg ha^{-1} was harvested under the conventional tillage treated with Stomp 330E. It was followed by the zero tillage under the same herbicide (1910 kg ha^{-1}). The lowest yield (1534 kg ha^{-1}) was realized in zero tillage and the weedy check under rice-chickpea cropping system (Table 9). For forage based cropping system, the studies on weed management practices in forage crops is meagre. Although some of the information on herbicide based weed management(Table 10) for forage crops and economically best integrated weed management practices (Table 11) for different crops has been collected from research articles from India. There is always a greater recognition of weed science research in India now and then before. The great potential of herbicides were well known earlier and now due to variety of reasons. The increased urbanization, higher income and more importantly the implementation of the National Rural Employment Guarantee scheme which guarantees employment/wages to one adult in a household for 100 days in a year have put tremendous pressure on availability of labour to agricultural operations including weeding. On other side their popularity and adoption by the farmers was limited due to a lack of knowledge, availability and cost.Hence integration of possible options are available in our basket could be a potential solution for future weed management under different cropping system.

Table 9. Effect of tillage methods and herbicides on the Yield (kg ha^{-1}) of chickpea

Herbicide	Conventional Tillage	Zero tillage	Herbicide Means
Stomp 330 E	2095	1910	2002 a
Puma Super	1905	1738	1822 b
Weedy Check	1666	1534	1600 c
Tillage Means	1889 a	1727 b	

Table 10. Selective herbicides for different intercropping system

S.No.	Intercropping system	Herbicides
1.	Maize + cowpea	Pendimethalin; Butachlor (Reddy and Reddi, 2000)
2.	Sorghum + cowpea	Isoproturon (Balasubramanian and Subramanian, 1989);Pendimethalin (Anonymous,2013)
3.	Pearlmillet+ cowpea	Pendimethalin (Yaduraju, 2003)
4.	Maize + soybean	Pendimethalin Oxyfluorfen+pendimethalin (Yaduraju, 2003)
5.	Maize + groundnut	Pendimethalin @ 0.75 kg a.i/ha (Gupta, 1998)
6.	Napier Bajra hybrid+ cowpea	Pendimethalin @ 0.75 kg a.i/ha. (Anonymous, 2013)
7.	Guinea grass + cowpea	Pendimethalin @ 0.75 kg a.i/ha (Anonymous, 2013)

Table 11. Most economical Integrated Weed Management for certain crops in India

Crop	IWM	Reference
Blackgram	(i) Pendimethalin at 0.75 kg/ ha fb HW 45 DAS	(i) Kumar *et al.* (2006)
	(ii) Pendimethalin at 0.50 kg/ ha fb HW 60 DAS	(ii) Rathi *et al.* (2004)
	(iii) Trifluralin (PE) at 0.50 kg/ ha fb HW	(iii) Sardana *et al.* (2006)
Cowpea	Pendimethalin 0.75 kg/ha fb HW 35 DAS	Mathew *et al.* (1995)
Garlic	PE of oxyfluorfen (0.15 kg/ha) or pendimethalin(1.0 kg/ha) fb HW 40 DAS	Porwal (1995)
Groundnut	Pendimethalin or alachlor 1 kg/ ha fb HW 30 DAS	Itnal *et al.* (1993)
Indian mustard	(i) Pendimethalin (PE) at 0.50 kg/ ha or fluchloralin at0.50 kg/ ha each fb HW 30 DAS	(i) Singh *et al.* (1999)
	(ii) Fluchloralin at 0.75 kg/ ha fb HW 25 DAS	(ii) Singh (2006)
Onion	(i) Pendimethalin at 1.5 kg/ ha fb HW 60 DAT	(i) Rameshwar *et al.* (2002)
	(ii) Oxyfluorfen applied at 0.25 kg/ha fb HW 40 DAT	(ii) Nandal and Singh (2002)
	(iii) Oxyfluorfen at 0.15 kg/ ha fb HW 35 DAT	(iii) Kolhe (2001)
	(iv) Fluchloralin or pendimethalin at 0.9 kg/ha fb HW 40 DAT	(iv) Sukhadia *et al.* (2002)
Okra	Stale seed bed with glyphosate application integrated with eucalyptus mulchingSowing at 20 cm apart with two	Ameena *et al.* (2006)
Dwarfpea	HW fb pendimethalin at 1 kg/ haPendimethalin (1.0 kg/ha) or	Tewari *et al.* (2003)
Pigeonpea/ Groundnut	fluchloralin (1.0 kg/ha) each fb two HW 30 and 42 DAS	Vijaykumar *et al.* (1995)
Intercrop	Pendimethalin at 1.50 kg/ ha+HW 40 DAS	
Pigeonpea/ pearl millet intercrop	(i) Application of butachlor 1.0 kg/ha,anilofos 0.4 kg/ha along with closer planting	Shinde *et al.* (2003)

Rice-transplanted rice	(ii) Anilophos 0.6 kg/ha 7 DAT+HW 27 DAT Butachlor at 1.0 kg/ha fb one hand	(i) Gogoi *et al*. (2001) (ii) Singh and Kumar (1999)
Rice-dry-seeded rice	weeding at 30 DAS by local tool 'Kutla' 60 kg N/ha+	Singh and Singh (2001)
Sesame	fluchloralin at pre-planting@ 1.0 kg/ha fb HW 21 DAS	Singh *et al*. (2001c)
Soybean	(i) Butachlor (Pre-em) 1.5 kg/ha fb HW 30 DAS (ii) Rows spacing of 22.5 cm with alachlor at 1 kg/ha	(i) Chandrakar and Urkurkar (1993) (ii) Shekara and Nanjappa (1993)
Sugarcane	Metribuzin or atrazine at 1 kg/ha+trash mulching (3.5 t/ha) in between cane rows at 60 DAP	Singh *et al*. (2001b)
Wheat	Pendimethalin at 0.75 kg/ha+ HW30 DAS	Singh and Singh (2004)

References

Alagh Y K. 1988. Inaugural address on "Pesticides and Indian Agriculture." In: *National seminar on "Changing Pest Situation in the Current Agricultural Scenario of India."* June 14-16, New Delhi.

Amare1 T, Sharma J J and Zewdie K. 2014. Effect of Weed Control Methods on Weeds and Wheat (*TriticumAestivum*L.) Yield. *World Journal of Agricultural Research* **2**(3): 124-128.

Ameena M, Kumari V L G and George S.2006. Integrated management of purple nutsedge (Cyperus rotundus L.) in okra. *Indian Journal of Weed Science* **38**: 81-85.

Anonymous 2013. I.G.F.R.I. 2014. Annual report 2013-14. Indian Grassland and Fodder Research Institute, Jhansi- 284003, Uttarpradesh, India. pp 136.

Antralinaa M, Istina I N, Yuwariahc Y and Simarmatac T. 2015. Effect of difference weed control methods to yield of lowland rice in the SOBARI. *Procedia Food Science* **3**: 323 – 329.

Baskaran R and Kavimani R. 2014. Productivity of sunflower as influenced by tillage and weed management. *Indian Journal of Weed Science* **46**(2): 138–141.

Chandrakar B L and Urkurkar J S. 1993. Efficacy and economics of weed control in soybean (*Glycinemax* L.) under vertisols of Chattisgarh region. *Indian Journal of Weed Science* **25**: 32-35.

Chaudhary S U, Iqbal J and Muzzammil H. 2011. Weed management in chickpea grown under rice based cropping system of Punjab. *Crop and environment* **2**(1): 28-31.

Chauhan B S and Johnson D E. 2011. Row spacing and weed control timing affect yield of aerobic rice. *Field Crops Research* **121**(2): 226-231.

Chhodavadia S K, Mathukiya R K and Dobariya V K. 2013. Pre- and post-emergence herbicides for integrated weed management in summer green gram. *Indian Journal of Weed Science* **45**(2): 137–139.

Dhakar P M, Pawar W S, Jugele A N.2000. Economics of weed control in groundnut. *Journal of Soils and Crops* **10**: 307-308.

Dhar S, Das SK, Kumar S and Tripathi SB. 2006. Response of fodder sorghum to different weed-management techniques and nitrogen levels. *Indian Journal of Agronomy* **51**: 310-313.

Fahad S, Niel L, Rahman A, Chen C, Wu C, Saud S and Huang J. 2013. Comparative Efficacy of Different Herbicides for Weed Management and Yield Attributes in Wheat. *American Journal of Plant Sciences* **4**: 1241-1245.

Gogoi A K, Rajkhowa D J and Kandali R. 2001. Integrated weed control in rainy season rice under medium land situation. *Indian Journal of Weed Science* **33**: 18-21.

Gupta O P. 1998. Modern Weed Management, Agro Botanica Publishers, Bikaner, India. pp 488.

Hassan G, Naqibullah K and Haroon K. 2003. Effect of zero tillage and herbicides on the weed density and yield of chickpea under rice-based conditions of D.I. Khan. Pakistan *Journal of Weed Science Research* **9**(3&4): 193-200.

Itnal C J, Lingaraju B S and Kurdikeri C B. 1993. Effect of herbicides and cultural methods on weed control in irrigated groundnut. *Indian Journal of Weed Science* **25**: 27-31.

Khalid U, Inayat U, Said M K, Muhammad U K, Ghulam S and Muhammad A K. 2012. Integrated weed management through tillage and herbicides for wheat production in rice-wheat cropping system in northwesternpakistan. *Journal of Integrative Agriculture* **11**(6): 946-953.

Kolhe S S. 2001. Integrated weed management in onion (Allium cepa L.). *Indian Journal of Weed Science* **33**: 26-29.

Kumar P, Ompal S and Ahlawat I P S. 2014. Influence of different tillage and herbicide management on weed dynamics and wheat productivity under rice-wheat cropping system. *Journal of Agri Search* **1**(2): 80-85.

Kumar S, Angiras N N and Singh R. 2006. Effect of planting and weed control methods on weed growth and seed yield of Blackgram. *Indian Journal of Weed Science* **38**: 73-76.

Kumar Y, Saxena R, Gupta K C, Fagaria V D and Singh R. 2013. Yield attributes and yield of groundnut (*Arachis hypogaea* L.) as influenced by weed management practices in semi-arid region. *Journal of Crop and Weed* **9**(2): 185-189.

Kunti, Gaurav S and Singh A P. 2012. Weed management practices on growth and yield of winter season brinjal under Chhattisgarh plain conditions.*Indian Journal of Weed Science* **44**(1): 18–20.

Ladha J K, Pathak H and Gupta R K. 2007. Sustainability of the rice–wheat cropping system. *Journal of Crop Improvement* **19**(1-2): 125–136.

Macneish R S. 1964. The origin of new world cultivation. *Science of America* **211**: 29-37.

Marwat K B, Muhammad A K, Saima H, Khalid N and Abdul M K. 2011. Integrated weed management in wheat. *Pakistan Journal of Botany* **43**(1): 625-633.

Mathew G, Sreenivasan E and Mathew J. 1995. Integrated weed management in grain cowpea. *Indian Journal of Weed Science* **27**: 42-44.

Menhi Lal, Tripathi S N, Sinsinwar B S, Niranjan K P, Gupta S D and Arya R L. 1994. Forage production technology, IGFRI, Jhansi pp. 17-24.

Mishra J S and Singh V P. 2003. Interference of *Euphorbia geniculata* in soybean-chickpea cropping system. *Indian Journal of Weed Science* **35**: 225-227.

Moorthy B T, Mishra J S. and Dubey R P. 2003. Certain investigations on the parasitic weed Cuscuta infield crops. *Indian Journal of Weed Science* **35**: 214-216.

Nandal T R and Singh R. 2002. Integrated weed management in onion (Allium cepa L.) under Himachal Pradesh conditions. *Indian Journal of Weed Science* **34**: 72-75.

Pandey A K, Prakash V, Singh R D and Mani V P. 2002. Studies on crop-weed competition and weed dynamics in maize under mid-hill conditions of north-west Himalayas. *Indian Journal of Weed Science* **34**: 63-67.

Pandey A K, Prasad K, Singh P and Singh R D. 1998. Comparative yield loss assessment and crop-weed association in major winter crops of mid hills of north-west Himalayas. *Indian Journal of Weed Science* **30**: 54-57.

Patel B D, Patel V J, Patel J B and Patel R B. 2006. Effect of fertilizers and weed management practices on weed control in chickpea *(CicerarietinumL.)* under middle Gujarat conditions. *Indian Journal of Crop Science* **1**(1-2): 180-183.

Pedde K C, Gore A K and Chavan A S. 2013. Integrated weed management in chickpea. *Indian Journal of Weed Science* **45**(4): 299.

Porwal M K. 1995. Integrated weed management in garlic (*Allium sativum*) under vertisols. *Indian Journal of Weed Science* **27**: 16-18.

Rameshwar S C, Sharma G D and Rana S. 2002. Evaluation of herbicides for weed control and economics in onion (*Allium cepa* L.) under cold desert region of Himachal Pradesh. *Indian Journal of Weed Science* **34**: 68-71.

Rao A N. 2011. Integrated weed management in Rice. *Rice knowledge portal*, Directorate of rice research, Hyderabad.

Rao M V and Bhardwaj R B L. 1986. Weed menace in wheat field. *Indian Farming* **29**(4): 16-18.

Rathi J P S, Tiwari A N and Kumar M. 2004. Integrated weed management in Blackgram (*Vigna mungo* L.). *Indian Journal of Weed Science* **36**: 218-220.

Ravisankar N, Chandrasekaran B, Raja R, Din M and Ghoshal Chaudhuri S.2008. Influence of integrated weed-management practices on productivity and profitability of wet seeded rice (*Oryza sativa*). *Indian Journal of Agronomy* **53**(1): 57-61.

Reddi T Y and Reddi G H S. 2000. Principles of Agronomy, Kalyani Publishers, Ludhiana, India. pp 414-417.

Saqib M, Alis, Ijaz M, Latif M, Ahmad M, Akbar N and Ghaffar A. 2015. The influence of weed management on the growth and yield of direct seeded rice (*Oryza sativa* L.). *Open access* **4**: 1-15.

Sardana V, Singh S and Sheoran P. 2006. Efficacy and economics of weed management practices in Blackgram (*Vigna mungo* L.) under rainfed conditions. *Indian Journal of Weed Science* **38**: 77-80.

Senthil Kumar M. 2009. Effect of plant density and weed management practices on production potential of groundnut (*Arachis hypogeae* L.). *Indian Journal of Agricultural Research* **43**(1): 57-60.

Shekara B G and Nanjappa H V. 1993. Nutrients uptake by crop and weeds and economics of different weed control treatments in soybean. *Indian Journal of Weed Science* **25**: 40-43.

Shinde S H, Pawar V S, Suryawanshi G B, Ahire N R and Surve U S. 2003. Integrated weed management studies in pigeonpea + pearl millet intercropping (2: 2) system. *Indian Journal of Weed Science* **35**: 90-92.

Shubra S and Deepak N. 2015. Effect of pendimethalin on weed flora and yield component of brinjal under agro- climatic condition of tripura. *International Journal of Research in Engineering and Applied Sciences* **5**(3): 79-83.

Singh R and Singh B. 2004. Effect of irrigation time and weed management practices on weeds and wheat yield. *Indian Journal of Weed Science* **36**: 25-27.

Singh R, Singh B and Chauhan K N K. 1999. Productivity and economics of Indian mustard (*Brassica juncea*) as influenced by varieties and different weed control treatments. *Indian Journal of Weed Science* **31**: 138-141.

Singh R. 2006. Effect of cropping sequence, seed rate and weed management on weed growth and yield of Indian mustard in western Rajasthan. *Indian Journal of Weed Science* **38**: 69-72.

Singh S N., Singh R K and Singh B. 2001b. Herbicidal cum-integrated approach of weed management in spring planted sugarcane. *Indian Journal of Weed Science* **33**: 136-138.

Singh S P and Kumar R M. 1999. Efficacy of single and sequential application of herbicides on weed control in transplanted rice. *Indian Journal of Weed Science* **31**: 222-224.

Singh U, Hiremath, S M, Halikatti S I, Shashidhara G B and Patil P L. 2011. Evaluation of herbicides for weed control in rainfed transplanted chilli (*Capsicum annuum*L.). *Karnataka Journal of Agricultural Sciences* **24**(2): 125-128.

Singh V P and Singh G. 2001. Weed control studies in spring rice (Oryza sativa L.) under rainfed low valley situation of Uttaranchal. *Indian Journal of Weed Science* **33**: 52-55.

Singh V P, Singh G and Singh R K. 2001a. Integrated weed management in direct seeded spring sown rice under rainfed low valley situation of Uttaranchal. *Indian Journal of Weed Science* **33**: 63-66.

Singh V P, Singh G Singh R K, Singh S P, Kumar A, Sharma G,. Singh M K, Mortimer M and Johnson D E. 2005b. Effect of weed management and crop establishment methods on weed dynamics and grain yield of rice. *Indian Journal of Weed Science* **37**: 188-192.

Subramanian E, James Martin G and Balasubramanian R. 2006. Effect of integrated weed-management practices on growth and yield of wet seeded rice (*Oryza sativa*) and their residual effect on succeeding pulse crop. *Indian Journal of Agronomy* **51**(2): 93-96.

Sugandha V, Shamshul H, Alyemeni M N and Aqil A. 2012. Effects of herbicide applications in wheat fields is phytohormones application a remedy. *Plant Signaling&Behavior* **7**(5): 570–575.

Sukhadia N M, Ramani B B and Dudhatra M G. 2002.Response of onion (Allium cepa L.) to methods of sowing and weed management practices. *Indian Journal of Weed Science* **34**: 76-79.

Sunil Kumar and Shiva Dhar. 2008. Influence of different herbicides on weed suppression, forage yield and economics of berseem (*Trifolium alexandrium*). *Indian Journal of Agricultural Science* **78**(11): 954-956.

Tewari A N, Tiwari S N, Rathi J P S, Singh B and Tripathi A K. 2003. Effect of cultural and chemical methods on weed growth and grain yield of dwarf pea. *Indian Journal of Weed Science* **35**: 49-52.

Vijaykumar B, Reddy M N and Shivashankar M. 1995. Integrated weed management in pigeonpea-groundnut intercropping. *Indian Journal of Weed Science* **27**: 12-15.

Walia U S and Brar L S. 2001. Competitive ability of wild oats (*Avena ludoviciuna* Dur.) and broadleaf weeds with wheat in relation to crop density and nitrogen levels. *Indian Journal of Weed Science* **33**: 120-123.

Yaduraju N T. 2006. Herbicides resistant crops in weed management. In: *the extended summaries,Golden Jubilee National Symposium on Conservation Agriculture and Environment Octo., 26-28. Banaras Hindu University, Varanasi. pp*. 297-98.

□□□

Chapter 19

IPM in Different Field and Horticultural Crops Under Integrated Farming Systems

N.K. SHAH
ICAR - Indian Grassland and Fodder Research Institute, Jhansi-284003 (UP)

Introduction

Among various biotic factors insects play an important role in agro-ecosystems. Pests and disease may lead to depressed crop production, problems in seed production and even the abandonment of cultivated species. A large number of pests have been found associated with a variety of crops grown in different parts of the country. Due to industrialization and population pressure India has remote scope for horizontal expansion of arable land. The only alternative left is the vertical expansion through increased crop productivity. In the present situation one of the ways to augment crop production is to reduce the losses inflicted by various pests and diseases. Agricultural and horticultural crops are subjected to the depredations of pests and diseases. These pests and diseases hamper crop establishment, impair its quality and reduce the yields.

Owing to limited scope of horizontal expansion to augment crop production, the alternate is to move with vertical growth through development of high yielding varieties, adoption of improved agronomic practices recommended for crop production and minimizing the crop losses inflicted by pests and diseases. In the present situation one of the ways to augment crop production is to minimize the losses imposed by various insect pests and diseases which altogether tolls 25-30 per cent of yield losses. The sustainability of the production system largely depends upon overcoming biotic stress factors like insect pests, plant pathogens and plant parasitic nematodes. Several promising crop production systems with recommended agronomical and plant protection package of practices are available that confer higher yields and are incredibly valuable for higher production.

Management Strategies

Under prevailing systems of crop production chemical pesticides are generally discouraged. Chemicals although provide quick and satisfactory control of many pests and diseases but these are not recommended as a blanket because of low economic returns, residual toxicity and related health problem to livestock and human beings. These issues restrict the scope of pest control through chemicals and call for the management approaches of co-existence with pest, pathogens and bio-control agents. This requires a close understanding of both the organism and environment. Different methods of pest management have been experimented in various crops with varying degree of successes. The general progress made in this direction is described in the following method:

The cultural control methods are prophylactic in nature and are considered as the first ditch defense around which to build other control options (Coaker, 1987). Though, these methods are unlikely to reduce pest infestation to desirable levels but within the frame work of IPM systems these can very well be used to reduce overall levels of infestation.

Tillage Practices: - Tillage practices create conditions not suitable for pest survival, expose pests to their natural enemies, inflicts direct physical injury by exposing to sun and other abiotic factors. One or a combination of these factors reduces the population. Deep summer ploughing twice during May and June at an interval of 15 days reduced the population of *Meloidogyne* spp. and *Rotylenchulus reniformis* infecting cowpea and field bean (Ahmad *et al.*, 1996).

Time of Sowing: - The choice of sowing dates in relation to crop pests has one principal aim to reduce the period to a minimum over which infective agents meet to susceptible stage of the host. Favorable environments (temperature, moisture, light, soil reaction, age of the host) are essential for growth and pathogenicity of pests. In vector transmitting viruses, the peak population of the vectors is reached at a particular period. Cowpea crop sown between 3rd week of June and last week of July escapes incidences of leafhoppers and other defoliators (Shri Ram *et al.*, 1989), Lucerne sown early 3rd week of October escapes the incidence of jassids and lucerne weevil (Shri Ram *et al.*, 1989).

Inter Cropping: - Introduction of polycultures like mixed cropping, rows inter cropping, stripes inter cropping, rely inter cropping and alleys inter cropping in fodder production systems can help in reducing pests. Inter cropping of cowpea in sorghum reduces leaf hopper and defoliator damage in cowpea (Shri Ram, *et al.,* 1989). Intercropping of mustard in wheat and Barley was found to reduce number of cyst/ plant and nematode infestation in barley (Rajvanshi *et al.,* 2002).

Crop Rotation: - From a plant protection point of view a good crop rotation is that which reduces the pest load by reducing the nutrition of pest in every subsequent year or totally deprive them of food, reducing the initial infection by disease and ensures unfavorable stratum for the development of pests. This method may not be very successful against insects having strong power of dispersal yet is very effective against nematodes and soil borne plant pathogens.

Organic Amendments: - Organic amendments are traditionally used to improve soil structure and plant nutrition but there are numerous reports that their addition can also lead to control of pathogens. Incorporation of various kinds of crop residues or agricultural wastes like neem oil cake (15-20 q/ha), saw dust (30 q/ ha) or chopped *Parthenium* plant (50 q/ha) have been found to be effective in reducing nematode population to a considerable level (33-66%) besides improving the dry matter yield (45-78%) in cowpea, sorghum and bajra.

Host Plant Resistance

Spectacular successes have been achieved in protecting different crops from insect damage and consequently reduce losses by using one of the relatively less emphasized methods, the host plant resistance. It is now being increasingly realized that this method deserves more attention than it has received so far. In crops with relatively low economic returns like forages, the development of resistant varieties can be the most ideal method for minimizing losses due to pests and pathogens. The farmer/cultivator is provided with a built in insurance against pests and pathogens and thus he is not hassled about the additional input for management of pests and diseases. Potentially the most economic and effective method of pest management in forage crops is to use pest resistant varieties. The primary objective is to develop crop varieties resistant to one or more pest species along with other desirable yield characters.

Biological control

Biological control is effective, environmental friendly and combines well with cultural control and host plant resistance. Although the use of biocontrol agents have not been taken up in our country in a large scale. In forage crops a number of workers have obtained control of soil borne plant pathogens through nematodes (Barnes *et al,* 1981), fungi like *Trichoderma viridae, T. harzianum* (Bhaskar and Ahmad, 1997). Exploitation of entomopathogenic nematodes, *Steinernema carpocapsae* and *Heterorhabditis indica* indicated alfalfa grub (*Hypera postica*) population was significantly reduced in *H. indica* and *S. carpocapsae* treatment at a concentration of 1 billion Ijs/acre. A decrease of 72.10% and 49.66% grub population over untreated control were noticed in *H. indica* and *S. carpocapsae* treatment respectively (Shah and Azmi 2005; Shah *et al,* 2011).

Management through ITK's

Work carried out at IGFRI have shown that crude water extracts, essential oils and certain chemicals isolated from several higher plants like neem, karanj, eucalyptus and castor have shown their potential as an insecticide, fungicide and nematicide. Aqueous extracts of neem (*Melia azadirachta*), karanj (*Pongamia pinnata*) and castor have revealed promising results for the management of insect pests, diseases and nematodes in sorghum, maize, cowpea, berseem and lucerne. Under laboratory bioassay asafoetida was appraised to be a potential pesticide.

Integrated Pest Management (IPM)

Our past experience in the pest management indicated multifaceted problem associated with ecology, behavior, environmental condition and biology of the pest/pathogen and inherent limitations of each control method. It is now almost universally accepted that ultimate solution lies in an integrated control. The connotation of the approach is to utilize all the possible methods of control including maximum exploitation of naturally existing populations and regulating factors in as compatible manner as possible and keep the population levels below economic threshold.

Integrated pest management or IPM is a systematic approach to crop protection that uses increased information and improved decision-making paradigms to reduce procured inputs and improve economic, social and environmental conditions on the farm and in the society. Moreover, the concept emphasizes the integration of pest suppression technologies that include biological, chemical, legal and cultural controls.

Integrated pest management (IPM) is a pest management strategy that focuses on long-term prevention or suppression of pest problems with minimum impact on human health, the environment and non-target organisms. Preferred pest management techniques include: encouraging naturally occurring biological control agents, using alternate plant species or varieties that resist pests, selecting pesticides with lower toxicity to humans or non-target organisms; adoption of cultivating pruning, fertilizing or irrigation practices that reduce pest problems or changing the habitat to make it incompatible with pest development.

Thus Integrated pest management is a sustainable approach to manage pests/pathogen by combining biological, physical and chemical tools in a way that minimizes economic, health and environmental risks.

IPM benefits can be realized if the following steps are taken:

i) Identify problems correctly
ii) Determine the extent of the problem by sampling
iii) Critically assess the importance of the problem
iv) Evaluate and select appropriate management alternatives
v) Implement selected management actions in a timely manner
vi) Evaluate the effectiveness of control actions.

Key Components of Integrated Pest Management

Knowledge: Understanding the biology and ecology of the pest and the crop is essential. Information about interactions within agricultural ecosystems is also important. IPM draws on the fundamental knowledge of plant and pests.

Monitoring: Farmers can use relatively simple techniques to keep track of what pests are where. This information, combined with knowledge of pest life cycles, can enable farmers to implement control measures at the most effective times. Monitoring on a broader scale can also be used to predict pest outbreaks and to forewarn farmers to proceed for action.

Economic threshold: This takes into account the revenue losses resulting from pest damage and the costs of treatment to prevent the damage. Below the economic threshold, the presence of the pest is tolerated. Only when pest numbers increase above the economic threshold level the farmer can proceed for its management.

Adaptability: Farmers must keep informed about what is happening in their fields so that they can adapt their strategies to changing circumstances. Research scientists, too, must aim to keep at least one step ahead of the pest, which is also undoubtedly changing and adapting over time.

Techniques in IPM

A wide range of pest control techniques are available to farmers. Some of them are as old as agriculture itself – rotating a crop to avoid a build-up of host-specific pests. Some are new – in recent years, genetic engineering has opened up many possibilities in pest control that were unavailable to agriculturists earlier. The simple philosophy is that the control will be more effective and resistance will be less likely to build up, when a range of measures are deployed against a pest. Wherever possible, different pest control techniques should work together rather than against each other. In some cases, this can lead to synergy – where the combined effect of different techniques is greater than would be expected from simply adding the individual effects together. Integrated pest management involves the integrated use of four basic control techniques (physical control, biological control, Genetic modification and of course, chemical control.

Physical Control

Physical controls are those that can be carried out by the farmer to alter environmental factors in a way that reduces pest populations. A simple and common example of this is crop rotation, which is the practice of planting different crops each year in a given field. This interrupts the normal life cycle of some pests/pathogens by changing their environment to one in which their favourite host plant does not feature. It is a strategy that has been used successfully for years by farmers. Another physical control method sometimes called 'mating disruption' involves the use of sex pheromones. These chemicals are produced by female insects to attract males for mating. For many insects, scientists have been able to analyze the chemistry of the sex pheromones and reproduce them synthetically in the lab.

Biological Control

The principle behind biological pest control is that a given pest has enemies - predators, parasites or pathogens. By introducing or encouraging such enemies, the population of pest organisms should decline. It is not a new concept. The ancient Chinese encouraged ants in citrus orchards because they attacked many citrus pests.

Genetic Modification

Crop plants can be bred to be resistant to pests. Farmers have been doing this for centuries, selecting the seeds of those plants least affected by a pest for use in the next year's crop. This preferential selection is a form of genetic modification. With advancement in biotechnology and molecular biology, it is becoming increasingly easy to transfer resistance genes into a plant - this is called gene transformation or genetic engineering. An example of genetic engineering is the insecticide-producing Bt gene in cotton. Scientists took the gene from a bacterium and inserted it into the plant, consequently making the plant resistant to insect attack. Similarly, potato plants have been genetically modified to increase their resistance to potato leaf roll virus.

Chemical Control

The use of chemical pesticides often forms part of an integrated pest management strategy. The key is to use pesticides in a way that complements rather than hinders other elements in the strategy and which also limits negative environmental effects. It is important to understand the life cycle of a pest so that the pesticide can be applied when the pest is at its most vulnerable stage - the aim is to achieve maximum effect at minimum levels of pesticide.

IPM in India

Integrated pest management (IPM) is a decision making process that identifies and reduces risks from pests and pest management strategies. It coordinates the use of pest biology, environmental information and multiple pest management tactics to prevent unacceptable levels of pest damage by the most economical means. Several organizations are actively involved in developing IPM techniques for adoption by farmers. Consequently, comprehensive IPM programs are developed for rice, cotton, sugarcane, pulses, oilseeds, vegetables and forage production system.

Most of the work on IPM has been conducted in developed countries. In India integration of various components of IPM for insect pest management have been worked out at IGFRI for cowpea, lucerne and Chinese cabbage (Shri Ram and Gupta, 1989). Substantial efforts have been made to combine several methods to manage nematodes in forage crops also, though on limited experimental stage. For instance certain intensive rotation combined with fertility management (sorghum) bajra wheat-cowpea x 50 kg N (Urea) was reported to be mostly associated with lower nematode population (Chawla and Prasad, 1973). Experiments carried out at this Institute revealed that combination of either deep summer ploughing with saw dust; neem cake (furrow application), carbofuran, aldicarb, phorate (furrow and seed treatment) or nematicide in reduced doses with neem cake sufficiently reduce nematode populations.

Plant protection researches established the efficacy for botanicals and bio rational insecticides on various crops including forages. Various commercial formulations of azadirachtin (the active ingredient of neem), neem seed kernel extract (NSKE), neem seed oil suspension (NOS), insecticidal soap, narrow range mineral oil and pyrethrins were tested on immature and adult life stages of insects in laboratory and greenhouse experiments. The results indicate that bio rational insecticides may be integrated into organic systems for insect pest suppression along with the conserving of predator field populations. Scrupulous investigations have yet to be done to optimize application timing, frequency and sprayer delivery methods to increase pest insect coverage.

Intercropping of leguminous and non-leguminous forages may result in lower insect populations as both share different pest species. Establishing alfalfa or alfalfa/grass with oats as a companion crop can provide excellent early season weed control, soil erosion protection and added forage. In years when insects reach in outbreak densities foliar insecticide must be incorporated to attain a successful IPM for profitable yield.

Amalgamations of various botanicals were evaluated at this institute for their comparative bio-efficacy for the management of pests/diseases in summer sown cowpea + sorghum fodder production. Seed coating with neem seed kernel (NSK) powder (50g/kg seed) + sprays of neem seed kernel (NSK) extract (3%) provided maximum protection in which a reduction 45.81% in *A. soccata* incidence, 8.70% in *C. trachypterus* and *H. nigrorepletus* intensity, 56.06% in *C. graminicola*, 69.87% in *G. sorghi* incidence in sorghum, 40.95% in *P. signata* infestation and 40.86% in *C. sorghi* incidence in cowpea were recorded. This treatment also provided maximum protection against PPN, *T. vulgaris* (41.62% reduction) and there was an increase of 14.06% GFY over control (Shah *et al*, 2007). Similarly neem seed powder (50g/kg seed treatment) resulted significant control of root rot of berseem (Bhaskar *et al.*, 2003).

Various combinations of botanicals were evaluated for the management of pests/diseases in berseem mixed with mustard fodder production. Seed coating with NSK powder + sprays of NSK extract (3%) provided maximum protection (a reduction of 68.22% in the root rot intensity in berseem and a reduction of 70.28% aphid population in mustard) consequently an increase of 33.61% in GFY over control. This combination provided maximum protection against plant parasitic nematode (PPN), *Tylenchorhynchus vulgaris* (a reduction of 42% PPN) and harbored maximum number of beneficial soil micro-arthropods viz. *Folsomia sp., Brachystomella sp.* (Collembolans) and Prostigmeta, Mesostigmeta and Cryptostigmeta (Mites) (approximately triple the population of beneficial collembolans and mites in comparison with untreated check) (Shah *et al.*, 2011).

Various IPM modules were studied at IGFRI for the management of biological stress factors in an intensive forage production system (berseem mixed with mustard + hybrid Napier). The treatments were factorial combinations of sowing dates and environmentally safe foliar sprays. Of various sowing dates firstsowing date in middle of October was found to be most appropriate time of sowing. This sowing date in combination with seed treatment of thiomethoxam + pencycurone (0.5gm/kg seed) along with foliar spray of imidacloprid + tebuconazole (0.025% conc.) was found to be the most effective IPM module as this combination significantly reduced biotic pressure on the crop (58.06% reduction in the root rot intensity caused by *Sclerotinia trifoliorum* in berseem, 78.65% reduction in aphid population (*Lipaphis erysimi* Kalt) and 52.55% reduction in plant parasitic nematodes (*Tylenchorhynchus vulgaris*) and yielded maximum GFY (89.52t/ha, an increase of 17.05% over control). This was followed by seed treatment with thiram + carbendazim + carbofuran and spray of endosulfan (0.07%) + mancozeb (0.09%) (88.38t/ha i.e.15.99% increase), seed treatment with tricho XP (5gm/kg seed) + foliar spray of azacel 0.15%conc. (86.38t/ha i.e.14.6% increase) and seed treatment with N S P 50gm/kg seed + foliar spray of NSKE 3 % (80.27t/ha i.e.7.5% increase) (Shah *et al.,* 2011).

Attempts are also being made through All India Coordinated Research Project on Forage Crops to collect information on various factors like pest status, damage thresholds in various agro-climatic regions of the country.

Major Constraints to IPM in India

A major limitation is the lack of trained personnel. Many farmers are not trained adequately in augmentative biological control, leading to misunderstanding of its potential efficacy. Logistical problems such as improper timing and delays in shipment can alter the effectiveness of natural enemies. Farmers often believe that natural enemies do not work well and that low pest populations will cause losses. The use of biopesticides is limited due to moderate toxicity, slow action, host specificity and photo-instability as well as a higher cost. Many farmers are not yet aware of the proper usage and available suppliers of biocontrol agents and biopesticides. A number of botanicals such as karanj, mahua, nuxvomica, custard apple, ipomoea, garlic and tobacco have been found to be effective against insect pests and diseases; however in absence of detailed scientific data, except for neem, most of them are localized to rural pockets. IPM adoption is influenced by the cost versus efficacy of products, need for sophisticated information for decision making, ability to integrate new products and techniques into existing farm management practices and managerial skills.

Conclusive Remarks

It can be concluded that although significant progress have been made, however much remains to be done before existing and new IPM options can be used in an optimal manner; therefore there is a need for further research on some of the following aspects of IPM considered to be more critical and significant.

1. Research on the applied ecology through detail studies of population dynamics to provide the biological basis for determining appropriate management tactics for the economic suppression of pest species.
2. The compatibility of control measures is needed to be tested during the early stages of an IPM programme.
3. Identification and utilization of multiple pest and disease controlling agents (chemical, varieties etc.) capable of managing insects, nematodes and pathogens.

References

Ahmad, S.T., Pandey, K. C. and Bhaskar, R.B. 1996. Integrated pest management for increased forage production, *Indian Farming*, **45**: 34 - 37.

Barnes, G.L., Russell, C.C., Foster, W.D. and Mc New, R.W. 1981. *Aphelenchus avenae* – A potential biological control agent for root rot fungi, *Plant Disease*, **65**: 423-424.

Bhaskar, R.B., Hasan, N., Pandey, K.C. and Melkania, N.P. 2003. Management of root-rot disease complex of Berseem (*Trifolium alexandrinum* Linn.), *Forage Research.*, **29**: 84-87.

Coaker, T.H. 1987. *Integrated Pest Management* [M. Burn A.J., Coaker, T.H. and Jepson, P.C. (Ed.)], Academic Press, London, pp. 69-88.

Chawla, M.L. and Prasad, S.K. 1973. Multiple cropping and nematodes II. Effect of intensity of cropping on nematode populations, *Indian Journal of Nematology*, **3**: 75-77.

Shah, N. K. and Azmi, M.I. 2005. Virulence of *Heterorhabditis indica* to the grubs of lucerne weevil, *Hypera postica* (Coleoptera:Curculinidae)'. In: Proceedings of *National symposium on Recent Advances and Research Priorities in Indian Nematology*. pp.42.

Shah, N.K., Saxena, P., Azmi, M.I., Roy, S. and Tyagi, P. K. 2007. Synthesis of botanical pest management technology for intensive forage production, *Range management and Agroforestry*, **28**(2): 157-158.

Shah, N.K., Saxena, P., Azmi, M.I., Roy, S. and Tyagi, P.K. 2011.Botanical pest management in Berseem+ Mustard mixed forage crop production, *Journal of Eco-Friendly Agriculture*, **6**(1): 59-62.

Shah, N.K., Azmi, M.I. and Tyagi, P.K. 2011. Pathogenicity of Rhabditid nematodes (Nematoda: Heterorhabditidae and Steinernematidae) to the grubs of alfalfa weevil, *Hypera postica* (Coleoptera: Curculionidae). *Range Management and Agroforestry*, **32(**1): 64-67.

Shri Ram and Gupta, M.P. 1989. Integrated pest management in fodder cowpea (*Vigna unguiculata* (Linn.) Walp.) in India and its economics, *Tropical Pest Management*, **34**: 348-351.

Shri Ram., Gupta, M.P. and Patel, B.D. 1989. Pest management in fodder cowpea (*Vigna unguiculata* (Linn.) Wasp.) through mixed and intercropping in India, *Tropical Pest Management*, **35**: 345-347.

□□□

Chapter **20**

Disease Management in Different Field and Horticultural Crops Under IFS: Issues and Strategies

H.V. SINGH AND D. DEB

ICAR - Indian Grassland and Fodder Research Institute, Jhansi-284003 (UP)

Introduction

For the last few decades, agricultural scientists in the developing and underdeveloped countries were looking for an alternate agriculture system to increase the net income of poor and small farmers so that the economic condition of these marginal farmers can be improved. Integrated farming system (IFS) may helps and provides opportunity which will improve the economic condition of the marginal and small farmers up to certain level with the help of appropriate alternative innovation in the field of agricultural technologies. An integrated farming system represents multiple crops (cereals, legumes, tree crops, vegetables etc.) and multiple enterprises (animal farming, bee keeping, fish farming etc) in a single farm. It is widely accepted as a means of achieving sustainable agriculture to meet the multiple objectives of poverty reduction, food security and competitiveness and sustainability. In India, more than 70 percent of the total population of the country still live in villages and mainly depend on Agriculture and/or related enterprises. Marginal and small farmers constitute more than 84 percent of the 115 million operational holdings in India which are cultivating only 29 percent of the arable land. Most of them are resource poor and work in diverse, risk prone environment. Further, small farm holders remained deprived of benefits of advancement in the field of agriculture, major beneficiary being resourceful well capable farmers. The livelihoods of the small and marginal house hold families are the major concern. One of the key challenges of the 21st century is developing ways of producing sufficient amounts of food while protecting both environmental quality and the economic well-being of rural communities. Due to declining profitability as a result of high input demand in crop production, increasing cost of inputs, the farmers are considering farming as a secondary occupation which resulted large scale migration of rural mass to the nearby towns and cities. In view of the fast changing agricultural scenario, a drift has resulted from sustenance to commercial farming. To maximize the yield, farmers are using high-yielding varieties and hybrids with higher inputs of chemical fertilizers and pesticides to a large extent. The newly released hybrids, indiscriminate and injudicious use of fertilizers and pesticides have resulted in susceptibility to various diseases and insect pests. Besides this, a number of other problems such as soil, water and air pollution, residual toxicity in fruit and vegetables, resistance to insects and pathogens, mortality of parasites, predators and pollinators, and resurgence with outbreaks of secondary pests have also developed. With increasing environmental awareness, the focus has now shifted towards the search for viable alternatives of disease control methods. Different agricultural activities are being applying to get maximum harvest (Fig. 1).

Plant diseases are considered an important biotic constraint, which leads to significant crop losses worldwide. Integrated disease management (IDM), which combines biological, cultural, physical and chemical control strategies in a holistic way rather than using a single component strategy proved to be more effective and sustainable. Host plant resistance, natural plant products, biopesticides, natural enemies and agronomic practices offer a potential viable option for IDM (Kumar, 2014). In practice and in the majority of cropping systems today, emphasis is still being placed on a single technology. Nevertheless, the use of IDM strategy is gaining momentum, but in developing countries it often lacks the enabling environment for its, successful implementation. Wide adoption of IDM practices is a pre-requisite for achieving impact at the country level. Experience over the last few decades clearly showed that adoption and support for using participatory approaches help farmers improve their overall field management, including disease management, reducing costs and improving production efficiency. In this paper we will shed light on all the elements that require attention to achieve successful IDM adoption at the national level in developing countries. Development of Plant clinic in rural areas helps farmers avoid futile costs of self medication, dosing fields with erroneous fungicides (Paul, 2012). Registration and commercialization of super stains of Biological control agents by Central Insecticide Board (CIB) helps to the farmers (Singh,

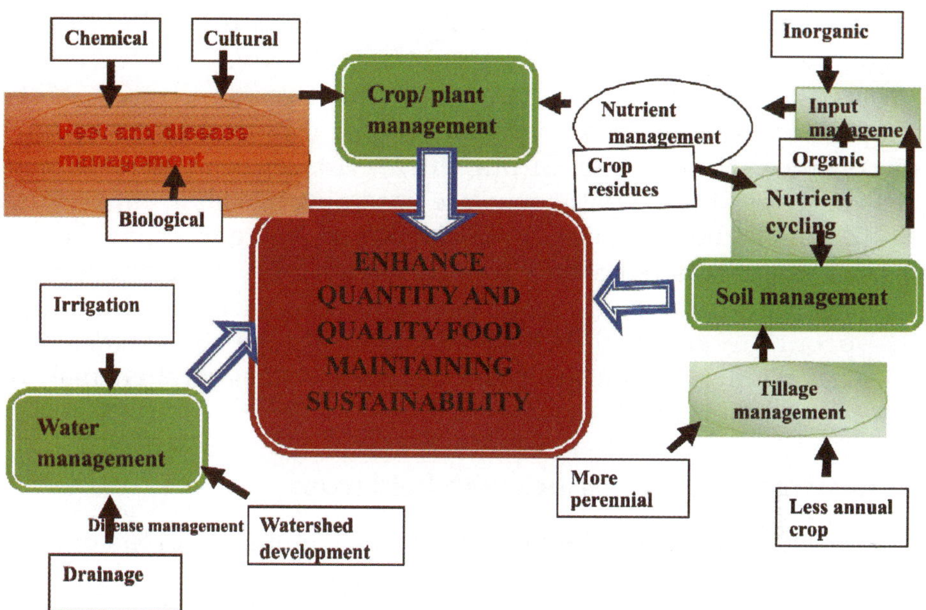

Fig. 1

2012). Mushroom (*Strobilurus tenacellus*) derived fungicides strobilurins also contributes in this direction (Thind, 2012).

Marginal and small farmers constitute more than 84 percent of the 115 million operational holdings in India. Most of them are resource poor and work in diverse, risk prone environment. Further, small farm holders remained deprived of benefits of advancement in the field of agriculture. The livelihoods of the small and marginal house hold families are the major concern. One of the key challenges of the 21st century is developing ways of producing sufficient amounts of food while protecting both environmental quality and the economic well-being of rural communities. The potentialities of each technology of the agricultural production system are not being fully harvested / utilized and make the system unviable. Traditionally different farming systems have evolved independently and being practiced by the farmers especially by poor and marginal farming communities without any rationale for utilizing the wastes and residues arising out of cropping/animals of their farm and other associated enterprises at farm resulting wastage of resources. Because of poverty, the small land holders are reluctant to adopt new technologies and/ or enterprises. Modern high input mono cropping based intensive agriculture has resulted in loss of biodiversity, out breaks of pest and diseases, degradation of soil and water, which ultimately led to stagnating agricultural production and productivity. Climate changes are also becoming a major factor in present scenario (Kumar, 2013). Crop yield loss, on field and after harvest due to diseases is vital importance (Kumar and Gupta, 2012). Insurgence of resistance against major disease pathogens are becoming alarming situation in cash crops and horticultural crops. Chemicals have reduced crop losses in many situations but even with a substantial use of chemicals pose numerous hazardous effects on soil biodiversity and environment. Inappropriate and furious use of chemicals led to increase and unnecessary disease outbreak and destruction of natural enemies of disease causing agents.

Disease management issues with field crops

Development activities in India mostly confined to urban or semi urban area while more than 65 per cent population in the country lives in rural area and mostly depend on agriculture and related avenues for their subsistence. Hunger and poverty persists because of many agricultural technologies specially plant protection methods not reach to them. Crop yield losses in field and after harvest are higher in these forming communities. Various types of direct and indirect losses caused by plant disease increased the production cost. The key tools of disease management are knowledge of disease and control measures with availability and accessibility (Kumar 2014). With the increasing area under hybrid varieties and intensive cultivation occurrence of diseases are more common with severity and more widespread (Thakur *et al.,* 2006). There has been considerable success in breeding for DM resistance using conventional pedigree breeding, and a large number of disease resistant hybrids have been developed and deployed (Khairwal *et al.,* 2004). This has contributed since the 1990s (Thakur *et al.,* 2006). In general, the long-term success of breeding for disease resistance is influenced by several factors that include: (i) the nature of the pathogen and diversity of virulence, (ii) availability, diversity and type of genetic resistance, (iii) screening method and selection environment, (iv) breeding methods, (v) utilization and deployment and (vi) monitoring resistance/virulence. Plant disease management is essential on the different field crops and particularly on vegetable farm to achieve optimal yield and profit. Plant pathogens infect plants with its ability to continue growing and potentially affecting the quality and yield of the crop plant. A comprehensive disease management approach is needed in order to minimize and limit the negative impact that diseases. Disease control options vary in strategy, timing, duration and efficacy, but when control options are combined in a holistic and integrated approach, the risk of disease-driven losses is minimized. The farmers with poor resource and unawareness with recent advances in agriculture are not using these technologies and besides the development of new technologies they are still deprive with progress. These are some of the issues which should be take care for developing disease management strategies of field crops:

1. Poor economic condition of large farming communities
2. Lack of adequate knowledge of pest and disease
3. Lack of alternative control options
4. Lack of access to biopesticides
5. Lack of accurate timing for fungicide applications
6. Lack of knowledge on proper use of new products.

Disease management issues with Horticultural crops

Small and marginal farms adopt diversified farming, despite very limited space and low sizes of holdings. However, the extent of diversification is relatively low on small and marginal farms, as compared to large farms. The proportion of area under non-food crops increased, though which is less in marginal to small farmers than large farmers. All most all marginal and small farmers are put higher proportion of their tiny plots to nonfood crops. Diseases destroy the plant in most active phage of crops especially in guava crop suffered with wilt, papaya with ring spot, banana with bubchy top citrus with canker and phytopththora aonla with color rot, apple with scab mango with malformation and many more examples of vegetable crops. For soil borne pathogens propagules (seeds of the pathogen) tend to decrease if its prime host is not planted in the same field. Horticultural crops are perennial in nature and disease infected parts may allow and develops the plant pathogens to infect other crops if suitable plant protection measure not adopted. Diseases trees are harbor the disease propagules and provide the inoculums source to other crops. With changing horticultural practices new diseases like phytoplasma, viruses and viroids are emerging. To protect horticultural crops from such diseases a greater vigilance and stringent quarantine measures are required. Significant gaps have been observed in the efficiency of proven protection technologies in between research farms and farmers fields. Factors responsible for low efficiency are variation in fungicides grade/quality, inappropriate/indiscriminate use, use of banned/spurious/expired fungicides and knowledge gaps in execution of integrated orchard management practices are the diseases management issues of today's agriculture. One of the evils faced by poor farmers of India is the spurious pesticides (Kumar, 2013). In rural areas traders forced selling of fungicides and their application by un aware farming communities is another issues.

Strategies of disease management under IFS

Plant pathology is challenging and an important science that deals with science of disease development and ability of managing disease. Farming communities able to get benefit from plant pathology by evolving and developing management techniques of emerging disease responded to the significant changes in agricultural practices. In the present day agriculture many new disease are developed due to changing agriculture and climate change. The diseases are less important earlier are becoming more important and causing hues loss to the crop. Under IFS an integration of crops and agricultural enterprises is followed, the scenario of disease development is change. A holistic approach is needed with local resource based plant protection measure which may be effective in diverse crop families and din different environment. The strategies under IFS should be location specific, crop specific, need based and local resource based. Among all the plant protection technologies so for used in protecting field crops and horticultural crops, IDM is the best solution and it should be modified and validated in particular site and crop. The component of IDM as under:-

1. Monitoring (scouting).
2. Forecasting.
3. Thresholds.
4. Choosing management methods.
5. Evaluation.

Strategic Preventative Measures

Many measures for preventative disease control have a long-term strategic character. Various aspects of the farm management and the long-term cropping system have impact on diseases of crops and thus should generally be considered in integrated disease management. Avoidance of pathogen sources in neighborhood of field and crop rotation may be one strategy. Abandoned orchards and orchards with high inoculums pressure should not be found in the neighborhood of an orchard. Maize stubble are often colonized by the same *Fusarium* spp. affecting wheat and such *Fusarium* spp. can survive and multiply on maize stubble for several years. Further, the negative impact of green revolution such as loss of genetic diversity of indigenous crops, mono cropping of selected crops (maize, rice, wheat) ignoring indigenous crops, increase dose of chemical fertilizers, development of pesticide resistant, emergence of new disease pests which were either nonexistent or present as minor pest become major pest. Indiscriminate use of pesticides that results killing of natural enemies of pests, beneficial soil micro organisms and its residual effect in food chain necessitates the development of integrated pest management concept to minimize and judicious application of pesticides are some of the corrective measures.

Prevention is better than cure. There is some of the preventive measure as follow:

- Avoidance of pathogen source
- Soil sanitation
- Crop rotation
- Tillage.

Preventive measures are planed with the planning and realization of a certain crop and locality. Typical measures are the choice of variety, seed quality, seeding time and crop structure. The preventive measure may be as under:

- Resistant variety
- Crop sanitation
- Healthy and improved seed
- Sowing time
- Crop structure
- Biological crop residue treatment.

Cultural measures

Cultural management manipulates the environment to make it more favorable for the plant and less favorable for the disease pathogens. Like all other control strategies, cultural control requires an understanding of the pathogens and the crop. Cultural control measures are usually applied at the most vulnerable stage of the pathogen life cycle and are generally preventive rather than curative actions. This work in under mentioned ways:

1. Prevent the pathogens from colonizing the crop.
2. Create adverse conditions that reduce survival of the pathogens.
3. Reduce the impact of pathogen damage.

This can be achieved by the adaptation of following methods:

- Adjust the planting to provide maximum drainage and air circulation.
- Plant disease-resistant or tolerant varieties or rootstocks.
- Remove overwintering sites *i.e.* crop residues and plant debris buried in soil.
- Prune to eliminate infected parts of plant.
- Proper irrigation schedules for specific crop.
- Manage fertilizer to limit the excessive plant growth.
- Follow soil sanitation.
- Select and use strong, healthy plants.

Successful implementation of IDM strategies

The ultimate aim of promoting IDM is to empower its users to engage in competitive and sustainable agricultural production with long-term positive impacts on poverty and environmental health. There are a number of the national programme and related activities that have direct impact on IDM adoption and scaling up, with special emphasis on the developing countries. Besides farmers' education in IDM principles, consumers, whether in rural or urban areas are major driving forces of the IDM market and could act as an important lobbying force for promoting food safety-related programme such as minimization of chemicals load on food commodities especially on fruits and vegetables. Awareness raising of policy makers by bringing them to the field and exposing them to the successes of IDM programmes and their impact on farmers and consumers has proven to be good strategy to get the national support to IDM programmes through policies, regulations, and funding. Land ownership and size have direct effects on the risk level that farmers would be ready to assume as well as on the motivation of farmers to invest in new technologies and adopt strategies with longer-term returns on investment. Crop and genetic diversity within the larger landscape, including gene deployment strategies, reduction of the number of varieties cultivated in agro-ecological areas, monoculture in time and space, management of field hedges harboring natural enemies or the management of alternate and secondary hosts of pathogens are all factors that would directly affect the effectiveness and successes of IDM strategies. Modern biotechnological tools especially to developed biocontrol agents could be used to minimized the diseases and to date about 26 microbes included in the scheduled to the Insecticide act (Mukhopadhyay, 2012). IFS adopted farmers are have more number of systems and especially for marginal who have poor resources cannot invest more on plant protection measure. Chemicals application is recommended only in extreme condition of disease. Commercialization of agriculture enhances the regular application of agrochemicals, while for sustainability application of traditional ITK with scientific skill is needed. Under IFS IDM is the best option and the component of that IDM should be chosen from locally available resources.

Conclusion

Indigenous disease management knowledge and traditional ecological knowledge derived through long experiences and perceptions accumulated by traditional farmers by interactions and with the nature and natural resources need to be effectively integrated and should used. Indigenous disease management knowledge well fitted in the traditional land use system which t need thorough validation in view of changing agricultural scenario of IFS through inorganic and organic farming methods. The success and sustainability of IDM strategy, especially with resource poor farmers greatly depends on their involvement in helping generate locally specifictec hniques and solutions suitable for their particular farming systems. These validated Indigenous traditional knowledge (ITK) may be incorporated as an integral component to the IDM concept for evolving better and effective disease management strategies in any of these farming systems. This systematic approach not only protects this fast disappearing ITK under the influence of modern agriculture but also preserve the indigenous disease management identity of farming communities. Therefore, this farming tradition should be harnesses, preserved, documented and developed as modern science. Training and awareness raising of farmers, disease survey teams, agricultural development officers, extension agents and policy makers remains to be an important factor for the successful implementation of IDM strategies. Training and awareness escalating programme of farmers, disease survey teams, agricultural development officers, extension agents and policy makers remains to be an important factor for the successful implementation of IDM strategies. Comprehensive eco-friendly approach for key diseases result of IFS should be focused as research priorities. Intensive research on soil health factors, chemical fertilizer application with declining soil organic carbon is needed. Research and technical issues should address the knowledge gaps in etiology, diagnostic techniques particularly for soil borne diseases and weather dependent outbreaks. Protection packages are largely chemical based against single pathogen resulting in resistance against agrochemicals especially against narrow spectrum based fungicides and resurgence of new pathogens which need to address. Fortification of the compost with bio-agents and nutrient enrichment will help in ensuring desirable soil health. Application of desirable microbial consortia in nursery production practices of fodder tree and perennial grasses should be standardized for enriching the myco-rhizosphere with to develop survival and vigor in saplings/ root slips against biotic and abiotic stresses.

Reference

Khairwal, I.S., Rai, K.N., Yadav, O.P. and Bhatnagar, S.K. 2004. Pearl millet cultivars. Mandor, Jodhpur, India: All India Coordinated Pearl Millet Improvement Project, Indian Council of Agricultural Research. 22 pp.

Kumar, S.and Gupta, O. 2012. Expanding dimentions of plant pathology. *JNKVV Res.J.* **46**(3): 286-296.

Kumar, S. 2013. Plant diseas management under Changing climate scenario. *J Mycol.Plant Pathol.* **42**(2): 149-154.

Kumar, S. 2014. Plant disease management in India: Advances and challenges. *Afr .J. Agaric Res.* **9**(15): 1207-1217.

Mukhopadhyay, A.N. 2012. Trichderma for plant disease management- a gift of God to humankind . *J. Mycol Plant Pathol.* **42**(1): 23.

Paul, Y.S. 2012. Role of plant clinics in global food security Souvinir 3rd Glo. Plant Pathol.Food Security, Udaipur, Rajasthan, P 34.

Singh, H. B. 2012. Commercialization of biocontrol agents. *J Mycol. Plant Pathol.* **42**(1): 25.

Thakur, R.P., Shetty ,H.S. and Khairwal, I.S. 2006. Pearl millet downy mildew research in ndia: progress and perspectives. *International Sorghum and Millets Newsletter* **47**: 125–130.

Thind, T.S. 2012. New generation fungicides for managing plant disease: advantage and limitation. J. *Mycol Plant Pathol.* **42**(1): 26.

□□□

Chapter 21

Post Harvest Processing and Value Addition of Horticultural Produce in IFS for Higher Profitability

SADHNA PANDEY

ICAR - Indian Grassland and Fodder Research Institute, Jhansi-284003 (UP)

Introduction

In India, more than 70 percent of the total population of the country still live in villages and mainly depend on agriculture and/or related enterprises. Marginal and small farmers constitute more than 84 percent of the 115 million operational holdings in India which are cultivating only 29 percent of the arable land. Most of them are resource poor and work in diverse, risk prone environment. Further, small farm holders remained deprived of benefits of advancement in the field of agriculture, major beneficiary being resourceful well endowed farmers. The livelihoods of the small and marginal house hold families are the major concern. The improper tapping of the potentialities of each component in the system make the system unviable. A multi-enterprise model based on an integrated farming system involving components of crops, fisheries, dairying, horticulture, vegetables, bee-keeping, poultry, duckery, gobar gas plant, solar heating system etc. may developed to provide regular income, employment and livelihood to small farmers. Horticultural crops are nutritionally rich mainly of essential proteins and vitamins which make human diet complete and help human being physically fit and mentally more sound. Agro and food processing industries can play a great role in terms of employment and income generation, poverty alleviation, export promotion and foreign exchange earning. Some of the process technologies and products have considerable potential for establishing fruits and vegetable industry in rural areas. Implementation of an efficient post-harvest system in any community must provide equitable benefit to all those involved in the system.

The horticulture sector in India is characterized by small, segregated farms with low per-hectare yields and huge post-harvest losses, owing to outdated practices. As per National Centre for Cold Chain Development, "The biggest wastage happens during the transportation of horticulture products from the farm gate to mandis and thereafter. Storage solutions can be provided only near the mandis, and this does not solve the problem. The answer lies in minimizing the wastage that happens during transportation." From a farm gate to a consumer, a horticulture product passed through seven different distribution channels, and in every step, there was a loss of five-seven per cent. Processing losses also abound. While China processed about 30 per cent of the food (fruits and vegetables) in 2009, the Indian food processing industry has been set a target of raising the level of processing perishable products from six per cent to 20 per cent by 2015. The Indian food processing industry is dominated by small and medium enterprises, which do not have the capacity to undertake large-scale processing of fruits and vegetables. There are several factors responsible for post harvest losses in horticultural produce at farmers' level need to be discussed in detail.

Critical Steps To Reduce Postharvest Losses

Pre-Harvest Sprays to Reduce Postharvest Losses

Pre-harvest sprays of chemicals may be applied to reduce postharvest losses in different fruits. Spray of Thiophenate methyl (0.05%) for Dashehari mango, 10 to 15 ppm giberellic acid (GA3) for on-tree storage of mango, Benomyl or Topsin-M or Carbendazin (0.05%) in Nagpur mandarin, 0.6% calcium chloride for grapes, bavistin (0.5%) for mango yield and 2% urea for increasing banana bunch weight.

Maturity Standards for Fruits and Vegetables

For both fruit and vegetable production the grower must know when the crop is ready to harvest. With field crops like wheat, the crop is said to be ripe when it is dry. With fruits and vegetables, first is mature, meaning that it is fully developed and will start to ripen. Constituents of ripe fruit crops depends on a variety of factors, a major one is the intended use of fruit. For example, for the fresh fruit market a table tomato is ripe when if it is fully red; there is a little value in green tomatoes for this purpose; although they will ripen off the vine. To the producer who ships bananas, the fruit is ripe when it is green.

Production and Harvesting

Vegetable production is highly specialized with specific practices for nearly every crop. Compared to most field crops vegetable are grown relatively in small acreages. Although a few large producers may account for a great portion of any crop nationwise. Commonly vegetable producers are highly specialized and have specialized equipment to meet the need of relatively few crops. Leasing land for the production of vegetable crops and producing contract crops are more common for vegetables than for other crops. Production is intense and competition for market is keen.

Producers face challenges of seedbed preparation, seeding pest management, fertilization, irrigation and harvesting as other farmer. To vegetable producers shipping and processing are also major concerns. Commonly producers must schedule harvest date to provide processing with a steady supply of raw materials for the processing facilities. Fruits and vegetable crops compared to most field crops may be grown on a much smaller scale. This does not show their unimportantness. These crops are important in the human diet as a source of essential vitamins, minerals and energy. Production and marketing require management of many diverse factors. Base principles must be understood as it is essential to effective management of fruit and vegetable production like growing, harvesting crops (fruit trees) tendering plants, including prunning during non bearing periods.

Harvesting the Produce

Due to improper harvest and post harvest handling about 22-40 % of the fruits are spoiled annually resulting in the loss of about 4000 crores. Efficient harvest of fruits and vegetable is one of the important unit operations to avoid these losses. At present, the fruits are harvested by traditional age old practices and methods. The methods and devices are most tedious and time consuming and cause damage to fruits and the tree branches. Such damages during harvesting amounts to 16 to 22 % of annual production. Several mannually operated improved fruit harvest have been developed and are in use which can be adopted for the purpose.

Pre-Cooling

Pre-cooling technology is extensively applied in the postharvest handling of horticultural produce. Pre cooling of mangoes to 12–15°C with 500 ppm Bavistin has been shown to increase their shelf life. Exposure of flowers to a temperature of 2–4°C prior to cold storage, results in prolonging their quality and shelf-life.

Grading

Systematic grading coupled with appropriate packaging and storage, will extend post harvest shelf life, wholesomeness, freshness, and quality, and will substantially reduce losses and marketing cost. Horticultural produce must be sorted and graded on the basis of parameters such as maturity, size, shape, color, weight, freedom from insects and pests, pesticide residues and ripeness. Mangoes are graded for export on a weight basis as: Class A (200–350 g), Class B (351–550 g) or Class C (551–800 g). Pineapple fruits are graded in accordance with their weight as follows: A (1.50 kg and over), B (1.10–1.50 kg), C (0.8–1.10 kg), D (800 g) and baby (550 g). Papaya fruits are graded to size typically in counts of 8, 10 and 12 in to 3.50 to 4 kg net weights. Vegetables like onion, potato, tomato, chillies, okra and french beans are graded on the basis of size, shape, weight and maturity stage.

Packaging and Transportation

Large quantities of mangoes, bananas, oranges and other fruits and vegetables are transported in open trucks. Window type conical bamboo baskets designed for stacking and aeration have been developed by the CFTRI, Mysore for transportation of produce by rail. The use of polyethylene film bags for wrapping whole bunches of bananas for transport, has been found to be most suitable for reducing wastage. The use of wooden crates having internal dimensions of 42x32x29 cm has also been recommended for the long distance transportation of bananas. Mandarins can either be individually wrapped in cling films or are packed in consumer packages such as plastic bags, plastic mesh, trays of moulded pulp, paper board, plastic, or foamed plastic. Losses in first grade tomatoes can be reduced from 15 to 30% by using upright cone baskets together with dry grass as a packaging material between the layers of fruits. Packing of tomatoes in sealed unventilated polyethylene provides a modified atmosphere which extends storage life. Printed plastic bags are used to reduce light transmission to potato tubers. Plastic oven ventilated bags of 25 and 50 kg of capacity are used for onions and potatoes.

Storage

Generally the quality of fruits and vegetable and their storage life are reduced by loss of moisture, decay and physiological breakdown. In India storage losses are high due to high temperature and humidity. These deteriorations are directly related to storage temperature, relative humidity, air circulation and gas composition. Storage at low temperatures immediately after harvest reduces the rate of respiration resulting in reduction of build up of respiration heat, thermal decomposition, microbial spoilage and also helps in retention of quality and freshness for a longer period. The storage life is governed by several factors e.g. variety, stage of maturity, rate of cooling, storage temperature, relative humidity, rate of accumulation of co, , per-packing air-distribution systems etc. In storage of fruits the cool stores play an important role. For short term storage of fruits and vegetables at farm level evaporative cooling system may be used.

Suitable value added products of fruits and vegetables for small scale production systems

Toffees

Fruit Toffee is a highly nutritious product as compared to sugar boiled confectionery. It is made from pulp of mango and other fruits along with certain ingredients. Small and cottage scale manufacture of fruit toffee provides potential avenues for self employment in the area where the fruits are available. Although fruit toffees are being made in the organized sector, there exists a vast potential for cottage scale production also.

Fruit Bars

Fruit bar is a concentrated fruit product meant for ready consumption. It has a good shelf life. Any variety of pulpy fruits, e.g. mango, guava, papaya, banana, apple etc. singly or in combination can be used for manufacture. Fruit bars are becoming increasingly popular due to good shelf life, taste, flavour and texture.

Fruit Jams and Jellies

These products are prepared by boiling the fruit pulp with sufficient quantity of sugar to a moderately thick consistency. The popular varieties of jam are pineapple, mango, mixed fruit, strawberry, grape, apricot and among jellies, guava and apple. The product is used as a bread spread and is also taken along with chapati, puri or similar products. Jams, jellies and marmalades share approximately 17% of the total processed fruit and vegetable products.

Improved Murabba Making

Murabba is one of the indigenous sweet preparations of the country. Murabba made from amla, bael, myrobalan, carrot, apple, mango, citrus peels are quite popular. The traditional method of the preparation requires a long processing time and does not ensure good keeping quality for the product often spoils due to miocrobial fermentation. The method has been improved to obtain murabba in a shorter period with good keeping quality, attractive translucent appearance and desirable texture.

Tutty Fruity

Fruits generally used for making preserves/ candies are apple, amla, papaya, mango, strawberry, raspberry, pear, cherry, etc. Among these raw papaya is largely used to make tutti-fruity used in bakery products, sweetmeats, ice creams, salads and pan.

The candied fruits and vegetables are quite popular food items. The consumption of these products is rapidly increasing.

Osmo-air dried Fruits

Osmo air-dried fruits are based on a novel approach towards dehydration. Slices of ber, pineapple, jackfruit, mango, etc. are processed in two stages. The first phase is the removal of most of the water using sugar syrup as an osmotic agent. The second phase is air drying where the moisture content is further reduced to about 15%. The osmo-air dehydrated product is near to the fresh fruit in terms of colour, flavour and texture. The product can be used in ready -to –eat type foods, ice cream, fruit salad, kheer, cakes and bakery products. Such osmo-air dried fruit based units can be set up in areas near fruit orchards to the benefit of people. The process is simple and involves operations like selection of fruits, cleaning, washing, peeling, curing and slicing/ dicing. The prepared fruit slices are steeped in sugar solution to remove water by osmosis. The slices are then drained, dried in a hot air drier and packed in flexible pouches.

Grape Raisins

Dried grapes are used in various food preparations and considered as a delicacy for direct consumption. Manufacture of raisins can be an important industry in the grape growing areas. An improved method has been developed which can easily be adopted. The process consists of washing, alkali treatment, sulphitation, drying in sun or in dehydraters.

Any grape variety with high sugar and low acid content can be used yielding a good quality product. No sophisticated equipment is needed and the unit can be installed in orchards. It can generate rural employment.

Dry Apricots

Apricot is an important fruit cultivated in the temperate regions. The dry fruit forms an important item of confection. The fully ripe fruits are harvested item of confection. The fully ripe fruits are harvested and placed in a wooden sulfuring chamber wherein yellow sulphur is burnt at the rate 4g/kg fruit. Sulfured fruits are dried in a solar drier for 5-7 days till the moisture content is about 17%. The dry fruits are packed in polythene bags for storage and marketing

Dehydrated Vegetables

Vegetables are seasonal and perishable. Dehydration is one of the methods to preserve them and make available throughout the year in hygienic conditions at reasonable cost. The dehydrated vegetables are easy to transport and cater to the needs of large catering establishments. They can be used in various preparations at any season of the year. Traditional sundrying is time consuming, less hygienic and climate dependent.

The process for controlled dehydration of vegetables consists of grading/ sorting, washing, peeling/ trimming, size reduction, blanching, Chemical treatment, dehydration and packing in unit can be established.

Anardana

The seed with adhering pulp (arils), when dried yields Anardana of commerce. The traditional method consists of manual extraction of arils followed by sundrying. It is unhygienic, labour intensive and slow.

The product cannot be stored beyond the following monsoon season since spoilage occurs due to discolouration, moisture ingress and insect infestation.

The improved processing technique consists of pre-cleaning, mechanised extraction of arils, solar drying and packaging. The mechanised aril extractor works on ¼ HP motor and can process 60 kg fruit/hr as compared to productivity of 5 kg/hr in manual operation.

Prepackaging of fruits and Vegetables

This simple technique involves cleaning, trimming, cutting of the fresh produce and packing the same in unit packages in polyethylene bags. Bean, carrot, brinjal, green chilli, root crops, leafy vegetables, and fruits like orange, lemon, banana, grape, and flowers like champaka, chrysanthemum can be prepackaged to obtain 1 to 2 times extension in shelf life in polyethylene bags under normal conditions without any refrigeration. The prepacked produce presents better consumer appeal, longer shelf life and has considerable handling advantages in transport and marketing. The major facility required at the farm level or in orchards is a packing shed.

Wax Emulsion for Fruit & Vegetables

A large number of units in tiny sector can be set up for improving the shelf life of fresh fruits and vegetables in villages where they are grown for marketing in the urban areas.

The wax emulsion is diluted with cold water and used for dipping fruits and vegetables. It enhances the shelf life, protects fruit from fungal attack, and reduces desiccation and weight loss during storage. The emulsion is harmless and imparts a gloss to fruits and vegetables. The process is quite simple and economical.

Anti fungal Paste

The banana bunches after harvests are cut into banana hands or clusters. The cut ends get infected. To check this spoilage, an antifungal paste has been developed. It is harmless and produces no undesirable odour or flavour.

The latex oozing out from the cut ends is washed in running water and wiped. The dry cut ends are treated with antifungal paste with the help of a paintbrush. The paste sticks well and dries in 20-40 minutes in the open air. It protects the cut ends of the bananas for a period of 15-17 days at room temperature and 21-32 days at 13°C while untreated cut ends show fungal rotting on 5th day room temperature and 10 day in cold storage. This antifungal paste application reduces infection during ripening also. For its manufacture the ingredients are mixed in a dough mixer to get a uniform paste. A food colour is also added to the paste to acertain the treatment of cut ends. The cost of the paste works out minimum economic unit is 25 kg paste/ batch of 90 min.

Improved Method of Mango Ripening

Mangoes are generally harvested when fully matured but green. The conventional method of ripening in hay has disadvantages like long ripening time, excessive handling and high degree of spoilage due to stem- end rot. The spoilage during ripening period is reported to be as high as 25 to 30%. In order to improve upon the ripening method, a simple technique has been worked out. It consists of dipping the fully matured but green mangoes in hot water at 48 to 54°C for 5 minutes, draining and keeping at room temperature till adhering surface water evaporates. Fruits are then packed in ventilated boxes/ crates. No hay or other packing material is needed. The ripening generally starts on 6th day of treatment and is complete on 12th day. The operation can be made continuous in fruit processing factories where large quantities of mangoes are handled. The technique has presently been standardized for Alphonso variety. The hot water treatment is found to reduce the spoilage to the extent of 50% and also helps in uniform ripening of fruits. The colour development in the flesh is better than conventionally ripened fruits.

Pickles and Chutneys

Pickles and chutney have a great importance in the Indian menu and have now become essential items in any feast and lunch. Pickling of fruits and vegetables is an old art. A large variety of these items are method of preparation varies. The basic method is salt curing of fruits and vegetables, acidifying, addition of vinegar / oil and the spices.

Instant Pickles

The traditional pickle making involves elaborate and time consuming process. A new concept has been worked out for making an instant pickle mix of lime and mango, which can easily be reconstituted, into a tasty pickle by the simple addition of oil and water. The dry mix has the features of convenience, better keeping quality, relatively low cost of production, reduced packaging and transportation cost. The ever expanding market for pickles, both in domestic and export sectors, can be exploited by adopting suitable market strategies.

The process is simple and consists of the selection of fully matured commodities for picking, washing, and cutting matured commodities for pickling, washing and cutting, brine curing at optimal conditions, drying, addition of dry spice mixes and packing in unit packages.

Potato Flour

Many ready –to -eat products are prepared from potato. Units based on potato products can easily be established in rural areas and the market can be exploited in urban and semi urban centres.Potato flour, granules and mash are used in the preparation of instant foods, soups etc. as binding materials and also for preparing kheer, tikki, chops, pakoda, cutlets, stuffed parotha, kofta and other products. Production of potato flour can be taken up easily using indigenous equipment's. The process involves peeling of potato, cutting, pre-treatment with salt and permitted preservatives, soaking, granulating and drying. The dried product is ground and packed. The flour obtained by this process can be easily reconstituted with boiling water to get the mashed potato and used for making a variety of products

Potato Chips and Wafer

Potatoes are grown extensively in the northern region of India. During the glut season, growers do not get remunerative prices. Therefore it is advisable to utilize the surplus produce by preparing potato chips under hygienic conditions.

The product based on developed sun drying technology is superior to the conventional one both in colour and quality. The process is simple and can be easily adopted at rural areas

Mushrooms

Mushrooms are fruit bodies of fungi that grow abundantly in the open fields, meadows, marsh edges, farm, yards etc. Some wild mushrooms are poisonous and narcotic but there are about 2000 edible spices too.

Technologies have been developed for commercial cultivation of mushrooms under controlled conditions. These are safe and readily available for the market. Mushrooms are rich sources of easily digestible protein and minerals, besides being low in calories and almost fat-free. Profitable mushroom cultivation can be taken up in the rural areas where agro-waste is easily available. Its cultivation can provide gainful employment. Mushrooms are highly perishable. The unmarketable surplus of fresh produce can be preserved and processed into value added products by canning, dehydration, brine preservation, conversion into pickles, soup and ketchup in small scale processing units near the site of production.

Papain

Papain is obtained from the latex of unripe green papaya fruit. It contains proteolytic (protein digesting) enzyme. Papain finds wide applications in different industries. It is used in tenderisation of meat, manufacture of protein hydrolysate, confectionery items (to prepare chewing gum), brewing industry (for cheese making). It is also used in the pharmaceutical, textile, tanning, aroma and perfume industries and for effluent treatments. Organised papaya Planatations can be taken up in the rural areas.

The process consists of tapping of latex from papaya fruits, mixing with potassium metabisulphite and storing under cold temperature. The latex is refined, vacuum dried and powdered. After testing and standardisation, it is packed and placed in rigid airtight containers.

Tomato Products

Tomato is extensive grown in India and used for the preparation of puree, paste, ketchup, sauce and ready- to- eat products. There is a good domestic and export market. Since the fast food sector is expanding rapidly the demand, particularly for tomato ketchup and sauces, is also increasing.

Constraints, which hampered the adoption of post-harvest technologies at farmers' level

Various researchers elicited the constraints, which hampered the adoption of post-harvest technologies. These prevailing constraints strictly control the adoption behavior (especially farmwomen, small and marginal farmer).

Sumathi, (1990)

i) **Farming constraints**

- lack of concrete or good floor,
- lack of space for storage,
- insufficient space for threshing and drying)

ii) **Social /cultural constraints**

- spreading dust threshing and winnowing,
- lack of skill,
- inadequate labour,
- lack of time,
- complexity of practice
- scared use of chemicals

iii) **Scientific / technological constraints**

- lack of conviction about advantage of post-harvest technology,
- lack of scientific knowledge about improved practices,
- non-availability of improved machineries,
- lack of technical guidance and training,
- non-availability of information regarding scientific storage methods
- non-availability of chemicals in time and with ease),

iv) **Economic constraints**

- lack of purchasing power,
- high cost of post-harvest machineries,
- loss of produce by using public transport and tractor,
- lower socio-economic condition
- high cost involved in constructing concrete floor)

Suggestions to improve the adoption of modern post-harvest technology among farmers

- Training need of the agro-processors to upgrade their knowledge and skills in modern post-harvest technology is strongly recommended.
- It was felt that provision of credit facilities could change the perceived constraints into enabling factors for adoption of modern post-harvest technology enterprises,
- A number of notable factors like socio-economic, contact with extension workers, provision of infrastructure and other institutional factors influence the adoption.
- Other factors also reported by authors' like personal factors including society membership education and household income
- Chamber et al. (1989) noticed that adoption of technologies has been enhanced by farmer-to-farmer extension network, on-farm demonstrations, and on-farm visits.
- Farmers through Self- Help Groups (SHGs) enabled the adoption of technologies and diffusion enhancement. SHGs have found to be reliable and efficient mode of technology transfer but require positive attitude of group members.
- It is need of the hour that agro-processors must be motivated to form the SHGs at cluster level. These groups should be linked with institutional credit system and markets.
- Transfer of technology through groups has found to be the effective means that requires minimum investment.
- Groups must be made aware of marketing strategies to get wider opportunities for their new products.
- Need based and skill oriented training should be organized to develop the skills among the group members.
- Training on farm implements and post-harvest technologies through demonstrations give better results.
- Availability of post-harvest processing equipment at competitive rates, popularization of improved methods / technologies and up-gradation of skills of the agro-processors will help in ameliorating the constraints and will promote adoption.

- This can be achieved by various approaches but formation of groups seems to be most appropriate in Indian context.
- Provision of extension of technologies by university coupled with other factors have had significant positive influence on decision to adopt and introduce cultural practices with improvement on the well being of participating farmers.
- The extension agencies can play an important role to educate the farmers in adoption of improved production technologies for their benefits and reversing their negative attitude.
- Increased knowledge and awareness level are generally considered prerequisites for adoption of new practices and technologies.
- Frequent visit of extension worker increased the efficiency / productivity of the farmers.
- The orientation of the groups' members must be changed from traditional business to the diversification of the agro-products.
- For wider publicity of the post-harvest technologies, agricultural programmes on modern post-harvest technologies and available government schemes should be prepared in local language and be broadcasted countryside.

References

Chikkasubbanna Venkatarayappa, 2006. INDIA (2), In: Postharvest management of Fruit and Vegetables in the Asia-Pacific Region, pp 143-151. Rolle Rosa S, (eds.), Asian Productivity Organization 1-2-10 Hirakawacho, Chiyoda-ku, Tokyo 102-0093, Japan and Food and Agriculture Organization of the United Nations Agricultural and Food Engineering Technologies Service Caracalla, Viale delle Terme di 00100 Rome, Italy.

Lisa Kitinoja, 2013. Innovative Small-scale Postharvest Technologies for reducing losses in Horticultural Crop. *Ethiopian Journal of Applied Sciences and Technology*. (Special Issue No.1): pp 9-15.

Meena, M.S., Prasad, M. and Singh, Rajbir 2009. Constraints Perceived by Rural Agro-Processors in Adopting Modern Post-Harvest Technologies. Indian Res. J. Ext. Edu. **9**(1), pp 1-5.

Pandey Sadhna, Pandey S. N. and Pandey P. H., 2011. *Post harvest management of horticultural crops*, pp1-453. Kalyani publications, B-1/1292, Rajendranagar, Ludhiana-141008, Punjab, India.

□□□

Chapter 22

Post Harvest Management of Forage Crops and Crop Residues

P.N. DWIVEDI AND P.K. PATHAK

KAR - Indian Grassland and Fodder Research Institute, Jhansi-284003

Introduction

Majority of livestock population in India subsists on vegetation of natural grasslands where grasses are cheap source of forage. Crop residues, straw and dry grasses form the basal roughage for feeding to livestock which are poor in quality being deficient in protein, available energy and minerals. Therefore, supplementation of green fodder and concentrate become essential for sustaining livestock productivity. However, there exists a huge gap between demand and supply of both green fodder as well as concentrate in our country. Moreover, green fodder is not available throughout the year. Therefore, it becomes imperative to search for alternate source of protein rich forage supplement. Supplementation of leaves of leguminous crops is an important and most practical feeding strategy for improving the feeding value of such poor quality roughages. Leaf meal being rich in CP, carotene and minerals could be an appropriate answer for providing nutritious diet to the livestock throughout the year. Supplementation of leaf meal can meet the requirement of livestock particularly during dry periods when availability of other protein rich forages is scarce. Thus it can bridge the gap between the demand and supply of the green fodder and concentrate. In India, traditionally also tree leaves particularly *Zizyphus nummularia* (Pala), Khejri (*Prosopis cineraria*), *Ficus* sp. and Neem leaves were lopped, dried and fed to different categories of livestock.

1. Post Harvest Technology of Forages

Handling and utilization of forage resources involves various unit operations starting from harvesting, collection and transportation, size reduction and processing (product formulation), drying and storage. However, material handling, transport and storage are the most important operations due to bulky nature of forage resources. Drying, size reduction and baling / densification / pelleting etc. are practiced for specific product formulation or when crop residues are wet or green fodder needs to be dried for conserving as hay or for adding in various products.

Forage Handling and Transport

Crop residues being voluminous in nature and have lower bulk density than grains need more space, labour, time in handling and farm operation causes more cost in the transportation. Traditionally the wheat straw in loose form is transported by trucks, tractor trolleys and by bullock carts depending upon the distance travelled.

Harvesting and Collection

Harvesting of crop/tree leaves is important in view of time of cutting, which influences the dry matter yield and quality. Cutting too early will give high quality meal but low yields and cutting too late will decrease nutrient content and increase fiber content but give higher dry matter yield. Among the leguminous fodder crops, *Stylosanthes seabrana* is first cut after 55 days whereas *S. scabra* and *S. guianensis* after 60 days and 80 days respectively for leaf meal preparation, depending upon soil fertility and water availability (Guodao *et al.,* 2004). Lucerne has been observed to provide 9 t dry matter yield per ha (Singh, 1988) and could also have a great potential for making meal.

The *Leucaena,* a leguminous fodder tree is cut first at 10–15 cm above soil level when it has grown to a height of 1.2 metres. Thereafter 4–6 cuts can be taken per year when the plant reaches a height of 1 metre. The approximate yield under Cuban farming conditions without irrigation or fertilization is 12 tonnes DM/ha with 22–25 % crude protein. Rajput *et al.* (2004) assessed a *Leucaena* field and plant base diameter above 2.5 cm (mean diameter 3.25 cm) was recommended to harvest for production of leaf meal. Yield of fresh *Leucaena* leaves was estimated as 12.10 t/ha/year by them.

Conservation and Drying

Conservation of herbage is accomplished by drying the crop and it is the single way to reduce moisture from the harvested plants. It could be through many ways including the simplest way of sun drying.

In sun drying the materials should be spread in a thin layer on a clean surface preferably concrete and dried for two days, then turned over and dried for another day. Alternatively, drying racks can be built using local materials such as bamboo to dry the materials above the soil surface and avoid spoilage. Indian Grassland and Fodder Research Institute, Jhansi has developed and evaluated two tier drying structure made up of Subabool *(Leucaena)* logs for shed/ventilated drying of green fodder and tree leaves. Use of racks can also hasten the drying process and improve quality through increased protein and β-carotene content. Though sun drying is a more acceptable and feasible alternative of many other drying methods, it is difficult to sun dry green forage crops/tree leaves, in the rainy season when the production is large. Also, extending the drying period diminishes the nutritional quality of the product. Hay production from stylo using sun drying resulted in loss of 15-40 % of the dry matter, mostly in the form of leaves (Amodu, 2004).

Heating air for drying reduces the drying time and maintains the quality of the product and is an essential operation in making leaf meal commercially. The high temperature drying was reported to be costly to dehydrate leaf meals due to expensive machinery used and the operational cost.

Densification of grasses/crop residues

Grasses/crop residues are densified with the help of high density forage baling machine developed by IGFRI, Jhansi. The machine is equipped with crank-piston drive mechanism and has a rectangular piston cross-section. It is operated by a 18.5 kW-3 phase, electric motor with V-belt and pinion-gear power transmission system. Density of the bales can be adjusted with the help of a spring tension mechanism provided at the outlet point of the machine. A force feed mechanism has been provided for the pre compression of the loose herbage in the hopper, which, in turn, increases the hay input inside the compression chamber. Feeding mechanism and tying of bales is manual. The capacity of machine is 200 to 600 kg/h depending upon material.

Value addition of wheat straw through molasses

For densification of wheat bhusa, hot molasses (10 to 20 %)and urea (2 %) was mixed thoroughly with bhusa and the admixture was densified thorugh IGFRI high density machine. The blocks were tied by 1.25 cm wide plastic tape from top to bottom and the steel wire was tied across the bale. The bales of molasses mixed wheat straw have the average density of 398.5 kg/m^3.

Adding urea

Beneficial effect of adding urea (2 to 4%) in wheat straw are well established in literature. Wheat stubbles are however baled by mixing 2 % urea with 20 % water. The tying of bales was done with steel wire rope. The average density of bales achieved was 200 kg/m^3.

Complete feed block

Complete feed blocks may also be developed in view to enhance nutritive value of forage through mixing with molasses and blending with leguminous fodder, concentrate mixture, minerals and vitamin additives. The composition of Complete Feed Blocks included wheat straw, molasses, dry leaves of berseem-, concentrate mixture - and recommended level of mineral mixture and vitamin additives.

2. Processing of Fodder

Fodder resources are subjected to various processing methods (Sharma *et al.*, 1995; Das *et al.*, 2009) such as chaffing, grinding, pelleting, chemical and biological treatments especially for their incorporation into complete feeds with the aim of enhancing palatability, intake and nutrient utilization with the ultimate objective of improved performance. The most common and widely adopted methods are processed in the form of reduced size (chopped or threshed), pellets, blocks etc. and are described below:

Size reduction

In the feed industry the size reduction is usually referred for grinding and producing powdery like material called meal. Commercial products must often meet stringent specification regarding the size and same time the shape of the particle they contain. Reducing the particle size increases the reactivity of solids, reduces the bulk of fibrous material for easier handling and is governed by the ultimate user (animal). For example smaller particle size is required for small animals and birds whereas larger particle size (straw) may be fed to the larger ruminants. Different mechanism involved in the size reduction process could be compression, impact, rubbing or attrition and cutting. In general compression is used for coarse reduction of hard solids, to give relatively few fines; impact gives coarse, medium or fine products; attrition yields very fine products from soft, non abrasive materials whereas cutting gives a definite particle size and sometimes a definite shape with few or no fines (McCabe *et al.*, 1985).

Forages and fodder crops, being fibrous material, could be reduced in their sizes mostly by using chaff cutter and concentrates by hammer mill type of grinder or beater type of pulveriser. Material passing through a 1-1.5 mm sieve after grinding dried *Stylosanthes* in China was considered as stylo leaf meal and ground material passing through 3 mm and 5 mm screen in India was considered as stylo leaf meal. The size of meal influences the digestibility (Dwivedi and Pathak 2004).

Pelleting of fodder

Feed Material consisting of subabool leaves, berseem hay, lucerne hay or other top feeds, mixed with wheat straw or dry grass is used for making feed pellets. Urea, molasses, concentrate mixture, minerals and vitamins are also added to enrich the feed pellets. All ingredients should be mixed thoroughly to make it a homogeneous mixture. Water is added to bring the moisture level of the mixture up to 55 - 60%. This is required to make the material flow freely through the pelleting machine. A power operated, auger type feed pelleting machine developed by Indian Grassland and Fodder Research Institute, Jhansi, is used for making feed pellets.

For making the feed pellets, the power is put on and the machine is operated. All well mixed material containing moisture from 55 to 60% is fed into the machine through feeding chute. The mixture gets ground and converted into dough while passing through auger and extruded out through the holes on the pressure plate in the form of cylindrical bars. These bars break into 10-15 cm long pieces due to their own weight which are known as "feed pellets". These feed pellets are then spread over the floor for drying which is required to bring down the moisture level up to 10-15 % for safe storage.

3. Storage

Storage of crop residues is different than grains in view of physical properties, lower density (70-120 kg straw/m^3), nature of crop, shape and size, location and ultimate use. Several structures have been traditionally used for storing these crop residues, grasses and fodders and may be classified into different groups. They may be permanent and temporary structures depending upon the constructional material used, underground or above the ground depending upon ultimate use of the product, on farm or at other places depending upon their marketing and use. However, most farmers prefer to construct their own storage structures depending on availability of infrastructure and economic considerations. The method of storing crop residues, grasses and fodders also differs depending upon their size and could either be stacked or stored in any permanent or temporary structures.

4. Feed Processing

Feed processing includes the treatment (physical, chemical, thermal) of a feed prior to consumption by animals. In general, the variability in processing effects is associated with the choice of equipment, with processing conditions as well as with the processing system, e.g. the combination and sequence of processing equipment . The commonly used process/equipments for production of animal feed are:

1. Cleaner
2. Destoner
3. Hammer mill / grinder or pulveriser
4. Mixer/Blender
5. Conditioner
6. Vertical pelletiser/Horizontal pelletiser
7. Extruder with pre-conditioner
8. Drier
9. Weighing and bagging.

Protein Quality

The properties of proteins in foods or food – systems have been classified: (1) sensory and kinaesthetic (e.g., flavour, odor , color , texture); (2) hydration, dispersibility, solubility and swelling ; (3) surface – active properties (required for e.g., emulsion and / or foam formation); (4) rheological properties , including gelation and texturization; (5)

other properties (e.g., Including the adhesions, cohesion, dough making, film and fibre formation during extrusion). In feed manufacturing especially the adhesive forces that protein may exert, are of interest. Protein can act as a binding agent between different feed particulates. Processing involves the combined effect of shear, heat, residence time and water resulting, among others in partial denaturation of the protein in the feed. As has been shown by Wood (1987) partial denaturation of the protein during processing may positively affect the hardness and durability of the feed pellets. Denaturation involves the breakdown of the (Spatial) three dimensional structure of the proteins (the secondary, tertiary and quaternary structures), thereby changing the bioactivity of the protein. Upon cooling protein reassociate and so bonds can be established between the different particles. No report could be found in which the effect of processing variable on feed manufacturing has been described on denaturation of protein and consecutive effect on physical quality.

Fibre

Fibre can be classified as water soluble part, which may cause a high viscosity, and a water insoluble part. The classification may be useful in explaining the effect of different fibre sources on pelleting characteristics.

Water-soluble fibre like glucans, arabinoxylans and pectin strongly raise the viscosity and may act as filler in the feed; the viscous material embeds the more coarse particles, thereby reducing the porosity in the feed. Subsequently, the structural integrity of the feed agglomerate is higher, resulting in higher durability and hardness of the pellets. During processing, process variables, such as residence time, moisture content, pressure and heat, alter the state of soluble fibre. To obtain the desired functional properties, the effect of time, pH and temperature are of major importance (Keller, 1983). Due to the complexity of the feed mash (blends of several raw materials that differ in composition and functional properties) and processing conditions (heat, pressure and time relationships), quantitative description of the effect of fibres, which forms with water a viscous solution /paste, on pellet quality are difficult to acquire.

The effect of insoluble plant fibres may be two fold. Plant fibres may prove beneficial in the pelleting process since they have the opportunity of entangling and folding between different particles or strands of fibre. On the other hand, due to their stiffness and elasticity, they may impair problems to the pellets-press operator since resilience characteristics of the material oppose good contact between particle fibres. Moreover, when large fibres are present within the pellet, they might induce a weak spot in the pellet. Large fibre particles may lead to homogeneity at which pellets are more likely to break. Increasing the residence time in the die of the pellet-press would diminish the effect due to the resilience the the material. This is in line with result, who found that an increase in hold time within a wafering machine resulted in the further stress decay of the materials under investigation, and subsequently increased durability of the formed wafer. found that expansion of wheat straw blocks after compression was lower after a longer residence time.

Fat

Added fat in the compound animal feed is known for its adverse effect on pellets hardness and pellets durability. Since most binding of feed particles incorporates water or, when involved, solubilise starches, protein, fibres, fat with its hydrophobic nature may interfere with binding properties of water-soluble components in the feed. Moreover added fat (and to lesser extent, fat enclosed in the matrix of cell walls) acts as a lubricant between particles and feed mash and the die-wall resulting in a lowering pelleting pressure. This will already cause lower pellet quality in many cases. Also, as a result of fat addition various properties of a raw material constituent might be influenced. For instance, gelatinization of starch can be inhibited in the presence of lipids or delayed to higher temperatures. This may directly lead to weaker pellets, vulnerable to deterioration in transport systems. Salmon (1985) showed that more fat in the pellet (0, 30, 60, 90 g kg^{-1}) significantly increased the amount of fines returned; 2.7, 7, 10.3, 15.6 % with no bentonite added. Richardson and Day (1976) showed that an increasing amount of fat added in mixer increased the amount of fines returned from the pellet mill and decreased the amount of mechanical energy (kWh $tonne^{-1}$) needed to convert 1 tonne of mash into pellets.

Fat enhances the production rates in pellet mills, primarily because of the lubricating effects of fats between the meal and die surface (Richardson and Day, 1976;). Because of low friction and pressure in the die decreased which in turn, would result in feed pellets with lower hardness and durability. However, some authors suggest that during blending, the feed mash with steam, natural oils and waxes would accumulate at the contacting sites of two particles and create a binding point either solid (waxes) or via liquid necking (oil or water), between particles upon cooling (Friedrich 1977). This could have positive effect on pellet hardness and durability because of immiscibility of water and fats, there would be an optimum in the amount of bonds established by waxes and fats on one side and water on the other side.

Pellet blinders

When physical quality of pellet is not sufficient to obtain a saleable product or it does not meet the manufacturer's quality standards, binder material can be incorporated to increase the physical quality of the pellets. For this purpose, different types of binders are available. The best known binders used in the animal feed production are bentonite (a clay mineral), carboxymethyl cellulose (CMC; modified cellulose) and lignosulphonates. The effect of binder material depends on the presence of water, which is required for a binding agent to become active.

Lignosulphonates are by product form the paper industry. During the process of delignifying of fibres, $NaHSO_3$ and or Na_2SO_3 is added to which the alkaline components $(SO_3)^{-2}$/ $(HSO_3)^{-}$ react with chemical bonds of the lignin to form water soluble ligno- sulphonates. Common inclusion levels range form 0.5 % to 3 %.

Bentonite as binder is used in feed technology to improve physical pellet quality. Its main target is acting as filler, thereby decreasing porosity in pelleted feed. In addition bentonite works as a lubricant in the die hole it decreases pressure and subsequently energy requirement of the pellet press. postulate the bentonite does not have any binding properties in itself. Especially with voluminous feed mashes as, for instance , fibre - rich diets, bentonite may prove advantageous.

Carboxy-methylcellulose (CMC) can also be used as binder in animal feed production. CMC gives a viscous solution/paste when mixed with water. Hydration occurs during conditioning. Subsequent compaction brings particulates close together and allows for the development of ionic attractions between the CMC and particulates.

Conclusions

The production of food grains is almost stagnant whereas the demand is on the rise. Under such a situation nothing can be wasted. In order to feed the livestock in a productive manner it is very important now to process /convert by-products, crop residues and agro-industrial wastes into animal feed. It needs a holistic approach involving animal nutritionists, engineers, bio-chemists and microbiologists to work together to process the various ingredients and agro-processing wastes and by-products into feed for livestock, poultry, piggery, fisheries sector.

□□□

Chapter 23

Dung Management Strategies for Integrated Farming System

S.B. MAITY, K.K. SINGH AND M.M. DAS

ICAR - Indian Grassland and Fodder Research Institute Jhansi-284003 (UP)

Introduction

Integration of various agricultural enterprises viz. cropping, animal husbandry, fishery etc strategically in the system to increase the income of farm is integrated farming system.

Livestock is an integral part of farming system. Recycling of available biomass is done from the farming system through animal component for the production of animal product like milk, meat, eggs or other products as per the integration. Dung is an important byproduct of livestock. It is used as manure or is converted to cakes for the use of fuel or mixed with clay for flooring and plastering of mud houses. It has great environmental value too, which may be negative and positive also. In negative, methane emission from manure have negative impact but on positive, the use of dung as fuel, manures, biofertilizers as well as use in Ayurveda. India is endowed with huge livestock population. In 2003, it had over 464 million livestock heads. This vast number is deployed primarily to produce milk and meat, fiber and draught power. However, in the process some byproducts or joint products are also produced. These are urine and dung. Dung is an important byproduct of livestock used for multiple purposes as per requirements. Manure contains on an average 0.35% nitrogen, 0.15% phosphoric acid and 0.20% potash (Anonymous,1966). In fact, only about 35, 65 and 75% of N, P, K, respectively, are absorbed by the first crop after application (Garg *et al.*, 1971) and rest available in subsequent cropping as residuals. Thus, dung manure has its role in farming system as recycling nutrients. Considering the importance of animal dung, in farming system the same has to be managed as available fertilizer resources for crop and other components. Availability will vary as per model of farming system. However, dung management knowledge is required for its efficient utilization to maximize the other returns.

Dung Production and Feasibility in Farming System

India is endowed with huge livestock population. In 2003, it was 464 million heads. Dung is an important by product of those livestock. Dung evacuation rate varied considerably across species, age groups and functional classification of bovines. Production of dung per animal per day largely depends on the quantity of feed intake and type of feed. Physiological and environmental factors also affects per animal per day dung production. Thus, it is difficult to give an exact data on it as shown in Table 2. Total wet dung production for the year 2003 was estimated at about 562 million tonnes. Species wise cattle had 60% and buffalo had 40% of total production. However, evacuation rate of buffalo was higher than cattle. Total dung production affected by season too as well as the quality of feed and forages.

In the integrated farming system, livestock component usually to be very small and manure production usually varied as per the number of livestock. When small animal and poultry introduced in the system, their droppings are difficult to collect. Therefore, the option for utilization of manure to the farmer is limited and to be judicious. Usually, farm yard manure utilized as biofertilizers, needs 20 tonnes per ha. area in irrigated land. One cow unit usually produce 6.63x365 kg wet dung per year or 24.2 tonnes wet dung that can produce 19-20 tones of farm yard manure after processing loss of moisture. Therefore, from one adult cow unit 20 tones farm yard production is possible if the process followed properly. Accordingly, assumption to be made as per integrated livestock unit in the system. Dung production in 2003 and its utilization for various purposes has been shown in Table 3.

Management of Dung for Biofertilizer Production

Dung collected in the process may be utilized biofertilizers as manure and vermicompost.

Farm Yard Manure (FYM)

This is the traditional manure and is mostly readily available to the farmers. Farm yard manure is a decomposed mixture of cattle dung and urine with straw and litter used as bedding material and residues from the fodder fed to the cattle. The waste material of cattle shed consisting of dung and urine soaked in the refuse of the shade is collected daily and placed in trenches about 6-7 m long, 1.5-2 m broad and 1 m deep. Each trench is filled up to a height of about 0.5 m above the ground level. The top of the heap is to be made dome shaped and plastered over with cow dung earth slurry. It becomes ready to apply after 3-4 months. It is possible to prepare by this process 7-8.5 cu.m of manure. (5-6 tonnes or 10-12 cart loads) per year per head of cattle. Well rotten farm yard manure contains 0.4 to 1.5 per cent N, 0.3-0.9 % P_2O_5 and 0.3-1.9% K_2O.

Table 1. Dung production by different categories of bovines, 2003

Categories	Average evacuation rate (kg/day)	Population (million)	Dung production (million tonnes)	% share in total dung produced
Cattle				
In milk	6.63	35.80	86.63	25.78
Dry	6.58	22.30	53.56	15.94
Adult male	4.46	57.60	93.77	27.91
Young stock	4.43	63.10	102.03	30.37
Total	178.80	335.99	59.79	
Buffalo				
In milk	8.35	33.30	101.49	44.91
Dry	8.49	13.90	43.07	19.06
Adult male	6.65	6.70	16.26	7.20
Young stock	4.43	40.30	65.16	28.83
Total	94.20	225.99	40.21	
Grand total	273.00	561.98	100.00	

Table 2. Dung production and utilization, 2003

Particulars	Values
1. Total production of dung (million tonnes) in 2003	562.0
2. The proportion of wet dung used for fuel (%)	36.81
3. Rate of weight loss during making of dung cake (%)	39.81
4. Other utilization of wet dung (%)	63.19

Methods of Preparation of FYM

a) **Pit Method**

This method is commercially adopted by farmers usually pits of 8m x 2m x 1m dimensions are prepared, which are filled in layers by the mixture of dung, urine and litter up to about 50 cm above ground level. The top is nicely filled, pits is covered by dry soil and then plastered by mud paste. One pit is needed every after two adult animals, the manure became ready after about 150 to 180 days of plastering. Usually 10-12 tones of FYM obtained/pit or every animal gives out about 5 to 6 tones of FYM /year (Yadav *et al.*, 2013).

b) **Trench Method**

This method is also called as Dr.C. N. Acharya method. In this method trenches of 6 to 8 m length, 1.5 to 2 m width, and 1 to 1.25 m depth are prepared. Mixture of dung and urine soaked litter is deposited in layers in these trenches until it becomes 50 cm above the ground layer. Now it is covered with 50 cm deep soil or wood ash- soil layer and then plastered by mud paste. Manure becomes ready for use after about 150 – 160 days of plastering.

c) **Heap Method**

Most commonly used by the farmers. Every day sweepings, cow dung and litter are collected and heaped at any fixed place. After about 6 to 9 months, the rotten manure is used. According to an estimate, about 30 to 35% N_2, 20 to 25% P_2O_5 & 4-6% potassium are lost during preparation of manure due to leaching, washing and volatilization.

Methods of FYM Application

1. FYM should be uniformly spread over soil surface and mixed thoroughly.
2. It should be applied 15-20 days before sowing or transplanting so that manure goes under ammonification and nitrification process.
3. Application of un-decomposed manure should not be applied.
4. Soil should have sufficient moisture at the time of application so that proper microbial activity takes place. Usual rate of

application ranges between 2-5 tones/ha for most of the crops. But it may show high as 50 to 100 tones/ha for vegetable, sugarcane etc. For best response it must be well powdered and it should be sieved, especially for use in nursery.

Vermicompost Production

Vermicompost is the product or process of composting using various earth worms like red wigglers, white wigglers and others to make a heterogeneous mixture of decomposing vegetable or food waste, bedding materials, and vermicast. Vermicast, also called worm castings, worm humus or worm manure, is the end-product of the breakdown of organic matter of an earth worm containing water-soluble nutrients, vermicompost is an excellent, nutrient-rich organic fertilizer and soil conditioner. This process of producing vermicompost is called vermi-composting.Vermicompost has been shown to be richer in many nutrients than compost produced by other composting methods. It also has outperformed a commercial plant medium with nutrients added, but needed adjustment for magnesium and pH. Unlike other compost, worm castings also contain worm mucus which helps prevent nutrients from washing away with the first watering and holds moisture better than plain soil.

There are two main methods of large-scale vermiculture. Some systems use a windrow, which consists of bedding materials for the earthworms to live in and acts as a large bin; organic material is added to it. Although the windrow has no physical barriers to prevent worms from escaping, in theory they should not due to an abundance of organic matter for them to feed on. Often windrows are used on a concrete surface to prevent predators from gaining access to the worm population. The second type of large-scale vermin-composting system is the raised bed or flow-through system. Here the worms are fed an inch of "worm chow" across the top of the bed, and an inch of castings are harvested from below by pulling a breaker bar across the large mesh screen which forms the base of the bed. Because red worms are surface dwellers constantly moving towards the new food source, the flow-through system eliminates the need to separate worms from the castings before packaging. Flow-through systems are well suited to indoor facilities, making them the preferred choice for operations in colder climates.

Small Scale Production

For vermin-composting at home, a large variety of bins are commercially available, or a variety of adapted containers may be used. They may be made of old plastic containers, wood, or metal containers. The design of a small bin made up of wood, metal sheet etc usually depends on where an individual wishes to store the bin and how they wish to feed the worms. Small pit may also be constructed for the purposes. Bins need holes or mesh for aeration. Some people add a spout or holes in the bottom for excess liquid to drain into a tray for collection. Worm compost bins made from plastic are ideal, but require more drainage than wooden ones because they are non-absorbent. However, wooden bins will eventually decay and need to be replaced.

Small-scale vermin-composting is well-suited to turn kitchen waste into high-quality soil amendment where space is limited. Worms can decompose organic matter without the additional human physical effort (turning the bin).

Worms for Vermi-composting

The worms used in vermicomposting are called redworms (*Eisenia foetida*), also know as red wigglers, manure worms, red hybrid or tiger worms. Usually red wiggler *Eisenia fetidae* is suitable for small to medium unit vermicomposting. They are surface dweller and together with symbiotic associated microbes are the ideal vectors for decomposing food waste. Other common earthworms such as *Lumbricus terrestris* are anecic(deep burrowing) species and hence unsuitable for use in a closed system.

Climate and Temperature

The most common worms used in composting systems, red worms feed most rapidly at temperatures of 15–25 °C (59-77 °F). They can survive at 10 °C (50 °F). Temperatures above 30 °C (86 °F) may harm them. This temperature range means that indoor vermin-composting with red worms is suitable in all but tropical climates. Other worms like *Perionyx excavates* suitable for warmer climates. If a worm bin is kept outside, it should be placed in a sheltered position away from direct sunlight and insulated against frost in winter.

Feed Waste Utilization

There are few food wastes that vermin-composting cannot compost, although meat and meat products waste and dairy products are likely to putrefy and in outdoor bins can attract vermi. Green waste should be added in moderation to avoid heating the bin.

Integration of Farming System

In integrated farming system, vermicompost unit could be laid as per the availability of dung and other waste. Usually farming system waste may be integrated and utilized for vermicompost production.

Process of Production in Farming System

- **Shed:** Shed could be prepared with wooden poles and structure giving thatched/ grass covered shed in 30ftx20ft structure (Pic 1). The central pole is of 12ft height and outer extremities of 6 ft height. Surface of floor should be 1ft higher than the surrounding areas (Maity *et al.*, 2009).

Pic 1: Low cost vermicompost shed for small farming system

- **Compost Pit:** Size of compost may be 1mt x 1mt x 0.75mt(ht), made up of brick wall 9 inches thickness, concrete floor and 5 inches pucca brick walled separation between the pits. Pit can accommodate 0.75 cft vermi-compost, for 1kg or 1000 earth worm (Pic 2).

Pic 2. Vermicompost pit

- Earth Worm Species: *Icenia foetida* (Pic 3).
- Inoculation rate: 1 kg or 1000 earth worm/ 1m3 compost or 1x1x0.75 mt half digested compost

Pic 3. Worm *I. foetida* for vermicompost

Pretreatment of Compost Materials

- Partial decomposition of cow dung was done for a period of 3 weeks in open area making its heap and sprinkling water in alternate days, covering a thin layer of dry grass.
- Partial decomposition of leaf, kitchen waste, grasses avoiding non bio degradable waste like polythene, plastics, stones or glasses.
- Used partly degraded organic waste as food materials of earth worms.
- Spread alternate layers of dung and leaf litters on bed/heap

Bed Preparation

- Spread 6-10 cm thick layer of decomposable organic waste such as grasses, leaves, sugarcane waste etc.
- Spread partially decomposed cattle dung, organic waste or biogass slurry on the top of 1st bedding layer in an inverted U shape till the central height reach 75 cm.
- Regular watering for 40-45 days to maintain 40-50% moisture till the compost is ready.

Inoculation

- Do not thrust worm culture into active bed. Wait for 2-3 week till the bed is cooled.
- Inoculate worm @ 1000/ 1 sq mt bed and cover it lightly with compost.
- Sprinkle water on surface to make the bed moist.
- Spread a gunny bag of 4 ft wide to cover bed.
- Maintain favorable temp. of 25-40° C

Table 3. Production of worm & compost

System	Area	compost	Worm initial	Worm at 35 days
Pit	0.75 m^3	1.5 qt	1.0 kg	3.1 kg
Platform	3.0 x .75 m^3	4.5 qt	3.0 kg	8.65 kg

Harvesting

- First lot may take 90 days. Subsequent cycle take only 35-40 days depending on density of worms.
- Stop watering before 2 days of harvesting. Allow earth to barrow at the bottom before taking out compost.
- Harvestable compost may be heaped in conical shape on the surface of bed.
- Collect compost from heap and sieve for packing

Vermicompost is ready for harvest when it contains few-to-no scraps of uneaten food or bedding. While harvesting, it's also a good idea to try to pick out as many eggs/cocoons as possible and return them to the bin. Eggs are small, lemon-shaped yellowish objects that can usually be seen pretty easily with the naked eye and picked out.

Benefits of using Vermicompost

Soil

- Improves soil aeration
- Enriches soil with micro-organisms (adding enzymes).
- Microbial activity in worm castings is 10 to 20 times higher than in the soil and organic matter that the worm ingests.
- Attracts deep-burrowing earthworms already present in the soil.
- Improves water holding capacity.

Plant growth

- Enhances germination, plant growth, and crop yield
- Improves root growth and structure
- Enriches soil with micro-organisms (adding plant hormones).

Environmental

- Helps to close the "metabolic gap" through recycling waste on-site.
- Large systems often use temperature control and mechanized harvesting, however other equipment is relatively simple and does not wear out quickly.

- Production reduces greenhouse gas emissions such as methane and nitric oxide.

As fertilizer

Vermicompost can be mixed directly into the soil, or steeped in water and made into a worm tea by mixing some vermicompost in water, bubbling in oxygen with a small air pump, and steeping for a number of hours or days.

The microbial activity of the compost is greater if it is aerated during this period. The resulting liquid is used as a fertilizer or sprayed on the plants.

Table 4. Quality of Dung and Compost

Compost	N%	P%	K%	Mg(ppm)	Fe(ppm)	C(ppm)
Dung	0.56	0.03	0.05	21	2.35	4.5
Digested Dung on 4th wk	0.67	1.12	0.04	35	4.7	5.2
Vermicompost on 4th wk	1.53	1.13	0.64	65	8.4	9.1
Vermicompost on 5th wk	1.68	1.18	0.66	66	8.6	9.4
FYM 3 m	.87	.04	.04	68	9.4	9.5

Moisture

If decomposition has become anaerobic, to restore healthy conditions and prevent the worms from dying, the smelly, excess waste water must be removed and the bin returned to a normal moisture level. To do this, first reduce addition of food scraps with high moisture content and second, add fresh, dry bedding such as shredded newspaper to your bin or pit, mixing it in well.

Worms Escaping

Worms generally stay in the bin, but may try to leave the bin when first introduced, or often after a rainstorm when outside humidity is high. Maintaining adequate conditions in the worm bin and putting a light over the bin when first introducing worms should eliminate this problem.

Small Scale Biogass Production in Farming System

Biogas is produced through the anaerobic digestion of organic waste such as animal and human, and household waste as well as agricultural by-products. Biogas typically refers to a gas produced by the biological breakdown of organic matter in the absence of oxygen. Organic waste such as dead plant and animal material, animal feces, and kitchen waste can be converted into a gaseous fuel called biogas. Biogas originates from biogenic material and is a type of biofuel.

Biogas comprises primarily methane (CH_4) and carbon dioxide (CO_2) and may have small amounts of hydrogen sulphide (H_2S), moisture and siloxanes(Mishra and Dikshit, 2004).

The gases methane, hydrogen, and carbon monoxide (CO) can be combusted or oxidized with oxygen. This energy release allows biogas to be used as a fuel. Biogas can be used as a fuel in any country for any heating purpose, such as cooking. It can also be used in anaerobic digesters where it is typically used in a gas engine to convert the energy in the gas into electricity and heat. In India, Pakistan and Bangladesh biogas produced from the anaerobic digestion of manure in small-scale digestion facilities is called gobar gas; it is estimated that such facilities exist in over two million households in India. In large farming system model small scale biogass plant may be integrated for efficient dung management.

Component of Biogass Plant

A complete biogass plant will have the gas generation subsystem (digester), feed and slurry handling sub system (inlet and outlet of slurry), gas distribution and utilization sub system etc. (George Otim *et al.,* 2010).

Integration in Farming System

In farming system, biogass plant may be recommended where, cow unit integrated not less than 10 ACU. If we consider dung per unit voided at least 10 kg/cow then production of dung would be 100 kg/day. Gas required for cooking and other processing in the system which will be equivalent to 6 member of a medium family size is 0.227x6 =1.35 m^3 gas. To operate 100 candle lamp (60 watts) =0.125m^3 per bhp-hr(Rai-2000). To operate 3 lamp 3 hrs per day, gas required 0.125x3= 1.13 m3 gas. Thus, gas required for cooking and lighting = 1.35 m^3 +1.13 m^3=2.45m^3 of gas. Take 2.5 m3 gas for small scale plant. Taking 1 kg dung produce 0.05m^3 of biogas (Duggal, 2002), this complies for 2.5m3=2.5/0.05=50 kg of dung. Thus, one 5-10 unit cow unit will sustain a small plant (Pic 4).

Pic 4. Small scale biogas plant

Modeling for the Management of Dung in Farming System

In integrated farming system, dung production depends on the number and type of livestock integrated as well as type of feed and fodder available from the system. Type of management also affects dung harvesting as grazing will propagate dung in field. As per the model, following system could be adopted for better dung management.

Table 5. Dung management in farming system

IFS Model	Area (ha)	Livestock	Poultry	Dung Production/day	Proposed management
Model 1	1 ha	1 cow/ buffalo	-	6.5-10 kg	75% FYM + 25% Vermicompost (Pit method)
Model 2	1 ha	1 cow/ buffalo	10 chick/ duck	6.5-10 kg + dropping	75% FYM + 25% Vermicompost + dropping in fish pond
Model 3	2 ha	5 cow/ buffalo unit	-	30-50 kg/day	75% FYM + 25% Vermicompost in platform
Model 4	2 ha	5 cow/ buffalo unit	10 chick/ duck	30-50 kg/day + dropping	75% FYM + 25% Vermicompost in platform + dropping in fish

					production
Model 5	2-4 ha	10 cow/ buff	10-20 chick/ duck	60-100 kg/day + droppings	25% FYM+ 25% Ver -micompost in platform + 50% in small biogas plant
Model 6	2-4 ha	10 cow/ buff + 10-20 goat	10-20 chick/ duck	60-100 kg/ day + 10-20 kg dung pallets + droppings	(25%Dung + 10-20 kg dung pallets) FYM+ 25% Vermicompost in platform + 50% in small biogas plant+ pond platform for poultry dropping in pond.

Thus modeling of livestock unit could be done as per availability of resources in farming system and plan of farming system.

Conclusion

In India, increasing pressure of human population triggered the increasing demand of food and dairy products. To cater the need, livestock population also facing pressure for more production. India's livestock sector, being largest in the world has increasingly come under international scrutiny for its greenhouse gas emission. Nonetheless, livestock also generate positive effects to environment like land savings due to dung used as domestic fuel or recycling of agricultural by products like animal feed or green house gases emission prevented by the use of manure in place of chemical fertilizer. Integration of livestock in farming system will have positive effect in that respect. Moreover, to maximization the net profit, different model of farming system to be practiced for better utilization of available resources.

References

Anonymous 1966. Hand Book of Agriculture p 105. New Delhi.

Duggal K N. 2002. *Elements of Environmental Engineering*. 6th Eds. Rajendra Ravindra Printers Pvt Ltd. New Delhi.

Garg A C, Idnani M A and Abraham T P. 1971. Organic Manure. Tech Bulletin (Agril) 32, ICAR. New Delhi.

George Otim, David Okaka and John Kayima, 2010. Design of Biogas plant for rural household in Uganda. *Proceedings of 2nd International Conference on Advances in Engineering and Technology*. Uganda.Image.slidesharecdn.com.

Maity S B, Kumar Anil, Tiwari S K, Singh Sultan, Nag S K, Rai A K and Kushwaha B P. 2009. Organic feeding system for milk and meat production- Vision and activities. Technical Bulletin, IGFRI Jhansi.

Mishra S N and Dikshit A K. 2004. *Environment and Livestock in India*. Manohar Publishers and Distributers. Ansari Road, New Delhi.

TNAU Agritech Portal, agritech.tnau.ac.in

Yadav A, Gupta R and Garg V K. 2013. Organic manure production from cow dung and biogas slurry by vermicompost under field conditions. *Int. j. recycling of org. waste in Agric.* **2**: 21.

□□□

Chapter 24

Biomass Recycling and Composting in Livestock based Integrated Farming System

SHIVA DHAR, R. K. PALTA AND J. P. SINGH

ICAR - Indian Agricultural Research Institute, New Delhi- 110 012

Introduction

In developing country like India, the economy depends on success and failure of agriculture. India has already enjoyed four decades of post green revolution. Although, extensive use of chemical fertilizer created several problems which deteriorated the soil and environment health, causes health hazard and insecurity of quality food, energy crisis, higher fertilizer cost. Such issues related to sustainability in agri-production system and ecological stability are renewed the interest of farmers and research workers in non-chemical sources of plant nutrients like biofertilizers, farmyard manure, green manure, composts etc. Human beings and cattle were also adversely affected due to the residues of these agrochemicals in food products (Kumar and Bohra, 2006). The organic manures like compost and vermicompost can improve food quality without compromising with food safety (Ersahin, 2011). Composts not only reduce the dependence on chemical fertilizers by supplying essential macronutrients, but also improve the soil organic matter, microbial biomass carbon (Sarangi and Lama, 2013), soil microbial activity (Dadhich *et al.*, 2012), inactivate the soil borne pathogen (Lodhn and Buramn, 2000), thereby encourage the growth, quality of crop, as well as sustain higher productivity (Sangeetha *et al.*, 2013

With the increased agricultural production due to green revolution, a huge amount of agro waste is also produced in the country. Disposal of such organic wastes from various sources like domestic, agriculture and industries has caused serious environmental hazards and economic problems. Generally, waste from the fruit and vegetable market is collected and dumped into the municipal landfills, causing a nuisance because of high biodegradability (Bouallagui *et al.*, 2004). This result in loss of potentially valuable nutrients (Baffi *et al.*, 2005). In India most common practices of waste processing are uncontrolled dumping which causes mainly water and soil pollution (Bouallagui *et al.*, 2004), whereas air pollution is caused due to burning of crop residue in the field. It leads to emission of GHGs and loss of valuable plant nutrient in the atmosphere (Jain *et al.*, 2014), that can be processed into fertilizer. In this context, recycling of agro waste through composting and vermicomposting represents the key technology for the production of organic manures and their use as nutrient supplements in soil to partially replace the chemical fertilizers or integrated use of both organic and inorganic sources of nutrients.

What agricultural biomass waste?

Waste generated from agricultural activities like farming of cereals and pulses, horticulture, floriculture, animal husbandry etc which may or may not be biodegradable. (ENVIS Newsletter, 2008). There are several types of agricultural waste like crop waste, food processing waste, animal waste and hazardous and toxic waste.

Table 1. Status of crop residue generated and residue burning in India

Crop	Dry residue generated Mt/yr	% residue burning	*Burned residueMt/yr
Paddy	192.82	40	77.13
Wheat	120.7	22	26.55
Maize	26.75	3	0.80
Jute	31.51	3	0.95
Cotton	90.86	8	7.27
Groundnut	11.44	1	0.11
Sugarcane	107.5	20	21.50
Rap. and mustard	17.28	1	0.17
Millets	21.57	2	0.43
Total	620.43		134.92

* estimation based on the % residue burning data (Jain *et al.*, 2014)

India produces about 500 million tons (Mt) of crop residues annually. Crop residues are natural resources with tremendous value to farmers. These residues are used as animal feed, composting, thatching for rural homes and fuel for domestic and industrial use.

About 25% of nitrogen, 25% phosphorus, 50% of sulphur and 75% of potassium uptake by cereal crops are retained in residues, making them valuable sources of nutrients. However, a large portion of the residues, about 140 Mt, is burned in field primarily to clear the field from straw and stubble after the harvest of the preceding crop. Primary crop types whose residues are typically burned include rice, wheat, cotton, maize, millet, sugarcane, jute, rapeseed-mustard and groundnut. Farmers in northwest India dispose a large part of rice straw by burning *in situ*.

Ministry of New and Renewable Energy (MNRE 2009), Govt. of India estimated that about 500 Mt of crop residue is generated every year. There is a large variability in crop residues generation and their use depending on the cropping intensity, productivity and crops grown in different states of India. Residue generation is highest in Uttar Pradesh (60 Mt) followed by Punjab (51 Mt) and Maharashtra (46 Mt).

Among different crops, cereals generate 352 Mt residue followed by fibres (66 Mt), oilseed (29 Mt), pulses (13 Mt) and sugarcane (12 Mt). The cereal crops (rice, wheat, maize, millets) contribute 70% while rice crop alone contributes 34% of crop residues. Wheat ranks second with 22% of residues whereas fibre crops contribute 13% of residues generated from all crops. Among fibres, cotton generates maximum (53 Mt) with 11% of crop residues. Coconut ranks second among fibre crops with 12 Mt of residue generation. Sugarcane residues comprising tops and leaves generates 12 Mt i.e., 2% of crop residues in India.

Generation of cereal residues is highest in Uttar Pradesh (53 Mt) followed by Punjab (44 Mt) and West Bengal (33 Mt). Maharashtra contributes maximum to the generation of residues of pulses (3 Mt) while residues from fibre crop is dominant in Andhra Pradesh (14 Mt). Gujarat and Rajasthan generate about 6 Mt each of residues from oilseed crops.

Table 2. Agro waste utilization pattern in India

Agro waste	Utilization
Rice husk ash and charcoal	Additive in cement mixes, glass manufacture, active carbon
Rice husk	Electricity generation
Banana peel and sugar fiber	Paper making pulp
Oil empty fruit bunch	Mulching and organic fertilizer
Oil palm stems, rubber wood	Soft wood furniture
Onion skin, G.N. husk	Heavy metal removal
Husk, Bagasse	Mushroom cultivation
Bagasse, banana fruit reject	Ethanol production, animal feed
Husk, straw, cow dung	Biogas production, electricity
Animal waste	Compost, fertilizer

Table 2 represents the different agro waste utilization pattern in India. From this table we can conclude that there are many economical uses of agro waste available in India other than composting. So we cannot say strongly that we can use such agro waste for compost making only but under certain circumstances we can use this waste for compost making which is explain as follows:

India produces about 133.57 and 86.74 Mt of actual dung through cattle and buffaloes respectively (estimated using formula given by Jain *et al.,* 2014). The states where the crop residue burning practices was observed intensively due to their disposal problem are the only states where production of dung is also higher. So if we utilize this crop residue and dung efficiently for compost making we can reduce GHGs problem upto certain extent and able to recycle plant nutrient to improve soil health with less investment.

The poultry industry has a potential to contribute in agro waste generation. The by-product generation categorized into two stages i.e. from production and dressing plant. From production stage the poultry litter and manure helps in surface dressing of agricultural land whereas the egg shell also have good source of calcium which is of special importance to earth worms health in vermicomposting. However, from dressing plant stage feather contributes about 7-8%. Apart from its use as decorative purpose feathers act as a bedding material as well as source of manure. Blood meal also forms 3.2-3.5% of poultry live weight. Blood meal also used for composting in western countries but its use in India is very less. It may be due to most of Indian population are vegetarian.

The slaughter house industry is an unorganized industry in India. At present there are no official norms for classification of slaughter houses. However, depending upon the type of animals slaughtered, the slaughter houses are classified. The local bodies must, therefore, take up modernization of slaughter houses and achieve the pollution control norms. Slaughtering of animals generates wastes consisting of non edible offal (like lungs, large intestines, various glands, animal tissues, organs, various body parts, etc.) stomach/intestinal contents, dung, sludge from waste water treatment, bones, etc. All these types of wastes are required to be disposed by adopting methods like rendering/controlled incineration/burial/composting/anaerobic digestion etc. The slaughter houses are normally controlled by local bodies, which should follow the standards prescribed, but due to non-existence of modernized slaughter houses, environmental pollution arising out of the slaughtering activities cannot be controlled.

In case of the different organic waste material, farm crop waste contains higher amount of NPK. Among the different animal waste cattle excreta contains more NPK which was followed by buffaloes (Table 3).

Table 3. Potentials of NPK from different organic wastes in India

Source	Plant nutrients (Mt)			
	N	P_2O_5	K_2O	Total
Cattle	2.977	0.793	1.332	5.102
Buffalo	0.745	0.276	0.487	1.508
Goat and sheep	0.214	0.063	0.020	0.297
Pig	0.044	0.027	0.029	0.100
Poultry	0.027	0.020	0.010	0.057
Other livestock	0.079	0.018	0.069	0.166
Farm crop waste	5.600	2.300	10.700	18.600

Fruit and Vegetable waste

At presently, India producing about 221 Mt of fruits, vegetables and perishables but India has capacity to store 23.6 Mt in 5,386 cold storages of which 80 % are used for potatoes. Hence, according to industry estimates, 25 to 30% of fruit and vegetable and 5-7% of food grains in India get wasted every year. Similarly, annually 10% of food grain loss due to rodents. Only 7% of food in India is processed. Therefore, fruit and vegetable producer subjected to loss of 25% worth of produce. The 40% post-harvest and processing loss (FAO Report, May 2011). If we utilize this waste for compost making we can recycle the nutrient that otherwise will go waste as fruit and vegetable waste contains higher amount of nutrients compared to other agro-waste. The total waste generated in the food grain market against total food grain arrival shown that waste generated from the different APMCs varies from 0.07% to 0.14%. These waste generated is quite low as food grain is not a perishable commodity as like fruit and vegetable but it has great potential of nutrients and can be recycle for returning the nutrients to soil.

The total waste generated in fruit and vegetables and its percentage compared to total arrival of fruit and vegetables in different APMCs. The % waste generated from these APMCs varies from 1 to 11.6% which is quite large amount. It was found that a huge quantum of fresh fruits and vegetables arrive and transacted daily in these Markets. As per traders survey, waste generated in fruits range from 2.6% to as high as 11.4 %. Wastage in Vegetables ranged from 3.15% to 12.6%. Markets like Azadpur generate approximately 4% wastages of both fruits and vegetables. Percent wastage reported by the Market Committee is comparatively quite low as against traders survey. This clearly indicates that a huge amount of food commodity wasted goes unreported. This also account for the lack of statistics and data on the quantum of food waste actually generated.

In some countries crop residues are used as a source of energy, animal feed, composting mushroom cultivation or even burned in field (Table 4). In China 37% of crop residues are directly combusted by farmers, 23% used for forage, 21% discarded or directly burnt in the field, 15% lost during collection, 4% for industry materials and 0.5% for biogas (Liu *et al.,* 2008). Thus burning of crop residues in the field is a major problem in China as well.

Table 4. Mode of crop residue management in other countries.

Mode of utilization	Country
Source of energy	Indonesia, Nepal, Thailand, Malaysia, Philippines, Indonesia, Nigeria
Composting	Philippines, Israel, China
Animal feed	Lebanon, Pakistan, Syria, Iraq, Israel, Tanzania, China, Africa
Mushroon cultivation	Vietnam
Burning	China, USA, Philippines, Indonesia

Why composting?

India has already enjoyed four decades of post green revolution and that to due to use of high yielding varieties and increased use of fertilizer. Although, extensive use of chemical fertilizer created several problems which deteriorated the soil and environment health, causes health hazard and insecurity of quality food, energy crisis, higher fertilizer cost. Such issues related to sustainability in agri-production system and ecological stability are renewed the interest of farmers and research workers in non-chemical sources of plant nutrients like biofertilizers, farmyard manure, green manure, composts etc. With the increased agricultural production due to green revolution, a huge amount of agro waste is also produced in the country. Disposal of such organic wastes from various sources like domestic, agriculture and industries has caused serious environmental hazards and economic problems. Generally, waste from the fruit and vegetable market is collected and dumped into the municipal landfills, causing a nuisance because of high biodegradability (Bouallagui *et al.*, 2004). This result in loss of potentially valuable nutrients (Baffi *et al.*, 2005). In India most common practices of waste processing are uncontrolled dumping which causes mainly water and soil pollution (Bouallagui *et al.*, 2004), whereas air pollution is caused due to burning of crop residue in the field. It leads to emission of GHGs and loss of valuable plant nutrient in the atmosphere (Jain *et al.*, 2014), that can be processed into fertilizer. In this context, recycling of agro waste through composting and vermicomposting represents the key technology for the production of organic manures and their use as nutrient supplements in soil to partially replace the chemical fertilizers or integrated use of both organic and inorganic sources of nutrients.

Benefits of Compost

Environment benefits

1. Water and soil conservation (erosion)
2. Protects groundwater quality
3. Minimizes odors from agricultural areas
4. Avoids methane production and leachate formation in landfills by diverting
5. Organics from landfills into compost
6. Drastically reduces the need for pesticides and fertilizers
7. Off-farm materials can be brought in and added to manure to make compost
8. Composted manure weights about one-fourth as much as raw manure per ton.

Agricultural benefits

1. Long-term stable organic matter source
2. Soil pH in favourable range
3. Regenerate poor soils
4. Suppresses certain plant diseases and parasites and kills weed seeds
5. Increases yield
6. Increases length and concentration of roots in some crops
7. Increases SNC and WHC of sandy soils and water infiltration of clay soils
8. Reduces fertilizer requirements
9. Restores soil structure; compost is a soil inoculants
10. Increases earthworm populations in soil
11. Provides slow, gradual release of nutrients
12. Provides opportunity for extra income.

What Are the Benefits to the Food Industry?

1. Reduces solid waste disposal fees
2. Educates consumers on the benefits of food waste composting
3. Markets your establishment as environmentally conscious
4. Markets your establishment as one that assists local farmers and the community
5. Reduces the need for more landfill space.

Agricultural use of composts remains low for several reasons?

1. Transportation problem as it is bulky in nature
2. The low nutrient value compared to chemical fertilizers
3. The nutrient composition variable compared to chemical fertilizers
4. Concerns regarding potential levels of heavy metals
5. Heavy application of composts to agricultural soils has been found to result in salt, nutrient, or heavy metal accumulation and may adversely affect plant growth, soil organisms, water quality, and animal and human health.

Urban composting

Urban composting is a emerging thrust area and the compost generated from these areas can be used for nutrition gardens and urban agriculture. This activity may be useful to produce better tasting, more nutritious food without the harmful effects of chemical pesticides and fertilizers. Urban agriculture supports more sustainable production of food by trying to eliminate the use of harmful pesticides and fertilisers. As most of the garbage generated in big cities goes to dumping grounds, this approach turns waste into resource and save the cities from turning into a garbage-dump. It enables recycling of a large percentage of this biodegradable garbage at the source itself. Thus, it saves the cost of transportation of food from far off places, incurring huge energy and money for their transportation. The energy used to transport food would be greatly reduced if urban agriculture could provide the cities with locally grown food reducing carbon footprint. These activities provide employment in the areas such as gardening, nurseries, consultancy, and contract for waste recycling, marketing and selling of organic produce.

Methods of composting

A. Traditional methods

Anaerobic Decomposition

The Indian Bangalore method (FAO, 1980)

This method of composting was developed at Bangalore in India by Acharya in1939. The method is basically recommended when night soil and refuse are used for preparing the compost. The method overcomes many of the disadvantages of the Indore method such as problem of heap protection from adverse weather, nutrient losses due to high winds / strong sun rays, frequent turning requirements, fly nuisance etc. but the time involved in production of a finished compost is much longer. The method is suitable for areas with scanty rainfall.

Aerobic decomposition

The Indian Indore method (FAO, 1980)

An important advance in the practice of composting was made at Indore in India by Howard during the period 1924 to 1926. The traditional procedure was systematized into a method of composting now known as the 'Indore method'. The raw materials used are mixed plant residues, animal dung and urine, earth, wood ash and water. All organic material wastes available on a farm such as weeds, stalks, stems, fallen leaves, prunings, chaff, fodder leftovers and so on, are collected and stacked in a pile. It may be prepared by pit method or heap method.

Large Scale Aeration : Windrow Composting (NRAES, 1992)

Turned Wind-rows

Windrow composting consists of placing the mixture of raw materials in long narrow piles or windrows which are agitated or turned on a regular basis. The turning operation mixes the composting materials and enhances passive aeration. Typically the windrows are initially from 3 feet high for dense materials like manures to 12 feet high for fluffy materials like leaves.

Passively aerated windrows

The method, passively aerated windrow system, eliminates the need for turning by supplying air to the composting materials through perforated pipes embedded in each windrow. The pipe ends are open. Air flows into the pipes and through the windrow because of the chimney effect created as the hot gases rise upward out of the windrow.

B. Rapid Composting Methods

While traditional composting procedures take as long as 4-8 months to produce finished compost, rapid composting methods offer possibilities for reducing the processing period up to three weeks.

1. The Berkley Rapid Composting Method (Raabe, 2001)

   Shredding and Frequent Turnings

2. North Dakota State University Hot Composting (Smith, 1995)

   Use of Mineral Nitrogen Activator

3. EM based Quick Compost Production Process (Hiraoka, 2002)

   Use of Effective Micro-organisms (EM)

4. Rapid Composting Technology (Virginia, 1997).

   Use of Cellulolytic Cultures.

Aerobic composting

In IARI the aerobic composting process adopted. A majority of the composting processes involve aerobic composting, wherein arrangements are made for exposing the composting material to oxygen. In this process aerobic microorganisms break down organic matter and produce carbon dioxide, ammonia, water, heat and humus. The intermediate compounds that may be produced are further oxidized and there is no risk of phytotoxicity. The heat generated destroys many human or plant pathogens and kills weed seeds. There is loss of more nutrients specially N as ammonia. Aerobic composting processes are preferred over anaerobic processes. Initially (first 2-3 days) the temperature in the compost windrow is around 20-45°C and mesophilic organisms multiply rapidly, mostly on easily decomposable organic compounds such as sugars and amino acids.

They generate heat by their own metabolism and raise the temperature to 50-70°C, where their own activities become suppressed. Thermophilic fungi and bacteria then dominate and continue the process of composting and raise the temperature of compost to 65°C or higher. This heat kills pathogen and weed seeds. Active composting stage is followed by a curing stage and the pit's temperature declines gradually. At this stage another group of thermophilic fungi start to grow. These fungi decompose cellulose and hemicellulose. Curing of the compost provides a safety net against the risks of using immature compost such as N hunger, oxygen deficiency and toxic effects of organic acids on plants.

Wind rows (crop residue and cowdung)

Microbial consortium application

IARI Wind-row composting

Wind-row composting involves placing the mixture of raw materials and cowdung (in the ratio of 5:1) in long narrow piles called wind rows, that are agitated or turned on a regular basis. The height of the wind-rows depends upon the raw materials. It is 0.9 m for dense materials and 3.6 m for light voluminous materials such as dry leaves. The width varies from 3 to 6 m. The size, shape and spacing of wind-rows depend upon the turning mechanism. Bucket loaders with a long-reach can permit high wind-rows, while turning machines permit low, wide wind-rows. There are a number of specialized machines for turning wind-rows that reduce time and labour involved considerably. Some of these machines attach to farm tractors or front-end loaders, while others are self-propelled. The microbial consortium mainly consisting cultures of *Aspergillus awamori*, *Aspergillus nidulans*, *Phanerochaete chrysosporium* and *Trichoderma viride*, is used before turning and mixing of the raw material. Optimum level of moisture in the heap of compost is maintained by frequent sprinkling of water.

Addition of water for moisture

Turning and Mixing of compost

The active composting stage generally lasts 3-9 weeks depending upon the nature of materials and frequency of turning. Where 3 weeks is the goal, the wind-rows require turning once or twice a day during the first week and every 3-5 days thereafter. In general all type of crop residue takes about 60-75 days. Such material needs 3-4 turning, first at the beginning and subsequent turning at 15 days interval. By using the above method the Biomass Utilization Unit is preparing about 5000 tonnes of good quality compost annually.

Loder machine

Sieving machine

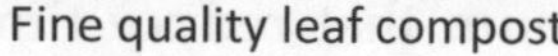

Fine quality leaf compost

Costumer purchsing Vermicompost

Twigs & wood Shredder machine

Shredded material

Compost Turner cum Mixer

Crop Residue mixed Farm yard Manure (CRFYM)

As the names indicate this manure mostly uses crop residues and cattle dung and urine as the raw material. At Biomass Utilization Unit, Indian Agricultural Research Institute, New Delhi, CRFYM is prepared using approximately 80% crop residues and 20% fresh cattle dung and urine on weight basis. This manure is prepared by wind-row composting involves placing the mixture of raw materials in long narrow piles called wind rows, that are agitated or turned on a regular basis. The height of the wind-rows depends upon the raw materials. It is about 1.2-1.5m for dense materials and 3.0- 3.6 m for light voluminous materials such as dry leaves and stover of crop residues. The width varies from 2.5 to 3.0 m. Bucket loaders with a long-reach used to make wind-rows of desired size and shape. A specialized machine is used known as 'Turning-cum-Mixing Machine' for turning wind-rows and mixing microbial consortium that reduce time and labour involved considerably. The liquid microbial consortium (diluted 1:100 or 1:50 dilution with water) is sprayed on wind-row at the time of first turning.

This composting process involves aerobic process, exposing the composting material to oxygen. In this process aerobic microorganisms break down organic matter and produce carbon dioxide, ammonia, water, heat and humus. The intermediate compounds that may be produced are further oxidized and there is no risk of phytotoxicity. The heat generated destroys many plant pathogens and kills weed seeds. Initially (first 2-3 days) the temperature wind-rows is around 20-45°C and mesophilic organisms multiply rapidly, mostly on easily decomposable organic compounds such as sugars and amino acids. They generate heat by their own metabolism and raise the temperature to 50-70°C, where their own activities become suppressed. Thermophilic fungi and bacteria then dominate and continue the process of composting and raise the temperature of compost to 65°C or higher. This heat kills pathogen and weed seeds. Active composting stage is followed by a curing stage and the wind-row's temperature declines gradually. At this stage another group of thermophilic fungi start to grow. These fungi decompose cellulose and hemicellulose. Curing of the compost provides a safety net against the risks of using immature compost such as N hunger, oxygen deficiency and toxic effects of organic acids on plants. The active composting stage generally lasts 3-6 weeks depending upon the nature of materials and frequency of turning. Where 9 weeks is the goal, the wind-rows require turning once or twice during the first week and every 15 days thereafter. The process of composting takes about 60-65 days. The C:N ratio of CRFYM produced by the wind-rows method varies from 15.4 to 20.5 depending upon the type of raw material used. It contains 4.8 to 7.0 % humus, 0.4 to 1.0 percent N, 0.2 to 0.4 per cent P and about 1.0 percent K depending upon the type of crop residue. It also contains secondary nutrients like Ca 0.6 to 0.95% and Mg 0.7 to 0.9%, and micro-nutrients like Fe 135-158 ppm, Mn 65-75 ppm, Zn 15-22 ppm and Cu 3-5 ppm.

CRFYM is an excellent soil amendment, which adds to balance of nutrients, while contributing valuable organic material to the soil. The latter is needed for the growth of soil microorganisms, to improve water-holding capacity, soil structure, for pH buffering and the organic complexing of nutrients - making them more available for uptake by plants. The main role of the CRFYM is to add organic matter and to increase humus content of soil.

Advantages of CRFYM include

1. Coarse CRFYM on the surface reduces the impact of the falling rain drops and thus reduces surface run-off and erosion.
2. Easily decomposable CRFYM leads to synthesis of complex organic substances that bind soil particles into structural units, called aggregates, which help to maintain a favourable condition of aeration and permeability.
3. Organic matter from CRFYM increases water-holding capacity of soil.
4. Organic matter adds and serves as a reservoir of essential plant nutrients which are released in harmony with the needs of the plants.
5. CRFYM produces organic acids that help in dissolving phosphorus and micronutrients in soils and increases their availability.
6. CRFYM increases exchange capacity as well as buffering capacity of the soils.
7. CRFYM serves as a source of food and energy for the growth of soil microorganisms as well as soil meso and macro fauna (ants, termites, earthworms etc.).
8. CRFYM used as mulch reduces water evaporation losses.
9. CRFYM used as mulch also helps in maintaining soil temperature.
10. The nutrients from manure are released to the plants slowly and steadily. The benefits will last for more than one season.

Conclusion

Crop residue, animal waste and food and agricultural commodity are wasted through the food supply chain at various stages starting from agricultural production till consumption but that can be processed into organic manure through composting. Enrichment of compost with vermi-wash, additives and microbial inoculants increase its efficiency. Off-farm materials can be brought in and added to manure to make compost which improves soil health. To recycle agro waste generated from various sectors agriculture, efforts needed at household as well as community level. IARI model of composting is an unique model to convert the agricultural waste into different types of valuable composts and it can be prepared at small scale, thus a feasible activity for integrated farming system.

References

Baffi, C.D., Abate, M.T., Silva S., Beneditti A., Nassisi A., Genevini, P.L and Adani, F. 2005. A comparison of chemical, thermal and biological approach to evaluate compost stability. By Geophysical Research Abstracts. 7: 09116. European Geosciences Union.

Bouallagui, H., Torrijos, M., Godon, J.J., Moletta, R., Cheikh, R.B., Touhami, Y., Delgenes, J.P., Hamdi, M. 2004. Two-phase anaerobic digestion of fruit and vegetable wastes: bioreactors performance. *Biochem. Eng.* J. **21**: 93-197.

Dadhich, S.K., Pandey, A.K., Prasanna, R., Nain and Kaushik, B.K. 2012.Optimizing crop residue-based composts for enhancing soil fertility and crop yield of rice. *Indian Journal of Agricultural Sciences* **82**(1): 85–88.

ENVIS Newsletter- 2008. *Agricultural waste management*. **4**: 4.

Ersahin, Y. 2011. The use of vermicompost products to control plant diseases and Pests. In: Karaca A (ed) Biology of Earthworms, Soil Biology. Springer-Verlag Berlin Heidelberg.

FAO 1980. A Manual of Rural Composting. FAO/UNDP Regional Project RAS/ 75/004 Field document 15.

FAO.2011. FAOSTAT. Food and Agriculture Organization, the United Nations.

Hiraoka, H. 2002. Biofertilizer Production Plant, Myanmar (FAO/UNDP Project), Back to Office Report, Feb. 2002, FAO- RAP, Bangkok

Jain, N., Bhatia, A and Pathak, H. 2014. Emission of air pollutants from crop residue burning in India. *Aerosol and Air Quality Research* **14**: 422–430.

Kumar, A and Bohra, B. 2006. Green technology in relation to sustainable agriculture. *In*: Kumar, A., Dubey, P. (eds) Green technologies for sustainable agriculture. Daya Publishing House, Delhi.

Lodhn, S and Burman U. 2000. Efficacy of compost on nitrogen fixation, dry root rot (*Macrophomina phaseolina*) intensity and yield of legumes. *Indian Journal of Agricultural Sciences* **70**(12): 846-849.

Liu H, Jiang GM, Zhuang HY, Wang KJ (2008) Distribution, utilization structure and potential of biomass resources in rural China: With special references of crop residues. *Renewable and Sustainable Energy Reviews* **12**: 1402-1418.

Ministry of New and Renewable Energy Resources (2009) www.mnre.gov.in/relatedlinks/biomassrsources

NRAES 1992. On-farm composting (Ed. Rynk, Robert). Natural Resource, Agriculture, and Engineering Service, Cooperative Extension, Ithaca, New York.

Raabe, R. D. 2001. The Rapid Composting Method. Co-operative Extension, Division of Agriculture and Natural Resources, University of California.

Sangeetha, S.P., Balakrishnan, A., Devasenapathy, P. 2013. Influence of organic manures on yield and quality of rice (*Oryza sativa* L.) and blackgram (*Vigna mungo* L.) in rice-blackgram cropping sequence. *American Journal of Plant Sciences* **4**: 1151-1157.

Sarangi, S.K and Lama, T.D. 2013. Composting rice straw using earthworm (*Eudrilus eugeniae*) or fungal inoculants (*Trichoderma viridae*) and its utilization in rice (*Oryza sativa*)-groundnut (*Arachis hypogaea*) cropping system. *Indian Journal of Agronomy* **58**(2): 146-151.

Smith, R. C. 1995. Composting Practices. NDSU Extension Service, North Dakota State University of Agriculture and Applied Science, and USDA.

Virginia, C. C. 1997. Rapid Composting Technology in Philippines: its role in producing good quality organic fertilizers. Extension Bulletin, FFTC, Taiwan.

□□□

Chapter 25

Soil and Water Conservation Strategies in Farming Systems

D.R. PALSANIYA, AKRAM AHMED, T. KIRAN KUMAR AND S.K. RAI
ICAR- Indian Grassland and Fodder Research Institute, Jhansi (UP), India-284003

Soil and water are the two important natural resources essential for the existence of life. For optimum crop production we need good soil health and sufficient water resource of appropriate quality. The tropical and subtropical parts of World suffer from serious soil erosion caused by intense rainfall that occurs within a few months of the year, and the topography of the cultivated upper catchments. Soil erosion remains a major problem to sustainable agricultural development in these areas. In addition to reducing agricultural productivity on-site, erosion causes negative impacts off-site. Land degradation and water scarcity are the major Natural Resources Management (NRM) issues at present. India, the second most populous country after China comprises nearly 16% of the world population and 15% of the world livestock resource. However, its share in land and water resources is merely 2.46 and 4.00%, respectively (Fig.1). It shows the resource deficit and population pressure on our resources. The per capita water availability is low and likely to decrease further, as its demand is rising for various needs (Table 1). Irrigation sector, the major user of water in India is facing tough competition with other sectors like domestic and industry for water use.

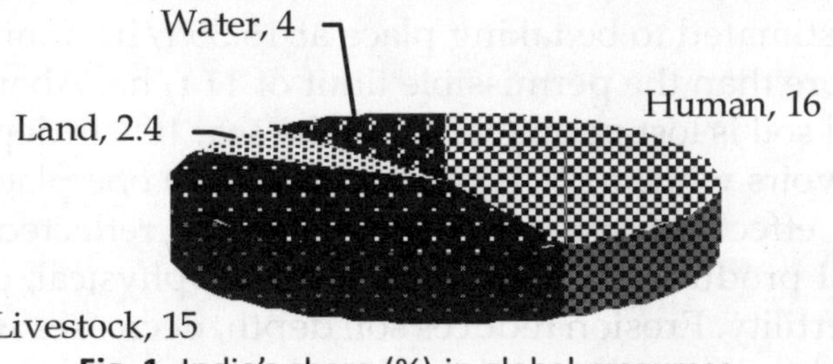

Fig. 1. India's share (%) in global resources.

Further, out of 400 m ha.m precipitation in the country, nearly 115 m ha.m is lost as run off resulting in drought in the catchments and floods in down stream areas. Our ground water table is depleting at very alarming rate. Our water resources are dwindling in quantity, quality and equity and will soon emerge as a binding constraint on India's progress. There are fears that water related conflicts will intensify at local, regional, national and international level in the coming future. River water sharing related disputes are common not only in India but also in other parts of the world. Our interstate water rivalry is evident from various water sharing disputes like Cauvery disputes amongst Tamilnadu, Karnataka, Kerala and Pudduchery; Yamuna dispute among Punjab, Haryana, Delhi and U.P. and Sutluj among Himachal Pradesh, Punjab and Rajasthan. Similarly, India also have water sharing disputes with Pakistan, Bangladesh, Nepal and China and it is rightly said that water may be the main cause the third world war. So there is urgent need to not only conserve this precious resource but also to develop it through surface and ground water management strategies.

Table 1. Estimates of water requirement for India (M ha m)

Use	2000	2025
Irrigation	63.0	77.0
Domestic	3.3	5.2
Industrial	2.7	12.0
Energy	2.7	7.1
Other	3.5	3.7
Total	75.2	105.0

Same is the case with our land resource. Nearly 187.8 m ha (57.1% of 329 m ha) area in India suffer from one or other kind of degradation (Table 2). Soil erosion alone results in loss of over 6 billion tones of fertile top soil containing around 5.53 million tones of N,P and K . Soil erosion is estimated to be taking place at 16.35 t/ha annually in India which is more than the permissible limit of 11 t/ha. About 29% of the total eroded soil is lost permanently to the sea, 10% is deposited as silt in the reservoirs while 61 % is transported from one place to another. The on-site effects of soil erosion is generally reflected in terms of reduced soil productivity through the loss of physical, chemical and biological fertility. Erosion reduces soil depth, decreases water holding capacity and reduces chemical fertility through nutrient and organic matter loss. The main off-site impacts of soil erosion are silting of reservoirs and resultant reduction in its storage capacity, accumulation of heavy metals and other toxic substances like pesticides in the reservoirs and environmental pollution. Our land and water resource base is shrinking and its productive capacity is diminishing. This dwindling water and land resource has necessitated the adoption of appropriate measures to conserve and develop them for sustainable agriculture production to impart food, energy and environmental security to the growing population.

Table 2. Land degradation in India

Type of degradation	Area (M ha)	% of total area
Water erosion	148.9	45.3
Wind erosion	13.5	4.1
Chemical deterioration (Salinity etc.)	13.8	4.2
Water logging	11.6	3.5
Total	187.8	57.1

Complementarities

Soil and water conservation are very important aspects particularly in rainfed areas. It can be strongly argued that rainfed agriculture holds the key for addressing the challenges of poverty, equity, food security and inclusive growth. There are always strong links and complementarities between measures for soil conservation and measures for water conservation. Reductions in surface run off by structures or by changes in land management will reduce soil loss. All such practices will increase infiltration and thus, help in water conservation also. Thus, soil and water conservation goes hand-in-hand and can not be separated. Soil and water conservation can be achieved through engineering and agronomic measures. Agronomic practices alone can serve the purpose if land slope is less than 2%. However, if slope is over 2% than we have to go for engineering measures along with agronomic practices.

Agronomic measures for soil and moisture conservation

Current land management practices had some effect on the degree of soil erosion. We can minimize soil erosion by manipulating these agronomic land management practices. The basic underlying principle to soil and moisture conservation is conserve and harvest as much rainwater as possible and provide safe disposal to excess water through adequate drainage. Thus, drainage is also an important aspect of soil and moisture conservation. Agronomic measures for soil and moisture conservation include contour cultivation, cover crops, strip cropping, mixed/inter cropping, vegetative barriers, wind breaks and shelterbelts, crop rotations, mulching, addition of organic matter, green manuring , conservation tillage, dead furrows ridging , broad beads and furrows (BBF), micro catchments, catch pits, use of chemicals, vegetative bunds and barriers, alternative land use systems like agroforestry etc. These agrotechniques help in intercepting raindrops and reduce their beating action, reduce run off and help in better intake of water by soil due to improvement in organic matter and soil structure.

Selection of crops

- Close growing, spreading type and dense canopy cover crops
- Deep root system, soil binding, N-fixing
- Cover crops protect the land like an umbrella by developing good dense canopy for interception of rain drops so as to expose minimum soil surface for erosion
- Among the field crops, legumes like cowpea, greengram, groundnut, etc. give a good cover for the land during the rainy season.

Crop Geometry

The crop geometry should be manipulated in such a way that it results in closer spacing within the rows across the slope so as to provide greater resistance to overland flow to minimize erosion losses.

Contour cultivation

Contour cultivation consists of carrying out agricultural operations like ploughing, sowing and intercultural operations across the slope and very nearly on the contour. In contour cultivation, each ridge of plough furrow and each crop row act as an obstruction to run off and reduces the velocity of the over land flow and thereby retard soil erosion and encourage infiltration. The reduction in runoff and soil loss due to contour cultivation has been shown in Table 3.

Both at Dehradun and Ootacamund, contour cultivation reduced the runoff and soil losses considerably as compared to cultivation along the slope. Besides conserving soil and moisture, contour cultivation also increases crop yield, fertilizer responsiveness and reduces nutrient loss with runoff water.

Strip cropping

Strip cropping refers to growing of different crops in alternate strips across the slope in such a way that they serve as vegetative barriers to erosion. The long and narrow strips consist of close growing erosion resisting crops alternated with erect growing erosion permitting crops. The suitable erosion resisting crops are pulses like cowpea, greengram, groundnut, soybean etc. while erosion permitting crop are pearl millet, sorghum, maize etc. The strip of dense growing crops reduce the transporting and eroding power of water by forming obstruction to runoff and filter out the soil from the runoff and retains it in the field. The width of the strips depends on the slope of field, rainfall characteristics and type of crop grown. The width of the strips depend on the slope of field are discussed in the table 4. To achieve the best results, strip cropping should be practiced in combination with others agrotechniques like contour farming, crop rotation, addition of FYM etc.

Table 3. Effect of contours cultivation on runoff and soil loss (Reddy and Reddi, 1995)

Treatment	Runoff (mm)	Soil loss (t/ha)
Dehradun		
Along the slope	670	28.5
Contour cultivation	511	19.3
Ootacamund		
Along the slope	52	39.0
Contour cultivation	29	15.0

There are four types of strip cropping systems.

1. Contour strip cropping

   Contour strip cropping consists of growing alternate strips of erosion resisting and erosion permitting crops along the contour. Contour strip cropping reduces the length of the slope. The width of the strip will vary according to the slope and topography.

2. Field strip cropping

   In field strip cropping, the strips are laid across the slope in uniform width without taking into consideration the exact contour. Field strip cropping is suitable for lands having uniform slope and high infiltration rates.

3. Wind strip cropping

   In wind strip cropping, the crop strips are laid at right angles to the direction of the prevailing winds regardless of the contour. Here, the erosion resisting crops are tall growing while erosion permitting crops are low growing. The objective here is to control wind erosion rather than water erosion.

4. Buffer strip cropping

   Buffer strip cropping consists of planting permanent strips of perennial grasses, legumes or shrubs in areas that are either badly eroded or that do not fit for regular cultivation. These buffer strips are established on steep and highly eroded slopes in the field.

Table 4. Width of erosion resisting and erosion permitting strips

Slope (%)	Width of crop strip (m)		Ratio (R:P)
	Resisting	Permitting	
<1	9.0	45.0	1:5
1-2	6.0	24.0	1:4
2-3	5.0	15.0	1:3

Tillage

Various properties of the soil like bulk density, porosity, surface roughness, hardness of the pan, etc. are altered by the tillage practices. Conservation tillage wherein the soil is disturbed to a minimum possible extent and the crop residues are left on the field itself helps in reducing the soil losses and conserving the water as compared to conventional tillage in which the field is ploughed twice or thrice followed by harrowing and planking. Minimum and zero tillage are the forms of conservation tillage. In minimum tillage, tillage operation is reduced to minimum necessary for ensuring a good seed bed, rapid germination, optimum plant population and favourable growing conditions. Zero tillage is the extreme form of minimum tillage where primary tillage is completely omitted and secondary tillage is restricted to seed bed preparation in the row zone only. These form of tillage besides other advantages also reduce soil erosion, enhance infiltration due to vegetation present on the soil and channels formed by the decomposition of dead roots, less soil compaction by the reduced movement of heavy tillage vehicles and improved soil structure. In most of the cases, conservation tillage reduces the soil loss by 50% over conventional tillage.

Sub soiling

Presence of hand pan in the sub soil also induces soil loss due to reduced infiltration resulting in enhanced run off. Breaking of hard pans by sub soiling using chisel plough improves soil moisture regimes, facilitate root penetration to deeper layers and reduce run off and soil erosion.

Crop rotation

Crop rotation refers to recurrent succession of crops on the same piece of land in a given period of time to get maximum profit from least investment without impairing soil fertility. Thick growing and spreading type crops like legumes should be incorporated the crop rotations. Such crops protect the soil from beating action of rain drops, intercept run off, increase infiltration and in turn, minimize the soil loss. These crops also improve soil structure and fertility by nitrogen fixation and addition of organic matter. Crop rotation for an area can be chosen such as way that during the rainy period there is a thick vegetative cover over the soil surface. Crops like groundnut, soybean, greengram, cowpea etc. are commonly used in crop rotations.

Mixed and inter cropping

Mixed cropping refers to growing of two or more crops in the same field simultaneously without any row arrangement while in intercropping; two or more crops are grown simultaneously in definite row arrangements in the same field. Generally, legume is grown as one of the component. The component should have different growing habits like tall growing –short growing, deep root system-shallow root system, N-fixing - non N-fixing, early maturing -late maturing, wide spaced- close spaced, erosion permitting -erosion resisting , etc.

Traditionally, these systems are extensively followed in India. At Vasad, inter cropping of pearl millet and greengram significantly reduced the runoff and soil loss compared to sole cropping (Table 5).

Benefit of mixed and intercropping

- It provides insurance against complete failure of crops under adverse conditions. e.g. Drought kills only shallow rooted crops, frost to legumes and epidemic to one crop only.
- It improves or maintains the soil fertility through adding organic matter, fixing atmospheric N and mining nutrients from different layers.
- It balances the farmers demand or need for cereals, pluses, oil seeds and fodder.
- These system provide better cover on the land, good protection to soil from beating action of rain, bind soil particles, increase infiltration rate and reduce run off and soil loss.

Table 5. Soil and water loss under various cropping system at vasad

Cropping system	Runoff (mm)	Soil loss (t/ha)
Pearl millet-fallow	167.7	2.3
Peal millet + greengram	121.0	2.2
Beedi tobacco	205.3	4.8
Sunnhemp – Beedi tobacco	142.6	4.4

(Reddy and Reddi, 1995)

Cover crops

Cultivation of close growing cover crops is one of the agronomic measures to reduce erosion and improve soil moisture. Cover crops protect the land like an umbrella by developing good dense canopy for interception of rain drops so as to expose minimum soil surface for erosion, Among the field crops, legumes like cowpea, greengram, groundnut, etc. give a good cover for the land during the rainy season.

Mulch farming

Mulch is any material, organic or inorganic in nature applied on the soil surface to check evaporation and improve soil moisture.

Mulching also results in additional benefits like reduction in run off and soil loss, moderation of temperature, reduction in soil salinity, weed control and improvements in soil structure and fertility if mulch is organic in nature. It also increases infiltration rate.

Types of mulches

i. **Dust mulch**

   The loose surface soil created through loosening by intercultural operation in called dust or soil mulch. Dust mulch checks evaporation through breaking capillaries.

ii. **Straw mulch**

   Use of straw as mulch in called straw mulch.

iii. **Stubble mulch**

   If crop residues like stalks, Stover haulms, straw, etc. are left on the soil surface, it is called stubble mulch.

iv. **Plastic mulch**

   If plastic materials like PVC (poly vinyl chloride), polyethylene, etc. are used as mulch, it is called plastic mulch.

v. **Vertical mulch**

   It consists of digging narrow trenches or slots across the slope at intervals and placing crop residues or straw or organic matter in these trenches. It is more beneficial in black soils. These slots with residues form ridges across the slope and thus, increase infiltration and reduce run off and soil loss.

Mulching with plant material reduces soil loss up to 43 times compared to bare soil and 17 times compared to cropped soil without mulches (Reddy and Reddi, 1995). Any locally available material may be used for mulching. It may be saw dust, straw, paddy husk, groundnut shell, crop residues, paper, plastic, stones, leaves, corn cobs, manures, wood chips, etc. Studies conducted at Dehradun (Table 6) showed that mulching @ 4 t/ha reduced run off from 15-37% and soil loss from 5.4-18 t/ha in maize at 8% slope.

Table 6. Effect of mulching on run off, soil loss and grain yield of maize

Treatment	Run off (%)	Soil loss (t/ha)	Maize yield (kg/ha)
No mulch	36.9	17.8	2433
Mulch @ 2 t/ha	17.9	7.3	2317
Mulch @ 4 t/ha	14.7	5.4	2395

Vegetative bunds

Growing of plants like *Gliricidia* on farm bunds serves dual purpose of producing organic matter under field conditions and also helps in conserving soil through reduced soil erosion. Planting of Gliricidia strengthens the bunds. It is easy to propagate and establish. During rainy season cuttings can be planted on bunds in 10-15 cm deep holes or 3-4 month old seedlings can be planted on bunds at spacing of 0.5 m apart. *Gliricidia* plants can be well managed through pruning at appropriate times (before sowing of rainy and post rainy season crop). The pruned green biomass are cut into pieces and applied to the soil surface as mulch or incorporated into soil as green manure to improve the soil. *Gliricidia* has fast growth, N- rich leaves, tolerance to pruning, ability to coppice vigorously and good fodder value. It can be grown as rainfed plant even at places, which receives only 500 mm rainfall annually. A study at ICRISAT reveals that by adding the green biomass from *Gliricidia* plants planted on bunds at a spacing of 0.5 m apart for a length of 700-800 m will provide about 25-30 kg N/ha, and also improve physical properties of soil over a period of time.

Vegetative barriers

Dense vegetation is raised across the slope for making live bunds or vegetative hedge. The plant species suitable for raising live bunds/ hedge/barriers are Saccharum munja, Vertiveria zizanoides, Cenchrus glaucus etc. These vegetative barriers are cheaper and more permanent soil and water conservation measures than engineering ones and helps to reduce the length of the field slope, check run off velocity, control soil erosion, reduce silt load and improve the soil moisture.

Agrostology

Grasses are known to control soil erosion and improve soil structure largely due to their good canopy and excellent binding force of their roots.

Agroforestry

Agroforestry is a landuse system that helps to enhance productivity and ensure sustainability by combining annual food crop, grasses and woody perennials under the same management. Under agroforestry, food crops supply the dietary needs while the trees help stabilize slope, reduce erosion, enrich soil, increase infiltration rate and meet needs of fuel, fodder and timber. Silvipasture, hortipasture, alley cropping, etc. help in rehabilitation of degraded lands through natural resource conservation and development.

Wind breaks and Shelter belts

Any barrier created to slow down the wind effect is called as wind break while a belt of shrubs or trees maintained for the purpose of shelter from wind, sun, cold, snow drift is known as shelter belt. Wind breaks and shelter belts play an important role in soil and moisture conservation particularly from wind erosion. These should be planted at the right angles to the prevailing wind direction. The most suitable conformation of shelterbelts is the type resulting from planting of tall growing trees in the centre and the lower bushy vegetation on the outer rows of the shelter belts. Wind carrying the soil particles is obstructed by wind breaks and shelter belts and hence minimizes deposition of soil particles. In general, wind velocity is reduced on the wind ward side from 5 to 10 times of tree height from shelterbelt and on leeward side the protection is up to 30 times tree height. *Acacia, Prosopis, Albizia, Tamarindus, Eucalyptus* etc. are the suitable species for this purpose.

Addition of Organic Matter

Addition of organic matter in the form of FYM, compost wormicompost, green manuring, etc. not only improve soil fertility but also conserve soil and moisture. Organic matter improves soil structure, enhance soil aggregation and improve soil water holding capacity. It facilitates the penetration of rain water thus decreases run off and erosion. It also checks evaporation losses of water from soil.

Use of Chemicals

Certain chemicals enhance the stability of soil aggregates and make them more resistant to break down due to beating action of rain drops and thus resist erosion and enhance infiltration rate. Spraying of chemicals like poly vinyl alcohol @ 480 kg/ha and soil treated with bitumen enhance aggregate stability.

Land configuration and planting techniques

Improved land configuration and planting techniques also helps in soil and moisture conservation. These measures include ridge and furrow, dead furrow, broad beds and furrows, micro-catchments, catch pits , furrow irrigated raised bed system (FIRBS) in case of wheat, etc. which not only conserve soil and moisture but also enhance greater water use efficiency and have provisions of adequate drainage in case of excess rainfall.

Conservation furrow

Series of furrows are opened on contours or across the slope at 3-5 m apart depending on soil type, slope, rainfall and crop to be grown. Generally, the size of furrow is kept as 15 cm deep and 20 cm wide.

Broad Bed and Furrow (BBF) technique

ICRISAT has developed a broad bed and furrow (BBF) technique for black soils. BBF consists of broad raised beds with furrows of 30 cm width and 15 cm depth on either side of each bed. The width of the bed is 120 cm from centre of one furrow to the adjacent furrow. The beds are maintained permanently. The beds act as minibunds at a grade and help in reducing the velocity of surface run off and increase infiltration opportunity time. The excess water is removed through the furrows.

Farm pond

Farm Ponds can be defined as bodies of water created by constructing a dam or embankment across a water course or by excavating a pit.

Farm pond is used for:

- As a source of drinking water for livestock, irrigation and fish production

Components of Farm Pond

- Storage area, earthen dam, mechanical spillway and emergency spillway
- Earthen dam is used to obstruct and confine the runoff water.
- Mechanical spillway is used for letting out the excess water from the pond and also as an outlet for taking out the water for irrigation.
- Emergency spillway is to safeguard the earthen dam from over toppling when there are inflows higher than the designed values.

Data requirement

- Area of farm, Area for Pond, Avg. daily evaporation, Annual rainfall and Crop information

Design of Farm Pond

- Site selection, 2. Capacity of pond, 3. Design of embankment, 4. Design of mechanical. Spillway, 5. Design of emergency spillway and 6. Seepage control from bottom

Site selection

The site for the pond is to be selected keeping in view of the following considerations;

1. The site should be such that largest storage volume is available with the least amount of earth fill. A narrow section of the valley with steep side slopes is preferable.
2. Large areas of shallow water should be avoided as these will cause excessive evaporation losses and also cause water weeds to grow.
3. The site should not cause excessive seepage losses.
4. The pond should be located as near as possible to the area where the water will be sued. When the water is to be used for irrigation, gravity flow to the areas to be irrigated is preferable.

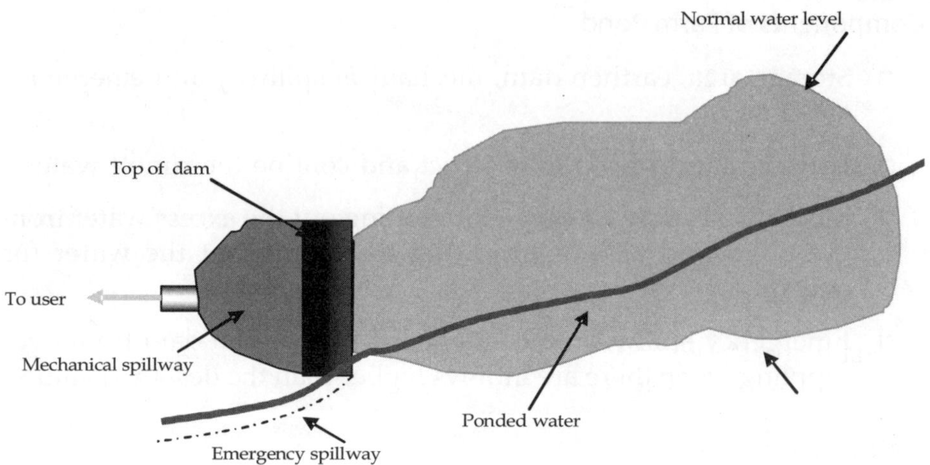

Capacity of pond

The capacity of the pond is determined from a contour survey of the site at which the pond is to be located. From the contour plan of the site, the capacity is calculated for different stages using the trapezoidal or Simpson's rule.

For this purpose, the area enclosed by each contour is measured using planimeter. According to the trapezoidal rule, the volume V between two contours at an interval H and having areas A1 and A2 is given by:

$$V = \frac{H}{2}(A1 + A2)$$

Using Simpson's rule the volume between any odd number of contours is given by:

$$V = \frac{H}{2}(Twice\ the\ area\ of\ odd\ contours + 4\ times\ area\ of\ even\ contours + area\ of\ the\ first\ and\ last\ contours)$$

Design of embankment

- Foundation condition: Coarse materials like gravel and sand provide good support but high seepage. Fine textured materials like silt and clay are impervious but not very much stable. Base width to be decided for better stability.
- Cross section: Decided based on the nature of the foundation material as well as fill material available at the site.

Seepage losses in ponds

The seepage losses from the ponds could be considerable depending upon the nature of the substrata at the pond site. The seepage losses will reduce with time due to the deposition of the finer particles brought in with the runoff water. At some locations this may take considerable time and as such lining of the pond becomes necessary. If the soil is well graded at the pond bottom (about 70% sand, 20% clay and 10% silt) excavating up to 30 cm and recompacting the soil will reduce the seepage losses. When proper soil types are not available at the site, other methods are to be used. Reduction in seepage rates achieved by mixing swelling clay materials, such as bentonite with the soil. In order to support the water column at least 10 to 15% sand should be present in the soil. Polythene sheets have been used as a lining material for ponds. A layer of soil of about 15 cm thick above the sheer is necessary to keep the sheet in place and avoid external damage. Mishra and Sharma (1994) they developed equation forcalculating the seepage loss for different situations and depth offarm pond for different capacity the capacity of the farm pond and side slope was decided accordingly.

Water harvesting through farm ponds and utilization for growing vegetable crops

In a research Subudhi *et al.* (2013) studied the water loss and economics of the lined ponds. The size of thepond is 7m top widths, 1m-bottom width, 3m heights, and 1:1side slope. The waterharvested in pond was reutilized for the pumpkin crop. Lined pond with soil cement (6:1) plaster of 8cm thickness gave highest Tomato yield of 4.8t/ha during kharif 2008-09 and radish root yield of 25.5 t/ha in rabi seasons of 2008-09. Thewater loss was 326 lit/day in lined pond and 24,000 lit/day in unlined pond. Thebenefit: costratio in lined pond was 3.04 as compared to 1.64 in unlined pond during 2008-09. Wesley et al. (2010) studied the effect of different lining material for water harvesting and supplemental irrigation through dug out ponds. The treatments included lining material, sodic soil; cement + sodic soil (1: 10); cement + soil (1: 8); cement + murrum (1: 6); cement + soil + murrum (1: 3: 5); cement + cement bricks; cement + sand + bricks and Kadapa slabs. Greater amount of runoff water was retained by cement + murrum as the lining material of the pond. The length of cracks developed in the cement + murrum lined farm pond was minimum as compared with other treatments. Four runoff events were recorded during the crop season and crop yields were increased by 120 per cent by giving two supplemental irrigation of 10 mm each for breaking the dry spell of 43 and 37 days during vegetative and pod development stages, respectively. The haulm yield increased by 50 per cent.

CONCLUSIONS

In farming system, depending upon the crop water requirement different in-situ and field water harvesting measures should be adopted. The crops which are water sensitive for them farm pond and furrow irrigation system is suitable. The crops which are not water sensitive, in-situ water harvesting measures such as bunding, mulching etc. should be adopted. Depending upon the climatic zone, proper farm pond size should be excavated. In arid and semi-arid zone, the farm pond size should be less in compared to humid zone. A considerable amount of water is lost through seepage from farm pond. So, proper cost-effective lining material should be used. An integrated, interdisciplinary, participatory and community based approach is required to manage our water and soil resources in a sustainable way focusing on the on and off-site impacts of soil and water erosion. Farmers should be educated to adopt the agronomic measures for soil and moisture conservation most of which are non monetary in nature and simple to practiced. Engineering measures should be adopted in conjunction with them if required.

References and further readings

B John Wesley, Swamy R and Reddy T Y. 2010. Rainwater harvesting and supplemental irrigationthrough farm-ponds and evaluation of lining materials, rainwater harvesting and reuse through farm ponds, CRIDA, Hyderabad.

Balasubramaniyam, P. and Palaniappan, S.P. 2002. Principles and practices of agronomy. Agrobios India, Jodhpur.

Mishra P K and Sharma S N. 1994.Theoretical consideration in design of farm ponds for minimizingevaporation and seepage losses. *Indian Journal of Dryland Agricultural Research & Development,* **9**(2): 114-120.

Murty, V.V.N. 2004. Land and water management engineering. Kalyani Publishers.

Reddy, T.Y and Reddy, G.H.S. 1995. Principles of agronomy. Kalyani Publishers.

Sharma, N.K. Agronomical measures for soil conservation and erosion control. Summer school on participatory integrated watershed management organized by CS&WCR&TI, Dehradun, 1-21 July, 2004.

Singh R P.2003. Rainwater management for sustainable dryland agriculture: An Indian experience. Indian Journal of Dryland Agricultural Research & Development, **18**(2): 180-186.

SubudhiC R and Senapati S C. 2013. Water harvesting through farm pond and utilization of conserved water for vegetable crops, Advanced Engineering Informatics, **57**: 13901-13905.

❑❑❑

Chapter 26

Watershed Based Integrated Approach for Resource Conservation and Livelihood Security

S.P. TIWARI

ICAR - Central Soil & Water Conservation Research and Training Institute, Research Centre, Datia (MP)

Introduction

Among the various factors responsible for slow down the agricultural rate of growth in the recent time, depletion and deterioration of natural resources, especially water and soil are considered most vital (Mangala Rai, 2009). Watershed based integrated approach broadly aims at scientific management of the natural resources like soil, water, plant, livestock, human and environment for increasing biomass production, employment generation, conservation of resources with due emphasis on landless, gender issues and biodiversity for over all development and livelihood security of rural poor's.The government of India has accorded highest priority to the holistic and sustainable development of the rainfed areas through integrated watershed development approach. Although watershed has been in existence since long time, it became popular in eighties when large numbers of model watershed under the aegis of National Watershed Programme were launched in the country. Compare to earlier programmes, the development of arable and non-arable areas were integrated in model watersheds. Encouraged with success of model watersheds, Ministry of Agriculture, Govt. of India in 1987-88, launched National Watershed Development Programme for Rainfed Areas (NWDPRA). Since then a massive national watershed programmes to improve and stabilize the agricultural production system for rainfed areas have been taken up in different agro-ecological zones.

Careful implementation of resource management technology on watershed basis is essential for the stabilization of primary production system for the sustenance of local population. This lecture note discusses on soil and water conservation technologies effective for integrated watershed management as evident through model watershed management projects implemented by ICAR-CSWCRTI, in different agro-climatic regions of the country.

Natural resources: problems and constraints

Soil, water and vegetation are the primary natural resources of the country. There rapid depletion posing a serious threat to the sustainability of production system in the country. The burgeoning population pressure coupled with unscientific land use practices and shrinking per capita land and water resources are largely responsible for over-exploitation and reduced productivity of natural resources. Due to increase in population per capita availability of arable land has steadily declined from 0.48 ha in 1950-51 to 0.15 ha in 2000-01 and is expected to reduce further to 0.12 ha by 2015-16 (Samra, 2005) Out of the total cultivated area in the country about 63% area is under rainfed condition. This rainfed sector contributes about 45% of the total food production; as compared to 55% production accruing from 37% irrigated area. Further it may be noted that 30% of the area (109 m ha) in the country, spread over in 99 districts is drought prone and agriculture in this area is non-sustainable due to critical water shortage and prevailing poor land and crop management systems.. Apart from this average forest vegetative cover stands at 20.55% against the recommended minimum cover of 33% in the country. Situation is much worse in the drought prone regions of the country like Bundelkhand (Table-1) where more than 70% population is engaged in agriculture and live stock rearing for their livelihood.

Table 1. State of forest cover in Bundelkhand region

Districts	Geographical Area	Forest cover (Year -2009 Assessment)			Total	Percent of TGA
		Very Dense	Moderate Dense	Open forest		
		Madhya Pradesh				
Chhatarpur	8,687	184	822	742	1,748	20.12
Damoh	7306	2	862	1741	2605	35.66

Datia	2691	0	78	79	157	5.83
Panna	7135	85	1501	1068	2654	37.20
Sagar	10252	2	1178	1726	2905	28.34
Tikamgarh	5048	1	93	309	403	7.98
		Uttar Pradesh				
Banda	4532	0	26	77	103	2.27
Chitrakoot	3092	0	358	203	561	18.14
Hamirpur	4282	0	66	108	174	4.06
Jalaun	4565	0	65	179	244	5.35
Jhansi	5024	0	33	167	200	3.98
Lalitpur	5039	0	128	442	570	11.31
Mahoba	2884	0	22	73	95	3.29
Total	70,537	274	5,232	6,914	12,419	17.61

Source: FSR-2009, Forest Survey of India

In view of the present state of natural resources, modern tool and procedures should be employed to manage and conserve natural resources for sustainable agriculture development.

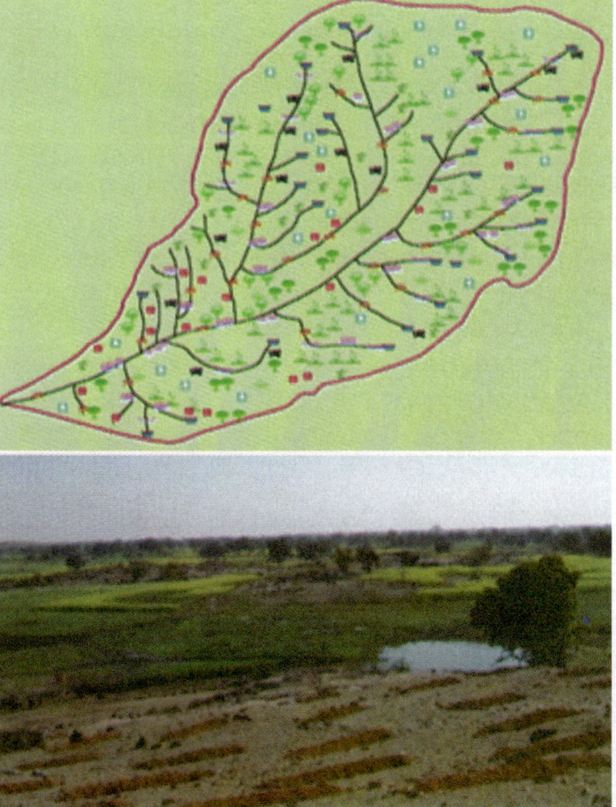

Definition and concept of watershed

Definition: The term "Watershed" strictly refers to the divide separating one drainage basin from another and defined as "Land area from which water drains to one given point". In this context Watershed is considered to be synonymous with "catchment" and "drainage basin". A watershed may be flat land or it may include hills or mountains or undulating terrain. The people and animals depend on the watershed are part of the watershed community and in turn influence the condition of watershed.

Concept: The concept of watershed management is not very new, as wide variety of programmes is being implemented by number of institutions in the country. However over last five decades there has been a paradigm shift in conceptualization and implementation of model watershed development in the country. Initial top-down approach of watershed development was mainly focused on protection oriented biophysical aspects. It has now been shifted towards bottom-up approach with due emphases on addressing the socio-economic and community participation issues in conjunction with bio-physical aspects.

How does a watershed affect the people?

A watershed affects the people in every walk of life. The sustained production of food, fuel, forage, fiber, fruits etc. depends on the management of resources like water, soil, vegetation and phenomena like floods and droughts are determined by the nature of watershed functioning. Watershed conditions affect and influence water quality, quantity and regime. These three water parameters are vital for (I) human and animal consumption, (ii) irrigation, (iii) industry, (iv) power generation, (v) irrigation, (vi) transportation and in many other ways. By manipulating different watershed characteristics the quality, quantity and regime can be regulated according to requirement.

Resource conservation technologies for watershed treatments

Rain water harvesting in conjunction with soil and water conservation techniques helps in augmenting water availability in the watershed. These technological interventions improve moisture availability in soil profile and enhance surface water availability in the watershed for protective/life saving/supplementary irrigation. Based on the nature, type and costs of soil and water conservation measures, these have been grouped in to three categories by Sharma and Mishra (1995) viz hardware land treatment, medium software land treatment and soft ware land treatment. While hardware land treatments are of permanent type executed for improvement of relief, physiographic and drainage line with the basic objective to check soil erosion, regulate runoff, reduce peak flow, medium software land treatments are usually of semi-permanent type and may last in 2-10 years. Inter-bund treatments like vegetative barriers, small section bunds, employed in large size fields where only conventional bunds are constructed along field boundaries are grouped in this category. These measures minimize the velocity of overland flow and needs major farmer's initiative with support of government departments. By and large soft ware treatments are temporary in nature and require making every year. Most of the agronomical measures such as contour farming, compartmental bunding, broad bed and furrows, ridge and furrow, dead furrows and mulching practices etc are grouped in this category and has wide acceptance due to favorable returns.

Development of water resources and utilization

Rain water harvesting (RWH) is a vital component of all integrated watershed management programmes (IWMPs) running in India and considered as a boon for livelihood and food security in rainfed areas. The principle of rain water harvesting is to conserve where it falls, according to need and geo-physical conditions. Rain water harvesting and management consists of ex-situ and in-situ harvesting technologies.

Ex-situ rain water harvesting

The ex-situ rain water harvesting technologies include dug out pond, storage tanks, *nala* bunding, gully control structures/check dams (weirs), water harvesting dam, percolation tanks and dams etc. These technologies are very much location specific and practices evolved in a given region has limited application in other region depending on similar soil-climatic condition. While working in Bundelkhand region our experiences showed that black Soils of the region are ideal for storage of runoff and construction of farm pond due to their low intake rate ranges from 1.6 to 2.1 mm/hr in case of Mar and 10 to 13 mm/hr in case of Kabar soil types. At the same time annual rainfall ranges from 750-1200 mm coupled with undulating topography provides good opportunities for water harvesting in the region. Depending on different land use conditions runoff in the region varies from 10-70 %. Sharda and Ojasvi (2005) estimated that in general water yield in the region is expected to the tune of 0.2 ha-m for every 5 ha catchment with flat lands and 0.3 to 0.5 ha–m for every 5 ha of catchment with sloping lands. Ephemeral streams in the region further offers good opportunity for storing water by constructing water harvesting structures called as stop dam.

The basic objective of *ex-situ* water harvesting includes creation of water availability in WHS for supplementary irrigation and ground water recharge in the watershed area. Rain water harvesting in watersheds thus enhances the overall water regime / surface and ground water availability and has direct impact on crops productivity and livelihood security.

Efficient utilization of harvested water

Since, runoff water harvesting structures are costly in nature; it is obligatory to utilize this harvested water more effectively for different purposes. Proper scheduling of supplemental irrigation to different crops may be one of the options to utilize the stored water in more efficient way in the watershed. Results of an experiment conducted on utilization of runoff harvested water from a farm pond constructed in red soils of district Datia (MP) revealed that one supplemental irrigation given at pod filling stage to soybean could enhance its grain yield by 42%. The short duration oil seed crop toria grown without supplemental irrigation failed due to moisture stress condition during reproductive stage of growth but one supplemental irrigation provided to *toria* at 30DAS could boost the yield by 4.28q/ha. Pre-sowing and Pre-sowing +one irrigation at 30DAS given to mustard increased grain yield by 2.6 and 12 times as compared to no-irrigation (Narayan, 2005). Efficient Utilization of harvested water for establishment of orchards, forest nursery, providing life saving irrigation etc by using modern irrigation methods and machinery may be other better options for increasing the resource use efficiency.

Impact of Water harvesting in model watersheds

Positive impact of runoff harvesting on surface/ground water improvement have been fairly documented in the watersheds. Samra (1997) summarized that integrated watershed management projects undertaken in different agro-climatic zones of the country could help in increasing the water table to the tune of 0.2 m to 2.0 m (Table-2). Substantial increase in yield of water is also reported by number of farmers in model watershed Jigna distt-Datia (MP) particularly by those located in the close vicinity/influence of the WHSs.

Table 2. Effect of watershed management strategies on ground water recharge in different regions of India.(Samra, 1997)

Watershed	Surface Storage capacity created (ham^{-1})	Observed rise in ground water table (m)*
Bajar- Ganiyar (Haryana)	79.0	2.0
Behdala (Himachal Pradesh)	18.0	1.0
Bunga (Haryana)	60.0	1.8
Chhajawa (Rajasthan)	20.0	2.0
Chinatekur (Andhra Pradesh)	5.6	0.8s
GR Halli (Karnataka)	4.0	1.5
Joladarasi (Karnataka)	4.0	0.2
Siha (Haryana)	42.2	2.0

**Difference between pre-project and post project water table*

In Jigna watershed, increased availability of irrigation water motivated the farmers to shift towards more remunerative crops and cropping pattern. Data collected from a user group of 25 farmers occupying 44.17 ha area in the influence zone of a WHS (check dam) at pre and post construction phase clearly suggested that majority of farmer has switched over from mustard to wheat crop during *rabi* season *due to increased availability of water in their well /tu be wells* (Table 3) . Due to this shift the total crop productivity of 44.17 ha land in terms of Wheat equivalent yield (WEY) become almost two fold from 997 quintal to 1835 quintal during rabi-season.

Table 3. Changes in crop pattern and productivity in terms of wheat equivalent yield (WEY) rabi-season (q/ha) in vicinity of a WHS at Jigna watershed, Distt- Datia (MP).

Before WHS					After WHS				
Crop	**Area**	**Yield (q/ha)**	**WEY (q/ha)****	**Total Productivity in terms of WEY (q)**	**crop**	**Area**	**Yield (q/ha)**	**WEY (q/ha)**	**Total Productivity in termsof WEY(q)**
Mustard (14)*	21.50	5.14	11.19	240.60	Wheat (19)	33.33	36.86	36.86	1228.50
Mustard +Gram (6)	11.50	4.00 +4.42	18.50	212.80	Wheat +Gram (1)	2.00	40.0 +3.75	48.30	96.60
Mustard +Wheat (4)	8.67	5.15 +36.25	47.46	411.48	Mustard +Wheat (1)	1.33	10.0 +40.0	61.79	82.18
Mustard +Gram + Wheat (1)	2.50	5.00 + 3.10 +35.00	52.75	131.87	Mustard + Gram + Wheat (4)	7.51	5.78 +3.63 +36.42	57.05	428.44
Total (25)	44.17			996.75	Total (25)	44.17			1835.72

*Figures in parenthesis are total number of farmers

**Based on minimum support price fixed by Govt of India for the year 2013-14 for Wheat @Rs 1400/q, Mustard @Rs 3050/q and Black gram @Rs 3100/q

Rain water harvesting and harnessing for livestock/animals

Rain water harvesting and harnessing primarily for stray cattle/ livestock is seldom given preference over irrigation purpose or invariably ignored in the watershed programmes. However this component emerged as a focal point during PRA conducted during project formulation of Jigna watershed located in Bundelkhand in the vague of consecutive drought situations occurred in recent past. Success story of this embakement cum dug out pond started since its filling with water in first year of its construction and its immediate benefit recognized by the down below farmers in their defunct wells which could start yielding water and become functional otherwise they were dry since long.

Table 4. Technical features of an embankment cum dug out pond constructed in Community pastures land at Jigna watershed, Datia (MP).

Particulars	Dimensions
Year	2010-11
Top of Pond	38 m x 28 m
Bottom of Pond	30 m x 20 m
Depth	2.0 m
Cost (Rs)	1,06,311
Dugout Capacity	1664 cum
Total Capacity	4493 cum
Catchment Area	4.5 ha
Surplus water arrangement	Natural

Out of the total live stock populations of about 6911 in the watershed, on an average approximately 10-15% livestock are more or less dependent for grazing in the panchayat and nearby private lands lying fallow adjacent to this newly constructed pond. Based on random count estimate > 1000 livestock including major and minor ruminants comes to this pond daily (Table 5) at different spells of time.

Table 5. Average Number of animals benefitted by the pond in community land .

Village	Farm Families	Cows	Buffaloes	Goat/ Sheep	Others (wild blue bulls, etc)	Beneficiary Communities
Jigna	58	250	150	700	>50	Bagla (25), Kumhar (5), Harijan (10), Yadav (8), Brahmin (10).

As per the farmer's perception due to construction of this pond (1) There is saving of about 3-4 hours a day otherwise spent in moving for search of water or forced to come back in village early. (2) Water in this pond keep grazing animals, nearby grazing area thus reduces pressure on crops etc.(3) increased number of animals and in milk production.

Monitoring of water availability in this pond was started since September, 2012 (i.e. after second filling) at fortnightly/monthly interval. Initially in the beginning years of construction, water dried in the pond during peak summer season (May-June) but drying days reduced over consecutive years. However during last summer season of 2014 water remain available throughout the summer period ranging from 1.2 to 1.5 m and no drying was recorded.

In-situ rain water harvesting techniques

Under rainfed conditions choice of technologies are going to vary from location to location due to high degree of complexity and diversity in situation. In -situ rain water harvesting is more feasible and practical preposition under most situations can be achieved increasing infiltration opportunity time by various on farm tillage, mulching, land configuration techniques. These are location specific and depend on rainfall intensity, slope and texture of soils etc.

Contour farming

Contour farming is one of the easiest and most effective low cost methods of controlling erosion, conserving moisture and improving crop yields. Carrying out farm operations viz ploughing, seeding planting and intercultural etc along the contour lines results in creation of furrows, which acts as miniature reservoirs to hold the excess runoff , dissipate the energy of flow and provide increased opportunity time for runoff absorption. Thus the erosive velocity of runoff is considerably

reduced and soil and nutrients are conserved and is reflected in higher crop yields. This practice is most effective on moderate slopes (2-6%) and on deep permeable soils, which are not prone to surface sealing. Contour farming alone cannot control runoff volume from higher sloppy lands which need bunding and grass waterways in natural drainage. It has been reviewed that seasonal runoff from the untreated catchment was reduced from 54% to nearly 40% of the rainfall where contour farming was practiced. The corresponding reduction in soil loss was 30t/ha to 20 t/ha (Mishra and Tripathi, 2013).

Contour and graded bunding

Contour bunding is adopted on relatively permeable soils. Surplusing arrangements must be provided for excess runoff generated from high intensity rainfall. On land having 2-6% slope contour and graded bund can be taken up, however between slopes range of 6-33% bench terracing should be adopted. Out of the several sizes and cross section of contour bunds evaluated at the research farm Datia Contour bunds, made of 0.55 m^2 cross section at 0.70 m VI found most effective in conserving more than 75% of the runoff, thus recommended for in-situ rain water conservation in the red soil of Bundelkhand region up to 6% field slope. Contour bunds are suitable for red soils in the region, being relatively permeable but not for black soils suffering with impeded drainage. On black soils often suffer logging condition graded bunds with 0.2% grade are suitable to prevent proper drainage and prevent water logging condition. Beneficial effect of contour bunding has been reported in agricultural watersheds (Ram Babu et al 1980, Gurmel Singh *et al.*, 1990).

Management of soil surface cover

Surface runoff and soil loss from cultivated fields can be reduced if the land has enough canopies during peak rainy season to protect the soil by intercepting the rainfall. Studies conducted at Datia clearly revealed that runoff from 2% sloppy cultivated fallow plot was 45 % of the monsoon rainfall and a good crop cover could bring it down to about 20 % thus suggested the importance of surface cover particularly during initial period of rainy season in reducing the erosion losses (Tiwari and Sharma,1994). Green gram, black gram and cowpea have been identified as promising cover crops for the region, which enhanced the soil fertility, reduced erosion losses and gave sustainable grain yields of 6.0-8.0 q/ha. In an another study conducted on sloppy red soils, groundnut also emerged as an efficient cover crop which recorded the minimum soil loss of the tune of 3.6 t/ha as against 8.6 t/ha under cultivated fallow plot with average yield of 7.9 t/ha

Green manuring for erosion control and fertility management

Green manuring, besides nutritional/fertility benefits to the land also act as a cover crop and minimize the erosion particularly during crucial period of rainy season. Results showed that on an average green manuring of sun hemp increased the yield of dwarf wheat by about 35% and water use efficiency (WUE) by 30% in red soils of the region (Narain,2005). The benefits of green manuring were found more pronounced in shallow and medium soil depth conditions. Practicing green manuring in the Bundelkhand region has greater relevance then in any other area with similar conditions due to peculiar soil, climatic and open grazing (*Anna pratha*) social problems, which forced farmers to keep their land fallow. Thus more than 70% cultivable area in the region remains fallow during rainy season, which further leads to higher soil and nutrients losses by runoff as well as by heavy infestation of weeds.

Tillage and Land preparation

Land configuration and tillage practices are important for increasing soil profile moisture storage, pulverize the deeper layers and creating better soil environment and ultimately minimize the adverse effect of shallow soil depth. Studies conducted at the Datia indicated that a plot having 30cm soil depth produced almost half of wheat grain (1.2 t/ha) as against the plot having 90cm soil depth (Narayan and Lal, 2006) clearly revealed the importance of soil profile depth in crop production. Bhusan *et al.* (1977) also reported a significant residual effect of deep tillage on yield of *toria* after maize harvest. Studies conducted at the research farm Datia revealed that deep tillage by mould board plough once in three years conserved 18 % more rain water and further with live mulching of sun hemp it enhanced to 34% and increased the sorghum yield by more than 35% over the traditional practice.

Deep tillage (20cm) practiced in sorghum resulted almost five fold increase (25.0 q/ha) in grain yield over zero tillage (4.7 q/ha). Runoff and soil loss occurred from deep tilled sorghum plot was of the order of 35.9% and 4.98% t/ha as against 43.4% and 3.08 t/ha from zero tilled sorghum plot, respectively. Growing of sun hemp in between two rows of sorghum for in situ turning was reported further helpful in reducing runoff (30.7%) and soil loss (3.87t/ha) and increasing grain yield (27.0 q/ha) indicated that practicing zero tillage is good from conservation point of view, but deep ploughing is a better choice for higher grain yields of sorghum and other cereals in the region. This may be attributed to the effective control of weeds through deep tillage, which in fact are great nuisance in the region particularly in rainy season.

Sub-soiler may be practiced on soils having hard layers beneath the plough zone. Tractor mounted sub-soiler (30cm deep) cracks the hard layer of shallow red soils, thereby increase the water intake in deeper layer, thus found beneficial improving root zone moisture regime and crop production. About 44.9% and 36.6% increase in case of sorghum grain and fodder yields respectively and about 83% in increase in case of pigeon pea grain yield was recorded due to use of sub-soiler as compared to the yields obtained using country plough Nitant and Singh (1992).

Intercrops for Contingent planning

Rainfed crops often face severe moisture stress during their growth period. Moisture deficiency may even lead to crop failure. In order to avert this situation a contingent plan to meet aberrant weather condition is necessary. As a contingency crop planning for rainfed condition, intercropping of soybean with castor in red soils and soybean with sesame in black soils found to be more remunerative than raising castor or sesame as the sole crop in Bundelkhand region. Further sorghum based intercropping study conducted with sesame, green gram, cluster bean and cowpea revealed that intercropping of 2 rows of sesame in between two rows of sorghum gave higher yield in terms of sorghum equivalent yield (41.4 q/ha) followed by one row of sesame (33.9 q/ha), two rows of green gram (30.6 q/ha) and one row of green gram (28.7 q/ha) (Narayan, 2005).

Vegetative barriers/live bunds

The effectiveness of a grass species, to develop as effective barrier, is only when it forms an adequate dense barrier across slope, which reduces run off velocity, increase infiltration opportunity time and deposits sediment behind the barrier. Vegetative barriers are closely spaced grass strips grown on contours across slope. These have been found effective in checking erosion and considered as an alternate to bund on mild slopes for soil and water conservation being relatively cheaper and pro-farmer. A study on 2% sloppy red soil revealed that vegetative barrier of *vetiveria zizaniodes* planted at 11 m horizontal spacing produced run off and soil loss of the order of 32% and 3.3 t/ha as against 50.0% and 8.8 t/ha respectively from cultivated fallow plot (Sharma and Santraj, 1995). In an another study *Vetiveria zizanioides* barrier planted at 11 m HI with Sorghum crop on 2% sloppy red soils increased yield of grain by 9.9 % (21.0 q/ha) than sorghum without barrier, in spite of the fact that there was reduction in the area of sorghum by about 5.4% in Sorghum+ barrier treatment, indicated that grass barrier has increased the yield of grain sorghum and produced less runoff (24.2%) and soil loss (1.74 t/ha) as compared to 47.2% and 4.6 t/ha respectively from the plot having sole crop of sorghum. Vegetative barrier of Marvel (*Dichanthium annulatum*) planted at 11 m spacing with up and down cultivation of pigeon pea, produced least runoff (32.4%), soil loss (2.7 tha-1) and N, P and K losses (5.9, 0.8 and 4.0 Kgha-1 respectively) against 65.0 % runoff, 13.7 tha-1 soil loss and 22.5 Kg N, 2.9 Kg P and 7.9 Kg K ha-1 losses from the plot without barrier (Tiwari and Kurothe, 2006).

Micro-irrigation systems for increasing WUE

The use of rain gun, a pressurized micro-irrigation sprinkler system found most effective irrigation technique to promote water use efficiently for increasing crops yield. Micro-irrigation through sprinklers (rain gun) for saving of precious irrigation water was demonstrated in wheat crop on farmers fields distributed over six villages of district Datia. The results indicated that there was 4 times saving in irrigation water given through sprinklers (rain gun) as against traditional method of irrigation. The water use efficiency under traditional practice of flood irrigation was ranged from 7.01 to 10.4 kg/ha-mm as against 17.4 to 21.7 under demonstrations with sprinkler irrigation through rain gun.

Development and utilization of common lands

Development of a good vegetation cover in community lands within the watershed is very important as these common lands are often major source of runoff and soil erosion. Rehabilitation of such lands should be initiated with closure treatment against biotic interference, followed by controlled grazing and afforestation by community mobilization. Soil and water conservation practices recommended on such non-arable lands include, contour and staggered trenching and contour furrows to support afforestation activities.

Closure treatment against biotic interference

Protection against biotic interference pays on denuded lands. A ten years closure/protection against biotic interference at the research farm Datia increased the tree population having DBH >15cm by about 6.2 fold irrespective of the species (A forest watershed protected at the research farm from biotic interference could develop almost 100% cover dominant with *Annogeises* and *Butea species* over a period of 10 years (Tiwari and Sharma,2000) and produced 2.3 % runoff only as against 12.8 % from unprotected forest watershed (Tiwari *et al.*, 1999). Silvi-pasture practices combined with closure to biotic interferences followed in IWDP watershed project also increased the density (number /100 sq m) of trees (3.75) , shrubs (128 and grasses 3350 as against 0.5, 89 and zero in the area having open access to grazing (Tiwari, 2005).

Staggered contour trenches

Depending on land slope and tree species to be planted trenches are dug up at horizontal spacing of 1-3 m up to 30% slope and designed to store 60-70 % runoff. Usually sizes of a trench are kept 3.0 x 0.3 x 0.3 m but as a thumb rule number of trenches are kept equal to number of trees. While Sharda *et al.* (1982) reported that contour trenches reduced the runoff from 25% to almost nil, Bhusan, *et al.* (1986) found that staggered contour trenches were very much effective in checking gully head extension in a ravine system.

Productivity enhancement of existing roots stock

Rejuvenation of existing root stock viz. *deshi ber* etc may be taken up extensively for increasing biomass production (fruits, fuel and fodder). Out of the three species of local ber species found in the area having different growth habits *Zyziphus mauriatiana* can be rejuvenated by patch budding technique at any time during June to September with assured success of > 60% whereas early budding in the month of June/ July in case of *Zizyphus numularia* . and during June month in case of *Z izyphus oinoploea* was found more successful (Tiwari, 1997).

Our experiences with a community rocky patch of 28 ha pasture area in Bajni watershed showed that combined effect of silvi-pasture techniques, mechanical measures (contour trenches, gully plugs, CPT/ stone wall and protection could help in increasing the forage yield from 0.09 t ha^{-1} to 1.97 t ha^{-1} and vegetation status cover from 24.7% in 1999 to 47.9% in 2001. Watershed Eco-Index (WEI) worked out to be 0.10 for the watershed in 2002 suggested that an additional 10% of watershed area has rehabilitated through green biomass cover from the community land.

References

Bhusan, L.S., Tiwari, A. K. and Mishra P. R. 1986. Contour trenching-An effective measure to control erosion on slopping lands. In: *Proc. National symposium on problems of arid and semi-arid regions and their remedial measures* (February 8-9, 1986). Jodhpur pp83-85.

Bhusan, L.S., Singh, G. and Anderson, D.T. 1977. Efficacy of deep tillage on moisture conservation and yield of rainfed crops. *Soil Cons digest* **5**(1): 8-14.

Gurmel Singh, Shastry, G. and Bharadwaj, S.P. 1990. Watershed responses to conservation measures under different climatic regions of India. *Indian Jour. Soil Conservation.* **18**(3): 16-22.

Mangala Rai (2009). Resource Conservation in Indian Agriculture. Souvenir, 4th World Congress on Conservation Agriculture, held from 4-7 February 2009, New Delhi. pp. 82.

Mishra, P.K. and Tripathi, K.P. (2013) Soil and water conservation Research for land management in India. Indian J. Dryland Agriculture Research and Development. **28**(1): 1-18.

Narayan, D. (2005) Resource conservation technologies for sustainable crop production in Bundelkhand. In (Ed: Sharda et al) Resource conserving technologies for social upliftment, IASWC, Dehradun India 256-260.

Narayan, D. and Lal, B. 2006. Effect of green manuring on soil properties and yield of wheat under different soil depth in alfisols of central India. *Bulletin National Institute of Ecology* **17**: 141-146.

Nitant, H.C. and Singh, P. (1992) Studies on tillage practices in red soils Annual report 1991-92 CSWCRTI, Dehradun: 134-135.

Ram Babu, Shrivastava, M. M. Shastry, G. and Puri D.N. 1980. Studies on hydrological behavior of small watersheds under different land uses. Annual report, CSWCRTI, Dehradun, 227 p.

Sharda V. N. and Ojasvi, P.R. 2005. Water harvesting through watershed management in different agro-ecological regions of India. Indian. J. Agril. Science **75**(12): 771-780.

Samra, J.S. 2005. Resource conservation and social upliftment through participatory watershed management. In (Ed: Sharda et al) Resource conserving technologies for social upliftment, IASWC, Dehradun India 555-566.

Sharma, S. and Mishra, P.K. 1995. Watershed management in dry land agriculture: Principles and practices. In Sustainable development of dry land agriculture in India. Scientific Publishers, Jodhpur pp 491-510.

Sharma A.K. and Santraj 1995. Runoff, soil and nutrients losses under different crop covers on 2% slope at Datia. Annual Report CSWCRTI, Research Centre, Datia: 11-17.

Sharda, V.N. 2005. Conservation technologies for sustainability. In (Ed: Sharda *et al.*) Resource conserving technologies for social upliftment, IASWC, Dehradun India 169-178.

Samra, J.S. 1997. Status of research on watershed management. CSWCRTI, Dehradun (UK).

Tiwari, S.P. and Kurothe, R.S. 2006. Effect of vegetative barriers on soil and nutrients losses at 2% slope on agricultural lands of reclaimed Mahi ravines. Indian J. soil conservation 34 (1): 37-41.

Tiwari, S.P. 2005. Soil fertility and vegetation: a case study of degraded community waste land developed through peoples participation under IWDP in Gujarat. In (Ed: Sharda et al) Resource conserving technologies for social upliftment, IASWC, Dehradun India 710-714.

Tiwari, R.K. 1997. Studies on rejuvenation of *deshi Ber* plants and their management. Annual Report CSWCRTI, Dehradun: 193-195.

Tiwari, R.K. and Sharma, A.K. 2000. Natural regeneration of trees. Annual report CSWCRTI, Research Centre Datia (MP) pp 7-8.

Tiwari, S.P. and Sharma, A.K. 1994. Runoff, soil and nutrients losses under different crop covers on 2% slope at Datia. Annual report CSWCRTI, Research Centre, Datia (MP) pp 10-15.

□□□

Chapter 27

Agroforestry for Livestock Based Farming System: Planning and Management

S.K. DHYANI

ICAR - Central Agroforestry Research Institute (CAFRI) Jhansi-284003 (UP)

Introduction

Sustainable management of natural resources is vital as agricultural development with inclusive growth and long term sustainability cannot thrive on a deteriorating natural resource base. The country today, confronted with widespread land degradation, ground water imbalances, impaired soil health and contamination of food and environmental pollution etc. The situation is getting further compounded with the recent climate change impacts on agriculture (Handa *et al.*, 2015). Livestock are the integral part of Indian agriculture and play an important role by augmenting farm income particularly in drought prone areas (Dhyani, 2003). Tree fodders are able to provide the fodder in terms of the amount and the nutrients. It has been estimated that fallen leaves, flowers, fruits, seeds, etc. available in various agroforestry systems can make up to more than 50% of the complete feed for ruminants and about 10% of the feed of non-ruminants. There is a need to utilize tree fodder/feed optimally and in combination with other commonly used fodder/feed.

Presently, India faces a critical imbalance in its natural resource base with about 18 per cent human and 15 per cent of livestock population of the world being supported only by 2.4 per cent of geographical area and 1.5 per cent of forest and pasture lands and 4.2 per cent of water resources. It is estimated that India will be the most populous country in the world by 2028, with a population of 1.4 billion by 2025 and 1.6 billion by 2050.

Thus, there will be an increase of about 33 per cent in population, but the per capita surface water availability and per capita utilizable surface water will decline by about 30 per cent by 2050 as compared to present (NRCAF, 2013).

Land use based alternative diversified production system such as crop diversification, integrated farming system and agroforestry can address some of these challenges. In India rainfed agro-ecologies contribute 60 per cent of the net sown area and 66 per cent of the livestock. Rainfed areas are complex, varied, fragile, under-invested, risky, and distress prone, however, there is greater scope for diversification in these areas than in the irrigated areas. Farming System as a concept takes into account the components of soil, water, crops, livestock, labour and their resources with the farming at the centre managing agricultural and related activities and non-farm avocations. The farm family functions within the limitations of its capability and resources, the socio-cultural settings and the interaction of these components with physical, biological and economic factors.

To design appropriate farming systems suited to diverse farming situations, farmers' participation beginning from planning till execution of R&D programmes will be inevitable and they would actually participate in research and extension efforts as foremost partners. In rainfed arid ecosystems, integration of livestock and horticultural crops with annual crops, especially their organic production, is expected to emerge as a major agroforestry system.

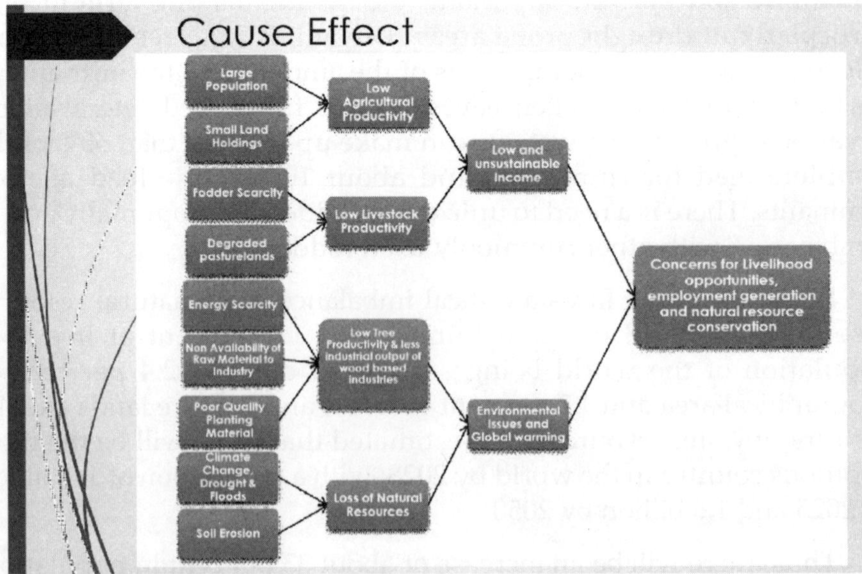

Fig. 1. Cause effect of challenges faced by agriculture (*Source:* Handa *et al.*, 2015).

Integrated Farming System (IFS) is considered to be powerful tool and holds the key for ensuring income, employment, livelihood and nutritional security in a sustainable mode for small and marginal farmers of the country who constitute more than two-third of the farming community. Integrated system meets the above goals through multiple use of natural resources such as land, water, nutrients and energy in a complimentary way thus giving scope for round the year income from various enterprises of the system. Research on IFS conducted over the decades, synthesis, philosophy and experiences ensures a monthly net income of Rs. 25000 ha^{-1} in irrigated and Rs 10,000 ha^{-1} in rainfed ecosystem can be obtained through appropriate combination of symbiotic enterprises.

Agroforestry has been defined variously but commonly it is a sustainable management system for land that increases overall production, combines agricultural crops, trees and forest plants and/ or animals simultaneously or sequentially and applies management practices that are compatible with the cultural patterns of local population. Agroforestry has potential for higher overall (biomass) productivity, soil fertility improvement, soil conservation, nutrient cycling and micro climate improvement. The carbon sequestration potential of an agroforestry system is generally greater than that of an annual system. Agroforestry has high potential for simultaneously satisfying three important objectives viz., protecting and stabilizing the ecosystems; producing a high level of output of economic goods; and improving income and basic materials to rural population. Besides, agroforestry is capable to conserve natural resources under different agro-climatic regions and is the only option to increase the forest cover in the country.

Need and Scope of Agroforestry

Agroforestry is an age old practice in India (Kumar *et al.,* 2012). At present, the farmers and land owners in different parts of the country integrate a variety of woody perennials in their crop and livestock production systems depending upon the agro-climatic conditions and local needs. Most of these practices are however, very location specific and their benefits have remained vastly under-exploited and un-extrapolated to other potential sites.

India has been at the forefront of agroforestry research (Dhyani *et al.,* 2008, Dhyani *et al.,* 2015). The Diagnosis and Design survey in different agro-climatic regions identified important agroforestry practices including livestock based systems in the country. Agroforestry systems are practiced in arable lands as well as on non-arable lands (Dhyani *et al.*, 2009). On arable lands growing of trees and woody perennials on terrace risers, terrace edges, field bunds, as intercrops and as alley cropping are common practices. Agroforestry practices for non-arable degraded lands such as bouldery riverbeds, torrents, landslide, shifting cultivation areas, waterlogged soils, control of desertification, mine spoil rehabilitation and treatment of saline and alkaline lands have been developed and demonstrated (Dhyani *et al.*, 2005). Agri-silviculture, agri-horti-silviculture, agri-horticulture, hortipastoral, silvipastoral and some other specialized combinations for specific conditions can be raised for livestock production with appropriate fodder species. The suitable forage, legumes and trees and shrubs for fodder purpose for different agro-climatic regions are presented in Table 1.

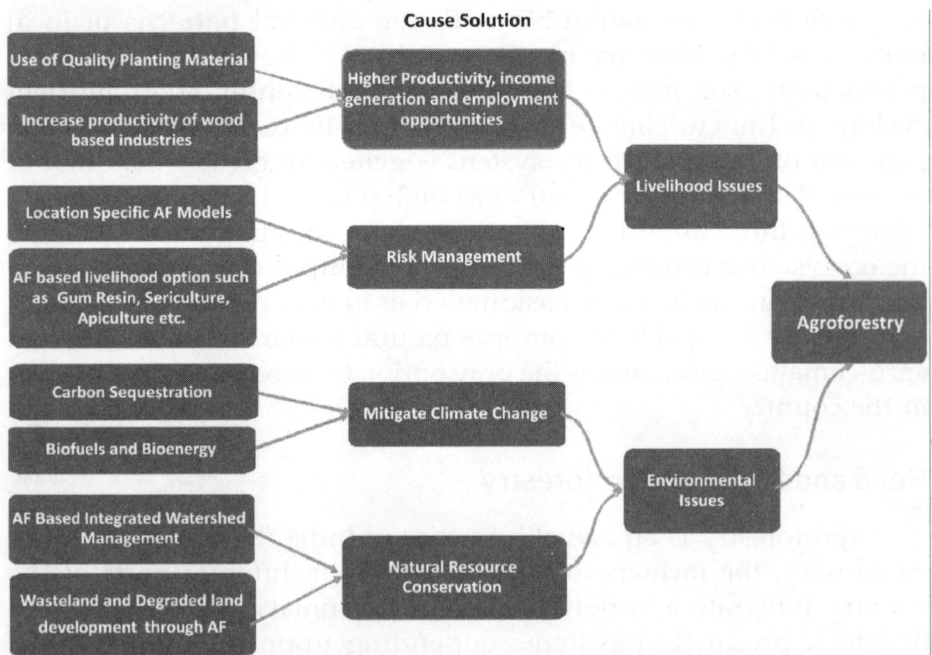

Fig. 2. Agroforestry as the solution to the challenges faced by agriculture (*Source:* Handa *et al.*, 2015).

Table 1. Soil Conservation Regions (SCR), their characteristics and suitable forage grasses, legumes and shrubs/trees for agroforestry[1].

SCR/Agroeco-logical region	State & Climate	Forage grasses	Forage legumes	Fodder shrubs/ trees
1	2	3	4	5
I. Glacier Region/1– W. Himalayas	Jammu & Kashmir (J&K),Cold arid	*Agrostis* spp., *Poa alpine, Trisetum spicatum*	*Medicago sativa*	*Hippophae rhamnoides, Ulmus wallichiana, Salix alba, Morus alba, Ailanthus*
II. Karewas Region/14–W. Himalayas	J&K, Himachal Pradesh, Punjab (North)-Cold semi-arid (dry) to humid and per humid	*Dactylis glomerata, Festuca rubra, Lolium perenne, Poa spp., Chrysopogon fulvus, Heteropogon contortus*	*Trifolium pretense, T.repens, Pueraria hisrsuta, Desmodium* spp.	*Quercus leucotrichophora, Robinia pseudoacacia, Grewia optiva, Fagus sylvatica, Celtis australis, Morus alba, Ficus, Melia*
III. Shiwalik Region/14-W. Himalayas	HP, Punjab, Uttarakhand, Semi-arid to humid and per humid	*Dactylis glomerata, Festuca rubra, Lolium perenne, Poa* spp., *Chrysopogon fulvus, Heteropogon contortus*	*Trifolium pretense, T. repens, Pueraria hirsute, Desmodium* spp.	*Q. leucotric-hophora, G. optiva, F. sylvatica, C.australis, Morus alba, Ficus spp., Bauhinia retusa, B. purpurea, Albizia* spp.
IV. Indo-gangetic Plain Region/4-N. Plains & Central Highlands. 9-N. Plains 13-E. Plains	UP, Bihar, Madhya Pradesh, Jharkhand, Chattisgarh, Sub-humic dry to sub-humic moist	*Brachiaria mutica, Bothriochloa intermedia, Cenchrus ciliaris, Chloris, Gayana, Cynodon dactylon, Dichhanthium annulatum,*	*Macroptelium atropurpureum, Stylosanthes hamata, S.scabra, Clitoria termatea, Arachis glabrata*	*Acacia nilotica, Albizia amara, A. lebbek, A.proc-era, A. stipulata, Bauhinia variegate, Dalbergia latifolia, D. sissoo, Hardwickia binata, Leucaena leucocephala,*

		Sehima nervosum, Paspalum notalum, Pennisetum pedicellatus		*Sesbania sesban, S. grandiflora, Ficus racemosa, Azadirachta indica, Pterocarpus marsupium*
V. Arid region- 2. W. Plains	Rajasthan, Haryana, Gujarat, Typic/arid to hyper-arid	*Cenchrus ciliaris, C. setigerus, L. sindicus, Panicum turgideum (Sand dune), Chloris gayama, Sporobolus marginatus (affected lands).*	*Cassia rotundifolia*	*Acacia nilotica, A. tortilis, Ailanthus excelsa, Dichrostachys cinerea, Prosopis cineraria, P. chilensis, Salvadora oleoides, S. persica, Ziziphus nummularia*
VI. West Alluvial Plain Region/2- Western Plains	Rajasthan (South-East), Gujarat, MP (North-West) Semi-arid dry to semi-arid moist	*Cenchrus ciliaris, C. setigerus L. sindicus* (sandy plains), *Panicum turgideum* (Sand dune), *Chloris gayama, Sporobolus marginatus* (Salt affected lands).	*Cassia rotundifolia*	*Acacia nilotica, A. tortilis, Ailanthus excelsa, Dichrostachys cinerea, Prosopis cineraria, P. chilonsis, Salvadora oleoides, S. persica, Ziziphus nummularia*
VII. Bihar Region/4- Northern Plains	UP, MP, Rajasthan, Semi-arid dry to moist	*Brachiria mutica, Bothriochloa intermedia, Cenchrus ciliaris, Chrysopogon fulvus, Chloris gayana, Cynodon dactylon, Dichhanthium*	*Macroptelium atropurpureum, Stylosanthes hamata, S. scabra, Clitoria ternatea, Arachis glabrata*	*Acacia nilotica, Albizia amara, A. lebbeck, A. procera, A. stipulata, B. variegata, Dalbergia latifolia, D. sissoo, Hardwickia binata, Leucaena leucocephala,*

		annulatum, Sehima nervosum, Paspalum notalum, Pennisetum pedicellatum		*Sesbania sesban, S. grandiflora, Ficus racemosa, Azadirachta indica, Pterocarpus*
VIII. Southern Malwa Region/10-Central Highlands	Gujarat, MP, Maharashtra, Semi-arid moist to sub-humic dry	*Andropogon gayanus, Pennisetum pedicellatum (red soil), Bothriochloea intermedia, Chrysopogon fulvus, Sehimi nervosum, Dichanthium annulatum*	*Atylosia scarabaeoides, Macroptehium atropurpureum, Stylosanthes hamata, S. scabra*	*Albizia amara, A. lebbeck, Anogeissus latifolia, A. pendula, Dichrostachys cinerea, Hardwickia binata, L. leucocephala*
IX. Plateau Region/6-Deccan Plateau7-Deccan Plateau Eastern Ghats	Maharashtra, Karnataka, Andhra Pradesh, Semi-arid dry to semi-arid moist	*Andropogon gayanus, Bothriochloea intermedia, Cenchrus setigerus, Chrysopogon fulvus, Dichanthium annulatum, Panicum maximum, Pennisetum pedicellatum*	*Atylosia scrabaeides, Macrotyloma axillarea, Macroptilium atropurpureum, Stylosanthes hamata, S. scabra*	*Acacia nilotica, A. lebbeck, A. procera, Anogeissus pendula, B. purpurea, B. retusa, Gliricidia sepium, Faidherbia albida, Holoptelia integrifolia, L. leucocephala, M. oleifera, Pterocarpus marsupium, Sesbania sesban, Terminalia arjuna*
X. Chalka Region/6-Deccan Plateau7-Deccan Plateau Eastern Ghats	Karnataka, Andhra Pradesh, Tamil Nadu, Arid (typic) to semi-arid moist	*Andropogon gayanus, Bothriochloea intermedia, Brachiaria decumbens, B. ruiziensis, Cynodon*	*Atylosia scrabaeoides, Macrrotyloma axillarea, Macroptilium atropurpureum, Stylosanthes hamata, S. scabra,*	*Acacia nilotica, Albizia lebbeck, A. procera, Anogeissus pendula, pendula, Bauhinia purpurea, B. retusa, Faidherbia*

		dactylon, Cenchrus setigerus, Dichanthium annulatum, Panicum maximum, Pennisetum pedicellatum	*Arachis, Clitoria tematea.*	*albida, Holoptelia integrifolia, L. leucocephala, Moringa oleifera, Pterocarpus marsupium, Sesbania* sp., *Terminalia arjuna, Ailanthus* sp., *Erythrina variegate, E. poeppigiana, Periserianthes falcataria, Tamarindus indica*
XI. Western Ghat Region/5-Central Highlands etc. 19-Western Ghats and Coastal Plain	Gujarat, Maharashtra, Kerala, Goa, Karnataka, Sub-humic to per humid	*Bothriochloea intermedia, Cynodon dactylon, Cenchrus ciliaris, Dichanthium annulatum, Panicum maximum, Pennisetum cladestinum, P. polystachyon, Setaria sphacelata*	*Arachis hagenbeckii, Clitoria tematea, Desmodium* spp., *Stylosanthes hamata, S. scabra*	*Acacia nilotica, Albizia lebbeck, A. procera, Anogeissus pendula, Bauhinia purpurea, B. retusa, Gliricidia sepium, Faidherbia albida, Holoptelia integrifolia, Leucaena leucocephala, Moringa oleifera, pterocarpus marsupium, Sesbania* sp., *Terminalia arjuna, Ailanthus malabarica, Erythrina variegate, Tamarindus indica*
XII. Central Eastern Upland/11-	MP, Orissa, AP, Sub-humid dry to	*Borthriochloa intermedia, Cynodon*	*Arachis hagenbackii, Stylosanthes*	*Bauhinia variegate, Dalbergia sissoo,*

Eastern Plateau	sub-humid moist	*dactylon, Dichanthium annulatum, Panicum maximum, Pennisetum pedicellatum, Setaria sphacelata*	*hamata*	*Leucaena leucocephala, Moringa oleifera*
XIII. Eastern Ghat Region/12-E. Plateau & E. Ghats 18-Eastern	Orissa, AP, Semi-arid moist to sub-humid dry	*Andropogon gayanus, Bothriochloa intermedia, Chrysopogon fulvus, Chloris Cynodon dactylon, Dichanthium annulatum Pennisetum pedicellatum, Urochloa mosambicensis*	*Atylosia scarabaeoides, Macroptelium atropurpureum, Stylosanthes hamata, S. guyinensis*	*Ailanthus malabarica, Artocarpus heterophyllus, A. lakoocha, Erythrina* spp., *Leucaena leucocephala, Moringa oleifera, Ficus* spp.
XIV. Diara Region/12-Eastern Plateau and Eastern Ghat	Bihar, Orissa, West Bengal (WB), Sub-humid dry to sub-humid moist	*Andropogon gayanus, Bothriochloa intermedia, Chrysopogon fulvus, Chloris gayana, Cynodon dactylon, Dichanthium annulatum, Pennisetum pedicellatum, Urochloa mosambicens*	*Atylosia scarabaeoides, Macroptelium atropurpureum, Stylosanthes hamata, S. guyinensis*	*Ailanthus malabarica, Artocarpus heterophyllus, A. lakoocha, Erythrina* spp., *Leucaena leucocephala, moringa oleifera, Ficus* spp.
XV. Sunderban & E. Valley Region/15-Bengal & Assam	Bihar (Eastern), WB, Assam, Nagaland, Meghalaya, Sub-humid	*Brachiana decumbens, B. mutica, Paspalum notatum, Setaria anceps*	*Desmodium heterophyllum, D. uncinatum, Vigna unguiculata, Phaseolus*	*Artocarpus heterophyllus, A. lakoocha, Ficus hookeri, Parkia roxburghii, Morus alba,*

Plains17-E. Himalayas	dry to per-humid		*calcaratus, Stylosanthes guyiensis, S. scabra*	*Sesbania sesban*
XVI. North Eastern Hill Region/16-Eastern Himalaya	Meghalaya, Manipur, Tripura, Mizoram, Humid to per-humid	*Goix lacryma-jobi, Pennisetum clandestinum, Tripsacum dactyloides*	*Desmodium* spp., *Centrosema pubescens, Stylosanthes guyiensis, S. scabra, Macroptelium atropurpureum*	*Dendrocalamus hamiltonii, Parkia roxburghii, Morus alba, Robinia pseudoacacia, Sesbania sesban*
XVII. Island Region/20-Islands of Andaman, Nicobar & Lakshdweep	Andaman & Nicobar, Lakshdweep, Humid to per-humid	*Andropogon gayanus, Cynodon dactylon, Cenchrus ciliaris, Pennisetum pedicellatum, P. polystachyon, Brachiana ruiziensis, Tripsacum laxum*	*Centrosema pubescences, Chiloria tematea, Macroptelium atropurpureum, Stylosanthes guyiensis, S. scabra*	*Bauhinia purpurea, Erythrina variegate, Gliricidia sepium, Leucaena leucocephala, Pithecellobium dules, Trema tomentosa*

[1]Modified from Dhyani (2003)

Agroforestry in the Indian Himalayas

Agroforestry is the most natural way of life in this region, considering the general physiography, vegetation, existing agricultural practices and the attitude of the people. The major benefit that agroforestry can bring to the sloping areas lies in its capacity to combine soil conservation with production function (Dhyani and Chauhan, 1994, Dhyani and Sharda, 2005). In the Himalayan region, agroforestry is being advocated for arable lands and non-arable degraded lands, bouldery riverbed lands, landslide and landslip stabilization, abandoned mine spoil area rehabilitation and as an alternative to shifting cultivation.

Much of this area can be reclaimed with agroforestry systems like silvi-pastoral for highly disturbed sites and horti-pastoral for moderate ones.

Fig. 3. Agroforestry soil basket (Source: Uthappa *et al.*, 2015).

The indigenous tribes like Lepcha and Limbu used to collect large cardamom from natural forests, which were later on domesticated. Alder (*Alnus nepalensis*) a non-leguminous nitrogen fixing tree is most abundant and preferred tree among the shade trees (Dhyani, 1998). The tree not only provides shade to arable crops, timber and fuelwood but also ameliorates the soil and protects it from erosion on hilly slopes. Besides alder, many trees growing on terraces or terrace risers are lopped for green fodder which is rich in crude protein and calcium. Different trees provide fodder at different times of the year sustaining the supply of fodder throughout the year. Among the introduced tree species, *Leucaena leucocephala* does well at low to mid hills and it could produce 9 t ha^{-1} dry foliage annually. Shade trees like *Albizia chinensis* has been reported to add 2.5 to 5.0 t ha^{-1} organic matter to the soil annually. Many species of bamboo, palms, coconut, arecanut and rattans (*Calamus* spp.) are cultivated widely in north-eastern areas as mixed or boundary plantations. Pasture in forests is also a common practice. In Sikkim on hilly areas, Nevaro (*Ficus auriculata*) based silvipastoral systems have shown production potential of 15.6-51.6 t ha^{-1} annually.

Agroforestry systems have been developed using local resources and conservation-based measures in the North Eastern Hill (NEH) region. Combining fine-root system of grasses and legumes, such as *Stylosanthes guyanensis, Panicum maximum, Setaria* etc. and deep-root system of fodder trees, such as alder (*Alnus nepalensis*) in a silvipastoral system stabilizes terrace risers and provides multiple outputs. Mulberry with frenchbean-groundnut followed by mustard is a profitable cropping system with mulberry, guava/lemon/pear and pineapple in paired rows and grasses on the bunds is an ideal system for silk production and additional income from fruits and cattle rearing (Dhyani *et al.,* 1996).

In the Western Himalayas, 60 to 70% requirement of the firewood and fodder is met from the trees/ MPTs growing along the bunds of agricultural terraces or scattered trees on the pasture lands. Hedge-row intercropping is feasible and important on sloping hilly lands when pruned biomass during cropping season can be used for fodder and fuelwood. Suitable grass, legume, shrubs and tree species suitable for fodder production in the region are presented in Table 1.

Indo-Gangetic Plains

This region contributes about 52 per cent to the national food grain output. Farmers use trees according to their need as well as suitability of the species. Many common trees such as *Azadirachta indica, Acacia nilotica, Dalbergia sissoo, Prosopis cineraria,* and others are found grown very frequently on farm lands particularly along crop-field boundaries for meeting the fodder requirements. In sub-montane zone of Punjab, farmers plant *Leucaena* either in block or boundary plantation. In south-west Punjab, farmers prefer to grow *Melia* and *Eucalyptus* as boundary plantation.

A sizable area in the Indo-Gangetic plains is salt-affected. A package of afforestation techniques has been developed by the Central Soil Salinity Research Institute (CSSRI), Karnal. To rehabilitate alkali soils, auger-hole technique is used for planting the saplings of salt tolerant species using FYM, gypsum, $ZnSO_4$ and insecticide powder for controlling termites. In long-term experiments it has been found that on highly alkali soil (pH>10) *Prosopis juliflora, Acacia nilotica* and *Tamarix articulata* can be grown with success. On soil of moderate alkalinity (pH 9-10), species like *Parkinsonia aculeata, Terminalia arjuna, Pithecellobium dulce,* aonla (*Emblica officinalis*), guava (*Psidium guajava*), ber (*Zizyphus mauritiana*), Karonda (*Carissa carandas*), Jamun (*Syzygium cumini*) and pomegranate (*Punica granatum*) can successfully be grown applying higher doses of gypsum and FYM. These species may be blended with forage grasses like *Leptochloa fusca, Brachiaria mutica* and *Chloris gayana.* On saline soils furrow-method of planting trees has been found successful. *Acacia nilotica, A. farnesiana, A. tortilis, Prosopis juliflora, Parkinsonia aculeata, Casuarina glauca, C. equisetifolia,* and *Tamarix troupii* have been found most suitable.

Seepage from unlined canals results in waterlogging and salinization. A wide belt of fodder trees such as *Syzygium cuminii, Pongamia pinnata, Terminalia arjuna, Dalbergia sissoo, Acacia nilotica* and other MPTS on both sides of the canal may be created to utilize the seepage water for biomass production. Perennial grasses *like Brachiaria mutica, Coix lachryma-jobi,* species of *Paspalum, Echinochloa colonum* and *Leptochloa fusca* may be grown in waterlogged areas for forage production. In ravine lands, trees like *Acacia nilotica, A. catechu, Dalbergia sissoo, Azadirachta indica* and *Pongamia pinnata* are most effective in association with forage grasses like *Dichanthium annulatum, Bothriochloa pertusa, Cenchrus ciliaris, Chrysopogon fulvus, Cynodon dactylon, Sehima nervosum* and *Panicum* spp. (Table 1).

Agroforestry in Humid and Sub-humid Region

In humid and sub-humid regions, *Albizia* spp., *Gmelina arborea* and *Gliricidia* are the important fodder trees for different agroforestry systems. In most of the region, tree component is used along with livestock and poultry component, whereas in irrigated areas fish component is also incorporated. Agrisilviculture is common in Jharkhand and Chattisgarh. *Acacia nilotica, Terminalia arjuna, Butea monosperma, Albizia* spp. are grown in Chattisgarh while *Zizyphus mauritiana, B. monosperma, Aegle marmelos, Mangifera indica, Schleichera oleosa* (Kusum) in Jharkhand. Homestead Agroforestry is also being practiced using *Gmelina arborea, Artocarpus heterophyllus, Madhuca latifolia, Zizyphus mauritiana* etc. In Odisha, agri-silviculture (*Cocos nucifera* for boundary plantation, block plantation of *Casuarina equisetifolia, Anacardium occidentale*) and homesteads are commonly practiced. In this region, *Leucaena leucocephala, Pongamia pinnata, Acacia nilotica* and *Dalbergia sissoo* are important component of silvipasture systems (Table 1).

Agroforestry in Arid and Semi-Arid Regions

Agroforestry is playing the greatest role in maintaining the resource base and increasing overall productivity in the arid and semi-arid regions. Agroforestry has an important role in reducing vulnerability, increasing resilience of farming systems and buffering households against climate related risks. It can also maintain or increase the supply of ecosystem services - water, soil health and biodiversity especially under climate change scenario that will continue to support social and industrial needs and sustainable functioning of ecosystem services (Dhyani, 2012).

The Central Agroforestry Research Institute, Jhansi developed and demonstrated a number of agroforestry systems in the farmers' field and three watersheds in the arid and semi arid regions. The important species suitable for agroforestry in the region are *Acacia nilotica, A. tortilis, Albizia lebbek, A. procera, Bauhinia variegata, B. purpurea, B. retusa, Dalbergia sissoo, Hardwickia binata, Leucaena leucocephala, Sesbania sesban, S. grandiflora, Ficus racemosa, Azadirachta indica, Ailanthus excelsa, Prosopis cineraria, P. chilensis, Salvadora oleoides, Gliricidia sepium, Moringa oleifera, Terminalia arjuna, Erythrina variegata, Anogeissus pendula, Tamarindus indica* and fruit trees guava, ber, bael and aonla, and biofuel species like *Jatropha* and *Pongamia.* Furthermore, *Ailanthus* and *Prosopis* based farming systems are important for livelihood security of the farming and rural community of semi-arid region of Gujarat and Rajasthan. These tree species not only meet the fodder requirement during lean period but also provide extra returns in cash due to industrial importance of *Ailanthus* as soft wood and *Prosopis* for higher calorific and nutritional value. *Acacia nilotica* has been identified and demonstrated as one of the most important tree species for tree based farming in rainfed regions of Madhya Pradesh and Chhatisgarh (Table 1).

Agroforestry Systems for Coastal and Island Regions

Low lying water logged marshy areas, flood plains, and ill-drained lands are the common features in the coastal areas swamps and river banks are occupied by the mangroves and associate halophytes. Other natural vegetation includes evergreen, semi-evergreen and deciduous forests on uplands and grazing lands in pockets. Plantation crops integrated with livestock and poultry and rice fields are main features of this region.

The systems and practices of agroforestry range from apparently 'simple' forms of shifting cultivation and farming in forests to sophisticated hedgerow intercropping systems, from systems involving sparse stands of trees on farm lands to high density, complex multistoreyed homegardens of lowland; and from systems in which trees play a predominantly 'service' role (e.g. shelterbelts) to those in which they provide the main commercial product (e.g. intercropping with plantation crops). Most of the systems are site-specific. These areas may be brought under multistoreyed cropping systems and forage grasses like hybrid napier (*Pennisetum purpureum*), kazungula (*Setaria anceps*) and guinea (*Panicum maximum*) and legumes like *Stylosanthes guianensis* may be grown as cover crops. Rubber and red oil palm plantations may also be integrated with forage crops particularly the legume covers.

Livestock based agroforestry systems: Planning and management

The adoption of livestock based agroforestry system depends on the climatic region, geographical features, demographic characteristics and socio-economic situation. In the humid areas, preference has been towards forest products dominating timber and wood lots. However, in the sub-humid tropics where deciduous trees and plantation trees allow sunlight penetration for growth and production of ground story, silviculture and agri-silviculture are commonly adopted. In the semi-arid tropics, where water is limiting, particularly during the dry season, agri-silviculture is less common, while silvipasture is more common. Since crop residues form the major portion of animal feed, the integration of livestock rearing in farming systems is common. Substantial research work on rehabilitation of degraded lands has been carried out in India which includes adoption of various agroforestry systems such as silvi-pasture, agri-silviculture, hortipastures, agri-silvi-pasture and agri-silvi-hortipasture in culturable and non-culturable lands and rehabilitation of degraded and mine spoil lands in a specific habitat (Dhyani *et al.*, 2005).

In agroforestry, aspects related to tree and species selection, planting geometry, perenniality of the system, land use as per the topography, soil and water conservation, and soil fertility is important. Tree-crop interactions are a key consideration when designing an agroforestry system and a layout for planting. Basic principles included is how to increase the overall value of the system, maximize complementarities, decrease or eliminate competition; and minimize crop displacement, through appropriate tree management. It has been possible to increase land productivity from less than 1.5 t $ha^{-1}y^{-1}$ to about 10 t $ha^{-1}y^{-1}$ (10-year rotation) by developing silvipasture systems. The additional forage availability through these systems is likely to reduce grazing pressure and thus have important environmental implications. A brief description of various agroforestry systems for livestock production which can be part of the farming systems is presented below.

Silvi-pastoral systems

Silvipastoral system which is mixed land - land use system for forage, livestock and wood production has paramount significance owing to forage supply from cultivated, marginal and degraded lands. The mixture of trees, grasses and legumes conserve soil and moisture. Establishment of successful silvi-pastures includes- selection of suitable species of trees, shrubs, grasses and legumes for different regions, adoption of measures for protecting the area from animals, suitable soil and water conservation measures and planting of tree and under storey components. Silvipasture system with trees such as *Acacia, Leucaena, Albizia, Melia,* and grasses *Cenchrus, Chrysopogon, Panicum, Pennisetum, Dichanthium* and legumes *Stylosanthes hamata* and *Macroptilium* are becoming very popular with the farmers (Table 1).

Horti-pasture system

Hortipasture system integrate livestock husbandry with fruit/ plantation crops with main aim to enhance biomass productivity of degraded lands and in turn helps in increasing the livestock production. It is an alternate land use system in arid and semiarid regions and may potentially support livelihood improvement through simultaneous production of fruit, fodder and firewood utilizing Class IV and V type of land. Animals also form part of hortipastoral system starting after three-year growth of the system. Hortipasture system supplies the protective food (fruit) for human being and fodder for animal and thus helps in bridging the wide gap between the supply and demand of fruit and fodder. *Anacardium occidentale, Annona squamosa, Emblica officinalis, Psidium guajava* and *Zizyphus mauritiana* along with grasses and *Stylosanthes* can be grown in wastelands under such hortipastoral systems. Production of grasses and legumes has been found to be between 2.5 and 6.0 t ha^{-1} which sustain livestock industry.

Agri-silviculture

In agri-silvicultural systems, MPTS (multi - purpose tree species) including fodder trees can be grown in association with field crops. Trees are normally pruned annually yielding fodder as well as fuelwood. In arid and semi-arid regions, hardy species like *Prosopis cineraria, Tecomella undulata* (Rohida), *Eucalyptus* sp., *Acacia tortilis, A. nilotica, Hardwickia binata, Azadirachta indica, Ailanthus excelsa, Zizyphus jujube,* etc. could be grown along with dry land crops. In Rajasthan, Gujarat, Haryana, Khejri and Rohida are grown along with crops of pearl millet /sorghum, moth, sesame, etc. without affecting crop production. The timber-cum-fodder trees produced fuel, fodder and timber besides the crop production. Such systems on arid and semi-arid areas have been considered highly economic in long run and also provide better employment opportunities.

Alley cropping

Alley cropping or hedgerow intercropping is an agroforestry practice in which perennial, preferably leguminous trees or shrubs are grown simultaneously with an arable crop. The trees, managed as hedgerows, are grown in wide rows and the crop is planted in the interspace or 'alley' between the tree rows. During the cropping phase the trees are pruned and the prunings used as green manure or mulch on the crop to improve the organic matter status of the soil and to provide nutrients, particularly nitrogen, to the crop. In the semi arid regions of India, the main benefit of alley cropping is fodder production during the dry season.

Forages on bunds

Grass, ground legumes, fodder shrubs and trees may be grown on the bunds, on the sloping land between terraces and along field boundaries. All of these constitute feed for dairy cattle. Farmers with marginal and small land holdings and livestock keepers with limited forage requirement can effectively grow forages on bunds. Potential forages for these niches are Napier - Bajra hybrid (irrigated areas); Tri-specific hybrid, Guinea grass and *Setaria* (rainfed areas) and fodder trees as per the agro-climates (Table 1).

Forage under plantation crops

Forage production under plantation crops has given encouraging results from different agro-climatic regions particularly in coastal belts of India. In Kerala, growing of forages under partial shade of plantation crops such as coconut, arecanut, etc. is a viable option. Guinea grass and other shade tolerant grasses like congosignal grass, signal grass provides unique opportunity for quality fodder production under the trees.

Trees in improved pasture

Improved pastures can be achieved by introducing fertilizer trees into grazing lands. This increases soil nitrogen, provides ground cover and keeps soils cooler, moister, less eroded and less compacted; allowing stressed systems to recover.

Forage from wastelands

In India, it is estimated that 120 million ha are degraded lands. Development of these lands for forage production will not only ensure enhanced supply of superior quality forage but also help in conserving the natural resources and recharging ground water, while improving the bio-diversity. There are many hardy grasses and legumes, shrubs and trees like *Panicum maximum*, *Cenchrus* species, *Dichanthium annulatum, Chrysopogon fulvus, Heterpogon contortus*, stylo, siratro, hedge lucerne, *Desmodium* spp., *Leucaena, Propsopis cineraria, Hardwikia binnata* etc. which can be grown on wastelands under rain fed situation. Shrubs and trees can be lopped regularly as tree fodder. Such tree species can be established on field bunds, home gardens and along farm boundaries in different agroclimates (Table 1).

Multi-storey systems

A typical mutli-storey represents an operational farm unit which integrates trees with field crops, livestock, poultry and/or fish and has the basic objective of ensuring sustained availability of multiple products for both consumption and sale. In Kerala such multistory are Homegardens which typically have a high degree of biodiversity as the result of generations of selection by farmers and natures responses to those choices. Depending on the space available (between coconut palms), one or two rows of multipurpose trees can be accommodated in the middle (spacing 1-2 m between plants). Tree management such as lopping, pollarding etc. is important to prevent any possible interspecific competition between the multi-purpose tree component and the coconut palms.

Fodder banks

Trees and shrubs with palatable leaves and / or pods are attractive to farmers as feed supplements for their livestock because they require little or no cash for inputs: they can be grown on boundaries as trees (often pollarded to reduce competition) or as hedges. They effectively do not compete for land as they are grown along boundaries, pathways - and along the contour to curb soil erosion. Managing fodder shrubs requires multiple skills including raising seedlings in a nursery, pruning trees, and feeding the leaves.

Choice of species and planting

Choice of MPTs for livestock based agroforestry is most important criteria. There may be many situations other than normal edaphic one where agroforestry practices are adopted. These may include soil salinity, acidity, marshy, waterlogged, calcareous and degraded lands. Salinity and acidity affects the germination and initial establishment of seedlings. Tolerance of MPTS and grasses and legumes differ remarkably. Some of the species adopt well under adverse growing environment. The following considerations should be kept while selecting the species under agroforestry.

- Suitability for growing under the prevalent agro-climatic conditions.
- Utility of trees [MPTS] for meeting the needs of fodder and other produce.
- Preference may be given to indigenous and fast-growing, and leguminous and other nitrogen-fixing species.

When the fodder trees are introduced in arable lands, then the primary concern is to keep up the production level of the arable crops. In that situation the following criteria are worth considering for planting trees under the agroforestry system:

- Non interference with arable crops;
- Easy establishment;
- Fast growth and short gestation period;
- Non-allelopathic effects on arable crops;
- Ability to fix atmospheric nitrogen;

- Easy decomposition of litter;
- Ability to withstand lopping; and
- Multiple use and high return.

Planting time & techniques

Under rainfed situation, planting should be done just before, or after the advent of first rains. Small seeds which must be planted close to the soil surface, need the moisture and warmer temperature which usually occur with the initial rains for germination. On prepared seedbeds, drilling seeds or planting in farrows is more effective than broadcasting because the seed is placed at the right soil depth. Planting arrangement should be made considering the particular space, the tree occupies in land use system and total number of trees of particular species is required to be planted. Planting density is higher in sole stands for single species. The density is adjusted in intercropping situation keeping at least 50-60 % of main crop. The plant to plant spacing of grasses is normally 50 cm in a row. There may be strip planting of grass and / or legumes at 100 cm apart.

Planting systems in different land use

On the basis of nature of components, common agroforestry systems where livestock can be integrated in the whole system partially or fully, fodder trees as MPTS, shrubs and grasses and legumes are invariably planted. In India, in intercropping systems, shrubs and trees are grown along with cash crops in the form of windrows or scattered among the cash crop plants. The shrubs and trees are lopped in regular interval to prevent overshading of the cash crops. In the undulating topography, some farmers grow shrubs instead of trees, on the bunds of embankment fields. *Sesbania grandiflora* is the shrub most commonly grown on the bunds of rice fields of eastern India, loppings of which are used as green manure and fodder as well. In the coastal areas coconut, arecanut, banana, jackfruit and mango are commonly grown in the field boundaries. In semi arid region, *Acacia* spp., *Ficus* spp. and in arid region *Prosopis cineraria* are the common species found in India supporting the fodder requirements of the livestock. In fenced systems, farmers generally use live fences (*Leucaena, Sesbania, Acacia*) to mark land boundaries and for livestock yards.

Livestock based agroforestry systems- some examples

Silvipasture provides a great opportunity for sustainable livestock production, besides having other added advantages. Silvipasture systems have been developed for different agroclimatic conditions. Some examples are given here.

- In most of the arid region of the north-western India, Khejri (*Prosopis cineraria*) based system is prominent. The second choice is *Zizyphus nummularia. Z. nummularia* is used for its leaves as fodder for camel and goats and berries for edible purposes. *Acacia tortilis, A. nilotica, A. senegal, A. leucophloea, Capparis decidua, Tecomella undulata, Salvadora persica and S. oleoides* are other common trees found on various grazing lands or as sand stabilizers (Table 1).
- *Lasiurus sindicus* and *Cenchrus ciliaris* are prominent grasses in grazing fields as well as in sand dunes. Besides above native species many woody perennials such as *Acacia albida, Hardwickia binata, Colophospermum mopane, Holoptelea integrifolia* and *Zizyphus rotundifolia* may form the constituents of silvi-pastoral or agrisilvicultural systems in these regions (Table 1).
- In Arid regions of Western India, introduction of compatible fuel, fodder and fruit trees in pastures as in agricultural fields acts as an insurance against frequent crop failures, and trees play a pivotal role towards peoples' survival and sustenance in such fragile ecosystems (CAZRI, Jodhpur).
- In Central India, *Dalbergia sissoo, Acacia nilotica, A. eburnea, A. leucophloea, A. catechu, Albizia lebbeck, Azadirachta indica, Butea monosperma, Pongamia pinnata, Holoptelea integrifolia, Balanites roxburghii* and *Dichrostachys cinerea* form important constituents of a silvi-pastoral system (CAFRI, Jhansi).
- In black cotton soil region very tall grasses supporting silvipasture system include *Heteropogon contortus, Chrysopogon fulvus, Themeda triandra, Iseilema laxum* and *Dichanthium annulatum* (IGFRI, Jhansi).
- A silvipasture system with bamboo and Anjan Grass is suitable for gullied lands in Mahi Ravines. The technology involves raising *Dendrocalamus strictus* (lathi baans) plants on gully bed and anjan grass on gully slopes and the interspaces of gully bed planted with bamboo. The grass yield from inter spaces of bamboo

planted on gully bed portion is about 10 t $ha^{-1}y^{-1}$ during initial years (CSWCRTI, Regional Centre at Vasad, Gujarat).

- In semi arid regions of Peninsular India, *Leucaena leucocephala* has been adopted as a common hedge-row-crop in many areas (CAFRI, Jhansi).

- The use of fodder trees as shelterbelts in areas that experience high wind or sand movement is well-established example of microclimate improvement that resulted in improved yields. Experimental shelterbelts of *Acacia nilotica* and *Dalbergia sissoo* were established over 120 km in square blocks at Jodhpur (Central Mechanized Farm of CAZRI, Jodhpur at Suratgarh in Bikaner District, Rajasthan).

- Silviapsture consisting of *Albizia amara* and *Leucaena leucocephala* as trees and *Dichrostachys cinerea* as shrub, *Chrysopogon fulvus* as grass, *Stylosanthes hamata* and *S. scabra* as pasture legumes and natural *Sehima – Heteropogon* grassland. The results indicate that under silvipastoral systems more number of sheep and goats can be reared with better performance as compared to natural grasslands due to availability of nutritious herbage from tree leaves and pasture legumes (CAFRI, Jhansi).

- Silvipasture with two tier (*Cenchrus ciliaris* + *Aialnthus excelsa*) and three tier (*C. ciliaris* + *D. nutan* + *A. excelsa*) systems were found very viable for small ruminants and results showed that 3 and 2 tier silvipastures were better than single *C. ciliaris* and rangeland for sheep production (CSWRI, Avikanagar, Rajasthan).

- In a silvipasture system comprising *L. leucocephala* + *A. tortilis* + *H. binata* + natural *Sehima – Heteropogon* grassland interspersed with native shrubs, the Barbari goat was not found suitable breed and local and Sirohi goats proved superior to it (IGFRI, Jhansi).

- In the Shiwalik foothills, silvipasture systems involving *Eulaliopsis binata, Saccharum munja* or *Vetivaria zizanoides* with *Acacia nilotica* were promising. In the southern hilly region, a silvipastoral combination of *Acacia melanoxylon* and *Digitaria decumbens* on steep slopes was promising for soil conservation and meeting the fodder requirement. In bouldery riverine land *Dalbergia sissoo* + *Chrysopogon fulvus; Acacia catechu* + *Eulaliopsis binata; Eucalyptus* hybrid+*Chrysopogon fulvus;* and *L. leucocephala* +Napier / *Panicum maximum* and a combination of *Bauhinia purpurea* or *Albizia lebbek* in association with *Eulaliopsis binata* was identified as the most promising silvipasture systems (CSWCRTI, Dehradun).

- From a silvipastoral systems in shalllow red chalka soils of Andhra Pradesh comprising *L. leucocephala* (K-8) and *Cenchrus ciliaris* and *Stylosanthes hamata* on an average, 6 t ha^{-1} y^{-1} DM yield was obtained from *S. hamata* and 3 t ha^{-1} y^{-1} from *C. ciliaris*, besides, 6 t ha^{-1} y^{-1} (biomass) from *L. leucocephala* (CRIDA, Hyderabad).
- A silvipastoral model comprising *Prosopis chilensis* and Karnal grass (*Leptochloa fusca*) together has proved highly successful for the development of community lands in Haryana. In this model, pahari kikar and Karnal grass are grown together for nearly 5 years. Later, when the surface soil is reclaimed, the grass is replaced by better forage such as shaftal (*Trifolium resupinatum*), berseem (*Trifolium alexandricum*) and senji (*Melilotus parviflora*). After 8-10 years of growing of trees and forage, the soil is reclaimed to an extent that a normal arable crop can be raised either after removing the trees or in association with trees (CSSRI, Karnal).
- Leaf fodder from the trees and shrubs is exceptionally rich in essential nutrients like crude protein, N-free extract, Ca, P, etc. Leaf fodder from *Leucaena, Sesbania grandiflora, Prosopis cineraria,* etc. has better nutrients than even Berseem and Lucerne. *Leucaena,* has very high nutritive value and is regarded as excellent leaf fodder species. The leaf fodder of *P. cineraria* (Khejri), *A. nilotica* (Babul), *Salvadora oleoides* (Pilujal), *Salvadora persica* (Kharajal) and *Zizyphus nummularia* (Bordi) was also quite palatable (good).
- Silvipastoral system comprising *Alnus nepalensis*, pineapple and forage crops like *Panicum maximum* or *Setaria sphacelata* coupled with *Stylosanthese guyanensis* in 1:1 ratio was found to be a sustainable agroforestry practice in soils having 30-60% slope in Meghalaya. Forage yield of 13.5 t ha^{-1} was obtained from the combination of Stylo and Setaria. In addition, 2.3 t ha^{-1} litters from *Alnus nepalensis* and 4000 fruits ha^{-1} from pineapple were obtained. This system also restored the fertility of the soils (Chauhan *et al.,* 1993).

Soil fertility enhancement and income generation

Plantation of compatible and desirable tree species particularly the Fertilizer/Nitrogen-fixing trees on farmland results in an improvement in soil fertility. There are several possible mechanisms of this as shown by Uthappa *et al.* (2015).

In India, nearly 120.72 million ha land or 37 per cent of the total geographical area is under one or the other forms of soil degradation (e.g., water erosion: 93 million ha, wind erosion: 11 million ha, salt affected soils: 6.74 million ha, and 16.53 million ha of open forest area; ICAR, 2010). Trees are known to grow in areas polluted by heavy metals and other hazardous industrial chemicals. In fact, there are trees which can absorb and tolerate such pollutants which not only reduce crop yields but also impair quality of crop produce. These areas can be brought under cultivation through biological amelioration.

The role of trees in soil conservation and erosion control is one of the most widely acclaimed and compelling reasons for including trees on farm lands prone to erosion hazards. The influence of trees on hydrological characteristics can extend from the microsite to the farm and regional levels. The primary objective of soil conservation is improved/maintained soil fertility. To achieve this, control of erosion, maintenance of organic matter and physical properties, organic matter addition, maintenance of nutrient is necessary. In this way agroforestry system constitute sustainable land use and helps to improve soils in a number of ways. Some of these beneficial effects are apparent in experiments carried out in different parts of the country. The results clearly indicate that in agroforestry system soil organic carbon and available nutrients are increased than growing sole tree or sole crop. The best example is from research conducted at CAZRI where increase in amount of N and P and other macro and micro-nutrients are reported in 14 -yr old plantation of Khejri (*Prosopis cineraria)* as compared to a bare site of *Prosopis juliflora* of the same age and forage species produce higher biomass under Khejri tree canopy due to a high fertility status.

Traditional agroforestry systems in north-eastern region indicated a spectacular increase in soil pH, organic-C, Ca, Mg, K, and P within 10-15 years of practice and the exchangeable Al, potential cause of infertility of these lands disappeared completely (Singh *et al.*, 1994). Similar results were obtained when multipurpose trees were evaluated in an extremely P-deficient acid Alfisol in Meghalaya (Dhyani *et al.*, 1994).

By alternate land use systems like silvipasture, agrisilviculture and agrihorticulture, the productive and protective benefits from watershed management (WSM) are considerably higher than the investment. Adoption of agroforestry based integrated soil and water conservation measures enhanced ET to 62% as compared to 53% in untreated (control) watershed receiving 815 mm annual average rainfall, reduced runoff (22% vs. 39%) along with increased base flow (4.5% vs. 1.2%) and groundwater recharge (11% vs. 7%) of total rainfall received as compared to untreated watershed (NRCAF, 2012). Economic Water productivity and total income increased from 3.7 to 5.9 m^{-3} and Rs. 17000 to 33000 $ha^{-1}yr^{-1}$ within six years, respectively. Moreover the interventions helped in reducing soil loss and N transport by 50% (Singh *et al.*, 2014).

Inclusion of trees and woody perennials on farm lands in the long run not only result in marked improvements in the physical conditions of the soil, but also provide economic returns. It has been found that the cost: benefit ratio under various silvipasture systems is more than 1.5. This is evident from the economic analysis of the silvipasture systems e.g. bamboo based (1.76) in Middle Gangetic Plain Region, *Hardwickia* based (2.13) in Eastern Plateau & Hill Region, *Albizia* based (1.56) in Western Plateau & Hill Region, and *Ziziphus nummularia* based (1.53) in Western Dry Zone (Planning Commission, 2001). A silvipasture system with bamboo and Anjan grass in gullied lands of Mahi ravines resulted in a cost benefit ratio of 1: 1.85 (CSWCRTI, Regional Centre at Vasad, Gujarat). IGFRI recently reported that Aonla- *Dicanthium* based horti-pasture has higher C:B ratio for hortipasture (1: 3.70) than pure pasture (1:1.85). Similarly, net income from fodder trees and other components of intensive integrated farming systems (IIFS) from the north-eastern region (Ngachan *et al.*, 2015) indicates the viability of the models (Table 2).

Table 2. Net income (Rs/ha/yr) for various units of IIFS in NEH region.

Models	Net Income (Rs/ha/yr)	B/C ratio (including labour)	B/C ratio (excluding labour)
Crop-fish- poultry-duck-MPTs-**hedgerow**	65,306	1.58	2.24
Crop-fish-poultry-**MPTs**	31,573	1.51	2.12
Crop-fish-goat-**bamboo-MPTs**	33,735	1.55	2.40
Crop-fish-pig-**hedgerow-MPTs**-vermiculture	34,276	1.42	1.86
Crop-fish-dairy-mushroom-liquid manure-broom - **MPTs** – vermiculture	1,21,634	1.76	2.48
Crop and fishery (without integration)	13,965	1.09	1.50

Carbon sequestration

Tree components in agroforestry systems can be significant sink of atmospheric carbon (C) due to their fast growth and high productivity. In agroforestry systems, tree component is managed, often intensively, by pruning for minimizing competition and maximize complementarity. The pruned materials are mostly non timber products. Such materials often are returned to the soil. In addition, the amount of biomass and therefore C that is harvested and exported from the system is relatively low in relation to the productivity of the tree. Therefore, unlike in tree plantations and other monoculture systems, agroforestry seems to have unique advantage in terms of C sequestration.

In India, evidence is now emerging that agroforestry systems are promising land use system to increase and conserve aboveground and soil C stocks to mitigate climate changes (Dhyani *et al.*, 2009). Average sequestration potential in agroforestry in India has been estimated to be 25 t C ha^{-1} over 96 million ha (Sathaye and Ravindranath, 1998). In another estimate agroforestry contributes 19.30% of total C stock under different land uses. The potential of agroforestry systems as C sink varies depending upon the species composition, age of trees, geographic location, local climatic factors and management regimes. The growing body of literature indicates that agroforestry systems has the potential to sequester large amounts of above and below ground C in addition to SOC enhancement, as compared to treeless farming systems (Ajit *et al.*, 2013).

Policy issues

Agroforestry has emerged as a major land use activity in the country after agriculture and forestry. However, with so much contribution and potential, the agroforestry on account of its dependency on multi-institutions and multi-disciplinary approach, has not gained momentum in a significant way. After years of discussion and deliberations at different quarters a consensus emerged that India needs urgently a policy on agroforestry (Dhyani and Handa, 2013). On 10th February, 2014 while inaugurating the 3rd World Congress on Agroforestry at Vigyan Bhawan, New Delhi, President of India Sh. Pranab Mukherjee launched the India's National Agroforestry Policy. The National Agroforestry Policy 2014 has the potential to substantially reduce poverty in rural India by creating employment in wood based industry, providing livelihood opportunities, help in conservation of natural resources and mitigation of c;limate change effects.

New Initiatives

There is no specific scheme or programme to promote agroforestry in the country. However, there are number of schemes of Government of India, state governments and other organizations where agroforestry is recognized as a component and being promoted as such. Some of the schemes and programmes in which agroforestry is a component are **Integrated Watershed Management Programme (IWMP)**, Mahatma Gandhi NREGA (MoRD), Soil Conservation in the Catchment of River Valley Project and Flood Prone River, National Horticulture Mission, National Bamboo Mission and National Mission on Medicinal Plants (DAC, MoA) to name a few. In XI and XIIth Plan, under Central sponsored **Tribal Sub Plan (TSP)**, funds have been allocated to different programmes including agroforestry for securing livelihood and creating employment opportunities for the tribal masses. Under the Scheme the tribal population provided with technical know-how along with inputs for adopting agroforestry for higher economic returns. Around 1.74 million ha forest land has been allotted to 4.66 million tribal and other forest dwellers under the aegis of FRA 2006. Agroforestry will play an important role to ensure regular income to the land allot tee. Thus, roughly on average nearly Rs. 2000 crores (or more) is being incurred annually on tree planting/agroforestry through number of Central Sector Schemes and programmes (Dhyani *et al.*, 2013).

Many states in the country have brought a number of innovative schemes for plantations. ***Lok-Vaniki***, an agroforestry program is successfully implemented in Madhya Pradesh for last few years. Similar innovative schemes are being implemented by a number of other states. However, it is not always that the area covered is private lands or farmer's field. In many situation the plantation are being raised in common lands and other areas as well. In recent years, BAIF (www.baif.org.in) in association with NABARD through its innovative Wadi (a horti-forestry orchard raised by tribal family on sloping uplands) programme implemented in about 60,000 ha in Gujarat, Maharashtra and Karnataka promoted agroforestry and livestock production is an important component in the Wadi. As per the Government of India initiative to encourage crop diversification in the earlier 'green revolution' states, Punjab wants to bring an additional area of 2 lakh ha under agroforestry to its present 1.3 lakh ha crop diversification strategy (CFAPPS, 2013). Haryana and Uttar Pradesh are also planning on similar lines. At Government of India level, a newly launched **Green India Mission** (http://moef.nic.in), 3.0 million ha of degraded lands and fallows are to be brought under agro-/social forestry within ten years. Agroforestry is being implemented presently as a sub-Mission under the National Mission for Sustainable Agriculture (DAC, Ministry of Agriculture, Govt. of India). Recently, Government of India launched a National Mission for Sustaining the Himalayan Eco-System (NMSHE) under National Action Plan on Climate Change (www.moef.nic.in/sites/default/files/Pg01-52_2.pdf). It attempts better understanding of the coupling between the Himalayan ecosystem and the climate factors and provide inputs for sustainable Himalayan development. In agriculture, it focuses to promote climate resilient farming models and validate technologies for hill farmers covering aspects related to crops, livestock, agroforestry, farming systems, horticulture and fisheries with a particular focus on jhum cultivation, soil and water conservation and water efficiency in relation to climate change. Similarly, schemes such as *Namami Gange* (for Ganga cleaning), boundary and bund plantation, and promotion of silvipsture in problem soils and other areas emphasize livestock production in non-conventional areas.

References

Ajit, Dhyani, S.K., Ramnewaj, Handa, A.K., Prasad, R., Alam, B., Rizvi, R.H., Gupta, G., Pandey, K.K., Jain, A. and Uma 2013. Modeling analysis of potential carbon sequestration under existing agroforestry systems in three districts of Indo-gangetic plains in India. Agroforest Syst DOI 10.1007/s10457-013-9625-x.

Dhyani, S.K. 1998. Alder-an essential component of agroforestry systems in the north-eastern hills of India. *Agroforestry Today (ICRAF)* **10**(4): 14-15.

Dhyani, S.K. (2003). Role of watershed management in improving forage production. In R.C. Jakhmola and R.K. Jain (Eds.), Sustainable Animal Production. p. 173-207. Pointer Publ. Jaipur, India.

Dhyani, S. K. and Chauhan, D.S. 1994. Agroforestry practices of North-eastern hill region of India. In: Agroforestry Traditions & Innovations (Ed. P.Narain *et al.*,) CSWCRTI, Deharadun, p. 19-23.

Dhyani, S.K. 2012. Agroforestry interventions in India: Focus on environmental services and livelihood security. *Indian J. Agroforestry* **13**(2): 1-9.

Dhyani, S.K. and Handa, A.K. 2013. India needs agroforestry policy urgently: issues and challenges. *Indian J. Agroforestry* **15**(2): 1-9.

Dhyani, S.K. and Sharda, V.N. 2005. Agroforestry systems as rural development options for the Indian Himalayas. *J. Tree Sciences* **24**(1): 1-19.

Dhyani, S.K., Handa, A.K. and Uma 2008. Twenty five years of agroforestry networking. AICRP on Agroforestry, NRC for Agroforestry, Jhansi, 92 pp.

Dhyani, S.K., Handa, A.K. and Uma 2013. Area under agroforestry in India: An Assessment for Present Status and Future Perspective. *Indian J. Agroforestry* **15**(1): 1-10.

Dhyani, S.K., Handa, A.K. and Uma 2015. Agroforestry Research in India: Present Status and Future Perspective. In: S K Dhyani, Ram Newaj, Badre Alam and Inder Dev Eds. Organized Agroforestry Research in India: Present Status and Way Forward. Boitech Book Publisher, Delhi.

Dhyani, S.K., Ram Newaj and Sharma, A.R. 2009. Agroforestry: its relation with agronomy, challenges and opportunities. *Indian J. Agronomy* **54**(3): 70-87.

Dhyani, S.K.; Chauhan, D.S.; Kumar, D.; Kushwaha, R.V. and Lepcha, S.T. 1996. Sericulture based agroforestry systems for hilly areas of north-east India. *Agroforestry Systems* **34**: 1-12.

Dhyani, S.K.; Sharda, V.N. and Samra, J.S. 2005. Agroforestry for sustainable management for soil, water and environment quality: looking back to think ahead. *Range Mgmt. & Agroforestry* **26**(1): 71-83.

ICAR 2010. Degraded and wastelands of India: status and spatial distribution. Indian Council of Agricultural Research, New Delhi, 158p.

Kumar, B.M.; Singh, A.K. and Dhyani, S.K. 2012. South Asian Agroforestry: Traditions, Transformations, and Prospects. In P.K.R. Nair and D. Garrity (eds.), *Agroforestry - The Future of Global Land Use*, Advances in Agroforestry 9, DOI 10.1007/978-94-007-4676-3_19, © Springer Science+Business Media Dordrecht.

Ngachan, S.V., Venkatesh, A. and Mohapatra, K.P. 2015. Agroforestry in Northeast India –prospects and scope. In: S K Dhyani, Ram Newaj, Badre Alam and Inder Dev Eds. Organized Agroforestry Research in India: Present Status and Way Forward. Boitech Book Publisher, Delhi.

NRCAF 2012. Garhkundar-Dabar watershed project- a successful model for Bundelkhand. Technical Bulletin 2/2012. National Research Centre for Agroforestry, Jhansi, p. 20p.

NRCAF 2013. NRCAF Vision 2050. National Research Centre for Agroforestry, Jhansi, p. 30.

Planning Commission 2001. Report of the Task Force on Greening India for Livelihood Security and Sustainable Development, Planning Commission, Govt. of India, New Delhi, 231 p.

Sathaye, J.A. and Ravindranath, N.H.1998. Climate change mitigation in the energy and forestry sectors of developing countries. *Annual Review of Energy and Environment* **23**: 387-437.

Singh, B.P.; Dhyani, S. K.; Chauhan, D.S. and Prasad, R.N. 1994. Effect of multipurpose tree species on chemical properties of an acid Alfisols in Meghalaya. *Indian J. Agric. Sciences,* **65**(5): 345-349.

Singh, R., Garg, K.K., Wani, S.P., Tewari, R.K. and Dhyani, S.K. 2014. Impact of water management interventions on hydrology and ecosystem services in Garhkundar-Dabar watershed of Bundelkhand region, Central India, *Journal of Hydrology* **509**: 132-149.

Handa, A. K., Dhyani, S.K. and Uma 2015. Three Decades of Agroforestry Research in India: retrospection for way forward. Journal of Research, PAU (communicated).

Uthappa A.R., Chavan, S.B.; Dhyani, S.K., Handa, A.K. and Ram Newaj 2015. Trees for soil health and sustainable agriculture: a win-win situation. Indian Farming (accepted).

□□□

Chapter 28

Agroforestry Approaches for Enhancing Soil Fertility in Farming Systems

RAM NEWAJ, S.B. CHAVAN, UTHAPPA, A.R.
ICAR-Central Agroforestry Research Institute, Jhansi-284003 (UP)

Introduction

Trees are part of human life, play crucial in almost all terrestrial ecosystems and provide a range of products and services to rural and urban people. As natural vegetation is cleared for agriculture and other types of development, the benefits that trees provide are best sustained by integrating trees into agriculturally productive landscapes - a practice known as agroforestry. In India, land degradation is global serious problem which can occur in various forms including decline in soil fertility, loss of organic matter, acidity/alkalinity/salinity, toxic chemicals in soils etc. According to global data, 24 billion tons of fertile soil disappear/year; due to drought and desertification each year 12 million ha are lost, where 20 million tons of grain could have been grown. Apart from this, 74% of poor (42% very poor and 32% moderately poor) are directly affected by land degradation globally (UNCCD report). Such type of degradation shows up gradually and may extend beyond yield loss and make land unfit for agriculture impacting farmers livelihoods. Monoculture of crops and varieties led serious problems such as depletion of biodiversity, deterioration of soil health, and the environment calls upon a critical concern. In addition, increasing levels of costly inputs like fertilizer, pesticides, etc. resulted not only in the escalation of cost of production but also jeopardized the buffering capacity of the agro-ecosystem. Under such context, sustaining soil's health and productivity has become the utmost need of the hour and a major shift in cultivation practices should be focused more for maintaining and improving soil resources which is crucial to our wellbeing.

In last sixty years, agriculture production has increased dramatically and even greater productivity may be expected in future as result of further advancement in technologies and innovations. Although it has improved the socio-economic status of farmers, but at the cost of environmental degradation. Earlier cropping systems were certainly mixture of desirable species used for food, fodder, fiber, etc., fulfilling the diverse needs of the community. Mixtures of species were chosen by farmers over the centuries to make judicious use of available natural resources and empirical choices of patterns were made among the best performing combinations observed, which were in harmony with nature.

It is considered that trees can sustain life by numerous services among that soil fertility enhancement is one. Trees are intergral part of agroforestry systems and contribute to environmental sustainability. Trees in agroforestry system enhance soil productivity through biological nitrogen fixation, adding litters (noon-woody litter, coarse woody litter and fine woody litter), turnover of fine roots, efficient nutrient cycling and deep capture of nutrients from soil. Even the trees that do not fix nitrogen can enhance physical, chemical and biological properties of soils by adding significant amount of above and belowground organic matter as well as releasing and recycling nutrients in tree bearing farmlands or crop lands. Intensification of cropping systems under changing climate scenario often depends on reducing the reliance of subsistence crop production, integration with livestock enterprises, greater crop diversification and agroforestry systems that provide higher economic value and foster soil conservation (Young, 1997). Maintenance and enhancement of soil fertility is vital for the global food security and environmental sustainability. Although India is self-sufficient in term of food production but country need to enhance food production, timber, fuel wood and fodder for further rise in human and livestock population. Maintaining and enhancing the soil fertility of farmlands to grow food grain as well as tree based product can help to meet the demand in future.

Soil fertility improvements

Agroforestry system constitutes sustainable land use and help to improve soils in a number of ways. Some of these beneficial effects are apparent in experiments carried out in different parts of the country. Tree crowns protect the soil from the impact of raindrops and create a specific microclimate in the particular area (Pradhan, 1973). Agroforestry practices play an important role in improving the fertility of the soils. Increase in amount of nitrogen and phosphorus and other macro and micro-nutrients are reported under the growth of 14 yr. old plantation of *Prosopis cineraria* as compared to a bare site of *Prosopis juliflora* of the same age (Shankarnarayan, 1984). It is reported that forage species produce higher biomass under *Prosopis cineraria* tree canopy due to a high fertility status (Singh and Lal 1969). There are evidences that soil under the trees were richer in organic matter and nutrients than those in the surrounds without tree cover. For *Acacia albida*, 50 to 100% increase in organic matter and nitrogen under canopy are known, together with increased water holding capacity (Felker, 1978).

In semi-arid India, it is also widely believed that shading by the tree is responsible for poor yields of associated crops, although the available evidence clearly indicates that the real problem is competition for moisture. It should, however, be recognized that alley cropping is very sound technology for improving the fertility of degraded low-activity clay soil in the tropics and the key to its success lies in the use of pruning as green manure or mulch (Ong *et al.*, 1991). An experimental evidence under agrisilvihorticultural system (aonla+ leucaena + blackgram) in rainfed conditions showed that after nine years 65 to 109.4% organic carbon was increased under below canopy and 28.1 to 62.5% under open canopy as compared to initial value (0.32%). The organic carbon per cent under the canopy of aonla is higher under the canopy than open canopy owing to complete falling of aonla leaves mostly limited to below tree canopy (Ram Newaj *et al.*, 1999). The influence of *Eucalyptus citriodora* plantation on the physical properties of the soil was observed due to the differential organic matter addition, root density, and root length and soil faunal activity. Contribution of 12 years old *Eucalyptus citriodora* plantation with tree density of 1600 tree ha^{-1} towards *in-situ* litter accumulation and nutrient concentration of leaves (Table 1) was carried out by Balamurugan *et al.* (2000).

Table 1. *In-situ* litter accumulation and percentage contribution of different vegetative components of total *in-situ* litter accumulation and nutrient conc. in leaf litter

Component wise *in-situ* litter accumulation (t ha^{-1})		**Nutrient conc. of leaves (%)**	
Leaves	3.25	N	0.58
Large twigs	0.75	P	0.04
Small twigs	1.25	K	0.41
Bark	0.5	Ca	1.55
Total	5.75	Mg	0.59

Source. Balamurugan *et al.*, 2000

In a silvipastoral study, the performance of different tree (*Albizia lebbek, Grewia optiva, Bauhinia purpurea* and *Leucaena leucocephala*) and grass species *Chrysopogon fulvus* and *Eulaliopsis binata* on changes in soil fertility over the 14 year period of experimentation from 1980 to 1993 revealed that soil pH decreased markedly from 7.3 in 1979 to 6.3 in 1993. There was a marginal decrease in organic carbon and available K content over the 14 year period. On the other hand, total N and available P increased appreciably by 50.0 and 74.5%, respectively over the initial status. The large improvements in soil organic carbon, total N and available K were noticed in the surface layer only (Table 3). The mean increase in 0-5 cm depth in 1993 over the 0-15 cm depth in 1979 was nearly two-fold in organic carbon and available K, and more than three-fold in total N. However, all these parameters were much lower in 5-15 cm depth that observed in the 0-15 cm depth at the beginning of the study. This was because the residues of leaf fall and other left over material remained lying on the soil surface and there was no incorporation into the deeper layers (Vishwanatham *et al.*, 1998).

Table 2. Changes in pH and nutrient status of soil over the 14-year period

Soil parameter	1979	1993		
	0-15cm	0-15 cm*	0-5 cm	5-15 cm
Soil pH	7.3	6.3	6.1	6.4
Organic carbon (%)	0.58	0.54	1.17	0.22
Total N (%)	0.082	0.123	0.296	0.035
Available P (ppm)	2.86	4.99	2.94	6.02
Available K (ppm)	44.0	40.6	77.8	22.0

(*Source:* Vishwanatham *et al.*, 1998)

Table 3. Soil improvement through trees

Trees	Available N(Kg/ha)	Available P(Kg/ha)	Available K(Kg/ha)	OC (%)	Reference
Poplar	190.52	17.25	162.94	1.22	Sarvade et al. (2014)
Eucalyptus camaldulensis	168.6	11.97	156.61	0.97	
Leucaena leucocephala	200.42	18.67	169.12	1.32	
Melia azedarach	168.26	16.14	163	1.07	
Albizia procera	186.26	16.52	165.97	0.54	Ram Newaj *et al.* (Unpublished)

Dalbergia sissoo	210.80	14.60	280.00	0.46
Hardwickia binata	221.57	15.83	281.34	0.56
Acacia auriculiformis	173.5	24.6	130.4	0.49 Das& Chaturvedi (2008)
Bambusa variegata	168.0	22.0	120.1	0.40
Bombax ceiba	169.2	22.2	122.3	0.41

The trees are able to maintain or improve soil health, which is evident from the high fertility status and closed nutrient cycling under natural forest. The processes by which trees improve soil fertility are as follows:

- Photosynthetic fixation of carbon and its transfer to the soil via litter and root decay;
- Nitrogen fixation by all leguminous trees and also by a few non-leguminous species (for example, alder and casuarinas);
- Improves nutrient retrieval by tree roots, including mycorrhizae from lower horizon;
- Erosion control by a combination of ground cover and barrier effects;
- Efficient uptake of nutrients from soil that would otherwise be lost by leaching;
- Soils under trees have a favorable structure and water- holding capacity through organic matter stabilization and conservation and root action;
- Exudation of growth promoting substance; and
- Pruning of trees synchronizes the time of release of nutrients from litter with crop demand

Biological nitrogen fixation to improve soil

Biological nitrogen fixation (BNF) constitutes a key nutrient input to agro ecosystems. The contribution of leguminous trees to building up N in degraded soils through BNF is well recognized as an important component of the ecosystem service of nutrient cycling (Barrios, 2007). There are significant differences in estimates of BNF in trees, ranging from high rates up to 472 kg N_2 ha^{-1} year $^{-1}$ in *L. leucocephala, Gliricidia sepium, C. calothyrsus* to low rates <50 kg N_2 ha^{-1} yr^{-1} in *Acacia melanoxylon* and *A. holoserica* (Giller, 2001). In another beneficial interaction mycorrhizal fungi associated with trees can help in taking up nutrient from deeper soil layers and increasing the availability of less mobile nutrients like phosphorus. Leguminous trees in agroforestry are considered as fertilizer tree for increasing nitrogen in soil through BNF. In case of legume, most of the nitrogen fixing tree species is from Mimosoideae and Caesalpinioideae, with fewer in Papilionoideae. Within these sub-families, 98, 60 and 30% of the tested mimosoids, papilionoids and caesalpiniods, respectively showed potential to fix atmospheric nitrogen (Table 4). Among the non-legumes, Alnus, Myrica and Casuarina are widely recommended in agroforestry system, which fix nitrogen in association with Frankia. The nitrogen fixing potential of *Casuarina equisetifolia* is 50-80 N kg ha"1 yr"1 and *Alnus nepalensis* is 29-117 N kg ha"1 yr"1 (Sharma and Kapoor, 2005).

Table 4. Nitrogen fixation of tree species in agroforestry

Species	N_2 fixation (Kg ha^{-1} yr^{-1})	Species	N_2 fixation (Kg ha^{-1} yr^{-1})
Acacia mearnsii	200	*Gliricidia sepium*	13
Casuarina equisetifolia	60-110	*Inga jinicuil*	35-40
Erythrina poeppigiana	60	*Leucaena leucocephala*	100-500
Acacia albida	20	*Sesbaniarostrata*	83-109

Maintenance or improvement of soil physical properties

The superior soil structure, porosity, moisture characteristics and erosion resistance under forest is well documented, as is their decline on forest clearance. Porosity is a key to many other physical properties: pores of 5-50 μm in diameter determine available water-holding capacity, whilst those over 250μ m are necessary for root penetration. There is much evidence of the influence of physical properties of tropical soils on crop growth, independent of nutrient or other effects. Trees help in breaking up of compact or in durated layers by roots. It has been observed that trees had significant impacts on the physical conditions of the sodic soil. The water holding capacity of the soil was found to increase with the reduction in bulk density of *E. tereticornis* plantation which could also be explained in terms of organic matter, humus and tree growth (Table 5). It was reported higher values of water holding capacity, sticky point, moisture and hydraulic permeability of soil under forest cover then those of soil under cultivation. The field capacity of the soil was significantly higher than that of water holding capacity, due to increase in micropores after 3 and 6 years old plantation (Mishra *et al.*, 2003).

Higher infiltration rate and soil permeability in 9 years old plantation may possibly be due to improved soil conditions. The increment in porosity, water holding capacity and field capacity and reduction in bulk density and $CaCO_3$ content under Eucalyptus plantation also increased the permeability of the soil. After afforestation the micropores increased in the soil due to network of roots. The tree roots and foliage of grasses are responsible for water infiltration into the substratum and thereby increasing the percolation of water within the soil, that's why the infiltration rate was high in plantation sites as compared to control sites.

Table 5. Physical properties of soil supporting *Eucalyptus tereticornis* plantations of 3 & 6 years age

Depth (cm)	Porosity (%)	BD(g/cm^3)	WHC(%)	FC(%)
Control				
0-30	41.31	1.63	44.07	24.06
30-60	42.22	1.66	44.60	25.81
60-90	35.41	1.72	44.29	24.07
3 years				
0-30	45.07	1.39	46.20	31.36
30-60	45.24	1.48	45.64	28.02
60-90	44.21	1.56	45.05	25.27
6 years				
0-30	47.56	1.27	48.04	33.16
30-60	45.03	1.38	47.66	27.39
60-90	38.45	1.45	47.00	25.45

Source: Mishra *et al.* (2003)

Agroforestry for controlling soil erosion

Erosion is now accepted as one of the main causes of soil degradation, including of physical, chemical and biological properties, all of which require attention. Mechanical means of soil erosion control measures are not cost effective and also cannot be well maintained; there is a greater interest in biological means of conservation. Here is greater emphasis on the effects of soil cover as a means of controlling erosion, as compared with checking runoff. It is accepted that we cannot stop the cultivation on sloppy lands in this region. The inclusion of trees in soil conservation and erosion control is one of the most widely acclaimed and compelling reasons for including trees on farmlands prone to erosion hazards.

Agroforestry plays a key role in keeping the soil resource productive which is one of the major sustainability issues. Closely spaced trees on slopes reduce soil erosion by water through two main processes: first as physical barrier of stem, low branches, superficial roots and leaf litter against running water and secondly as sites where water infiltrates faster because of generally better soil structure under trees than on adjacent land. Agroforestry systems on arable lands envisage growing of trees and woody perennials on terrace risers, terrace edges field bunds, as intercrops and as alley cropping in the shape of hedge row plantation integrating trees on the fields act as natural sump for nutrients from deeper layers of soil, add bio-fertilizer, conserve moisture and enhance productivity of the system.

The tree cover reduces erosion to low levels, primarily through the effect of the ground surface cover of litter and understorey vegetation, the protection afforded by the tree canopy being relatively slight. One of the major advantages of agroforestry in terms of improving or sustaining soil productivity is through its effect on soil conservation. Trees reduce the rate of erosion by:

- Protecting the soil from the impact of rain
- Transpiring large amounts of water, which counteracts very wet soil
- Binding soil to sloping land with their roots.

The Cardemon-alder based agroforestry system of N-E India provides excellent prove on soil erosion control (Table 6). The soil fertility is continuously maintained through nutrient cycling from the natural tree cover. Ecological sustainability is even greater with cardamom when the Himalayan alder is used as the shade trees, as this tree regenerates naturally on sites affected by landslides. Various data were collected for large cardamom and other crops based on 6 rainfall events. Soil loss (30 kg ha^{-1}) and overland flow (2.17% of precipitation) were low in large cardamom agroforestry compared to other agroforestry types whereas 477 kg ha^{-1} soil was lost and 9.55% of precipitation was lost as overland flow in the rainfed agricultural systems (Sharma *et al.*, 2007). Role of shelterbelt in arid zone is well known and maintains soil fertility by controlling soil erosion due to wind (Table 7). Trees like *Prosopis cineraria, Ziziphus nummularia, Z. mauritiana, Tecomela undulata, Hardwickea binata, Cassia siamea, Acacia tortilis* and many others play important role in production system. Many of these also act as shelterbelt for associate crops and also improve soil health.

Agroforestry for problematic soil

In our country 120 million ha land is subjected to various type of degradation out of which 9.38 mha suffered by sodicity and alkalinity. In sodic soils the survival of crops are very difficult. The reclamation of sodic soils, through chemical and mechanical mean is very costly, but under such condition trees may do well to sustain productivity. The establishment of permanent vegetative cover of suitable trees and crops is an option of great promise in problematic soils in view of growing demand for fuel wood, timber and fodder as well as environmental improvement.

Table 6. Runoff and soil loss under different land use systems of Sikkim

Land use	Runoff (L/ha)	Soil loss (kg/ha)
Agricultural field of maize	6427	122
Large cardamom Agroforestry	6989	66
Natural forest	5581	12
Barren land	8097	166

Source: Sharma *et al.*, 1992

Table 7. Effect of different type of shelterbelts on soil erosion in arid region

Type of shelterbelt	Total amount of soil loss (kg ha^{-1})		
	Year I	Year II	Mean
Prosopis juliflora	93.2	609.3	351.2
Cassia siamea	91.5	277.1	184.2
Acacia tortilis	106.0	494.1	300.0
Agricultural field without trees	262.7	831.0	546.8

Source: Harsh and Tewari, 1993

Singh (1994) extended site suitability studies to 20 alkali soil sites in Ganga-Yamuna Doab. Barren alkali soils, represented by Natric

Comborthids and Calciorthids, had pH of 10 to 10.5, ESP of 60 to 95, and EC between 1 and 43 dSm^{-1} and constituted mainly by carbonates and bicarbonates of sodium. *Prosopis juliflora* was found to grow in typic Natrutalfs with a maximum pH of 10.0, ESP of 70, and ECe at 12 dSm^{-1} in the rooting zone, even though the top 44 cm of soil had pH 10.3 ESP of 70, and ECe at 12 $dS\ m^{-1}$. Other planted species failed to grow in this soil. Along with some natural growth, *P. juliflora* and *A. nilotica* established on typic natrustalfs with an average pH of 9.5 ESP 50 and ECe 10 dSm^{-1}. Other species likes *Dalbergia sissoo, Pongamia pinnata, Albizia lebbeck, Terminalia arjuna, Butea monosperma, Capparis deciduas* and *Salvadora persica* did well on soil with a pH less than 9.1, ESP up to 44 and ECe up to 7.5 dSm^{-1}. *Casuarinas equisetifolia* and *Acacia nilotica* could grow well in soil with ESP 30.6, whereas *Pongamia pinnata* and *Dalbergia sissoo* survived only upto ESP of 15.2.

The rehabilitation of sodic soils, through chemical and mechanical mean are very costly, so that it is the beyond the reach of Indian farmers. Under such soil condition trees may do well with crops to sustain the productivity of soils (Khan *et al.*, 1995). Table 8 conclude that sodic soil can be permanently rehabilitated by the adoption Agroforestry system, to supplement the requirements of fuel wood, fodder and timber to the farmer's and to maintain friendly eco-system in the area. The establishment of permanent vegetative cover of suitable trees, and herbaceous crop is an option of great promises for the use of such lands in view of growing demand for fuel wood, timber and fodder as well as environmental improvement (Table 9).

Table 8. Sodic soils amelioration through multipurpose tree species

Tree species	Chemical properties (on 2001)				
	SoilpH	EC (mmhos)	OC (%)	Available P(kg/ha)	Available K(kg/ha)
Azadirachta indica	8.45	0.25	0.75	12.0	240
Dalbergia sissoo	8.6	0.20	0.61	14.4	210
Albizzia procera	8.3	0.25	0.45	9.0	125
Terminalia arjuna	8.7	0.25	0.58	10.8	175
Eucalyptus hyrid	7.2	0.20	0.53	9.0	140
Leucaena leucocephala	7.8	0.20	0.56	10.8	210
Acacia catechu	8.35	0.20	0.67	20.0	225
Acacia nilotica	8.52	0.20	0.85	13.4	165
Morus alba	8.38	0.20	0.61	14.4	210
Cacia siamea	8.35	0.20	0.85	23.4	165
1989 (planting time)	10.5	0.74	0.12	6.0	85.0

Khan *et al.*, 1995

Table 9. Recommended tree species for problematic area

Problematic Areas	Recommended tree species
Sodic soils (pH 8.62-10)	*Acacia nilotica, Casuarina equisetifolia, Prosopis juliflora, Prosopis cineraria, Tamrix articulate, Pithecellobium dulce, Salvadora persica, Terminalia arjuna, Pongamia pinnata, Sesbania sesban, Eucalyptus tereticornis, Butea monosperma, Acacia auriculiformis, Azadirachta indica, Poplars*
Wetland	*Salix babylonica, Salix xuchonensis, Alnus cremastogyne, Alnus trabeculosa, Paulownia tomentosa, Morus alba, Taxodium distichum and Taxodium scandens*
Waterlogged area	*Eucalyptus spp., Casuarina equisetifolia, Bambusa arundinacea, Tamarix, Lucerne, Syzygium cumini, Terminalia arjuna, Corymbia tessellaris, Barringtonia acutangula, Psidium gujava, Ficus glomareta*
Sand dunes stabilization	*Prosopis cineraria, Acacia tortilis, Acacia nilotica, Prosopis juliflora, Azadirachta indica, Tecomella undulata, Capparis deciduas, Acacia senegal, Ziziphus spp., Parkinsonia Aculeta and Ailanthus excelsa*
Coastal Area	*Anacardium occidentale, Ailanthus malabarica, Casuarina equisetifolia, Pongamia glabra, Sesbania aculeate, Salicornia bigelovii, Salvadora persica, S. oleoides, Terminalia catappa, Calophyllum inophyllum*, species of *Pandanus, Nypa fruticans, Borassus flabellifer*
Mining area	*Anthocephalus cadamba, Gmelina arborea, Acacia auriculiformis, Acacia mangium, Albizia lebbeck, Bamboo species, Grevillea robusta, Dalbergia sissoo, Annoguisus pendula and Albizia lebbeck, Leucaena leucocephala, Pongamia pinnata, Acacia catechu, Erythrina indica and Bombax ceiba*
Ravine area	*Acacia catechu, Dalbergia sissoo, Morus alba, Dendrocalamus strictus, Acacia nilotica, Azadirachta indica, Albizia spp., Holoptelia integrifolia*
Acid soils	*Pinus roxburghii, P. wallichiana, Celtis australis, Pinus, Cassia, Acacia mearnsii, A. decurrens, A. dealbata, A. pycnantha, Manilkara hexandra, Phoenix* spp. *A. auriculiformis, Pterocarpus marsupium, Shorea robusta, Hardwickia binnata, Terminalia* spp., *Tectona grandis, Gmelina arborea, Xylia xylocarpa, Bamboo* spp., *Alnus* spp.

Conclusion

Agroforestry is an ecologically most complex system of natural resources management for sustainable crop production and providing benefits to all those who use the land by integrating trees on farms and in the agricultural land scape. Tree based farming systems play an important role in maintaining the physico-biochemical properties of the soil thus sustaining the overall productivity. There is a huge task to check soil degradation throughout the country due to erroneous crop rotations being practiced in most parts of the country to achieve high productivity. In the light of present day situation, it can be said that agroforestry serves as 'island of fertility' in an agri-ecosystem where resources are very much limited. Thus, more attention should be focussed on conservation and maintenance of soil health for evergreen agriculture and livelihood security.

References

Balamurugan, J., Kumaraswamy, K. and Rajarayan, A. 2000. Effects of *Eucalyptus citriodora* on the physical and chemical properties of soil. *Journal of Indian Society of Soil Science* **48**(3): 491-495.

Barrios, E. (2007) Soil biota, ecosystem services and land productivity. *Ecological Economics* **64**: 269–85.

Das D. K and O. P. Chaturvedi. (2008). Root phytomass recovery and rooting characteristics of five agroforestry tree species in eastern India. Journal of *Tropical Forest Science* **20**(3): 156–166.

Felker, P. 1978. State of the art: *Acacia albida* as a complementary permanent intercrop with annual crops. Report to USAID. Riverside, California, USA: University of California, 133 p.

Giller, K.E. (2001) *Nitrogen fi xation in tropical cropping systems* (2nd Edn.) CAB International Publishing, Wallingford.

Harsh, L.N. and Tewari, J.C. 1993. Sand dune stabilization, shelterbelts and silvi-pastoral plantation in dry zones. In: Sen, A.K. and Kar Amal. (eds) Desertification in Thar, Sahara and Sahel Regions. Scientific Publishers, Jodhpur, pp 269-279.

Khan, S.A and Shukla, I.N. 2003. Suitable MPTS for sodic soils of central plain zone of Uttar Pradesh. *Indian J. Agroforestry.* **5**(1&2): 119-121.

Mishraa, A., S.D. Sharmaa, G.H. Khanb. (2003). Improvement in physical and chemical properties of sodic soil by 3, 6 and 9 years old plantation of Eucalyptus tereticornis: Biorejuvenation of sodic soil. Forest Ecology and Management **184**(1-3): 115–124.

NRCAF Annual Report 2013-2014. National Research Centre for Agroforestry, Jhansi pp 127.

Ong, C.K., Corlett, J.E., Singh, R.P. and Black, C.R. 1991. Above and below ground interactions in agroforestry systems. *Forest Ecology Management* **45**: 45-57.

Pradhan, I.P. 1973. Preliminary study of rainfall in interception through leaf litter. *Indian Forester* **99**: 440-443.

Ram Newaj, Shukla, S.K. and Yadav, R.S. 1998-99.Varietal evaluation of aonla (*Emblica officinalis*) in agrisilvihorticultural system under rainfed conditions on marginal lands. *Annual Report*, NRCAF, Jhansi, pp.44-46.

Sarvade, S., H. S. Mishra, Rajesh Kaushal, Sumit Chaturvedi and Salil Tewari Wheat (*Triticum estivum* L.) Yield and Soil Properties as Influenced by Different Agri-silviculture Systems of Terai Region, Northern India. *International Journal of Bioresource and Stress Management* 2014, **5**(3): 350-355

Shankarnarayan, K.A. 1984. Influence of *Prosopis cineraria* and *Acacia nilotica* on soil fertility and crop yield. In: Shankarnarayan K.A. (Ed.). Agroforestry in Arid and Semi-arid Zones, Central Arid Zone Research Institute (CAZRI) Publication No.24, Jodhpur (Raj.), India, pp.59-66.

Sharma, E., Sundriyal, R. C., Rai, S.C., Bhatt, Y.K., Rai, L.K., Sharma, R. and Rai, Y.K.1992. Integrated Watershed Management: A Case Study in Sikkim. G.B. Pant Institute of Himalaya Environment & Development, Koshi.

Sharma, P.K. and Kapoor, K.K.2005. Problems and Prospective of Nitrogen fixation in Agroforestry systems. Proc. Indian Natl.Sci.Acad. B71 pp 145-161.

Singh, K. N. 1994. Crop and agronomic management. In: *Salinity Management for Sustainable Agriculture*. D. L. N. Rao, N. T. Singh, R. K. Gupta and N. K. Tyagi (Eds.). Central Soil Salinity Research Institute, Karnal, pp. 124-144.

Singh, K.S. and Lal, P. 1969. Effect of *Prosopis cineraria* and *Acacia arabica* trees on soil fertility and profile characteristics. *Annals of Arid Zone* **8**: 33-36.

Vishwanatham, M.K., Samra, J.S. and Sharma, A.R. 1998. Biomass production of silvipasture systems on gravelly lands of Doon valley. *Bulletin* No.T-39/ D-26, CSWCRTI, Dehradun, pp.68-69.

Young, A. 1997. *Agroforestry for soil management* (2nd Edn). CAB International, Wallingford.

❑❑❑

Chapter 29

Integrated Farming System Research Under Climate Change Scenario

R.K. AGRAWAL AND J.B. SINGH

ICAR - Indian Grassland and Fodder Research Institute, Jhansi-284003

Abstract

Conventional agriculture has caused economic problems associated with increased costs of energy-based inputs and decreased farm incomes. Farming system is a resource management strategy to achieve economic and sustained agricultural production to meet diverse requirements of farm livelihood while preserving resource base and maintaining a high level of environment quality. It interacts adequately with environment without perturbing the ecological and socioeconomic balance on one hand and attempts to meet the common goals on the other. Advantages of integrated farming system include positive impacts on natural resources and other key inputs besides generation of regular employment for the farm family members. The realm of climate change that is projected to affect the agriculture system most is the phenomena of global warming. Agriculture and livestock sector is a source for the global climatic change by emitting green house gases (GHG) as well as a sink by providing ample scope to sequester carbon and reducing GHG emission by innovative management practices. Its negative impacts are projected to more severely felt on agriculture and allied livestock sector for their livelihoods. The inter relationship of integrated farming system and climate change is having different important aspects of study such as crop weather relationship during germination, establishment, vegetative and reproductive growth stages, yield and yield attribute as well as during post harvest period. Crop modeling studies on forage sorghum (C_4) and cowpea (C_3) showed increased dry matter biomass by 3% in sorghum but more prominent in cowpea by 46% under elevated CO_2 from 330 ppm to 770 ppm. The anticipated rise in temperature between 2.3 and 4.8°C over the entire country together with increased precipitation resulting from climate change is likely to aggravate the heat stress in dairy animals, adversely affecting their productive and reproductive performance. The feeding of urea treated straw improved the feed intake, and milk yield by 22.13 and 22.3 % than the untreated animals.

Key words: Integrated farming system, Elevated CO_2, Climate resilience, Straw treated urea.

Indian economy is largely driven by agriculture, which in turn depends upon the vagaries of monsoon. Successful crop production requires suitable combinations of several factors like soil, water, crop and variety etc. Conventional agriculture has caused economic problems associated with over production of crops, increased costs of energy-based inputs and decreased farm incomes. It has also produced ecological problems such as poor ecological diversity, soil and water pollution and soil erosion. The adoption of integrated systems of agricultural production involving lower inputs of fertilizer, pesticides and cultivations can reverse the trend. Such systems are dependent upon a good understanding of the nature of interactions between the four main components of such systems, which are fertilizers, pesticides, cultivations and rotations, and how these interactions influence crop yields and farm income. Integration of various agricultural enterprises *viz.*, cropping, animal husbandry, fishery, forestry etc. have great potentialities in the agricultural economy. These enterprises not only supplement the income of the farmers but also help in increasing the family labour employment. Farming system is a mix of farm enterprises such as crop, livestock, aquaculture, agro forestry and fruit crops to which farm family allocates its resources in order to efficiently manage the existing environment for the attainment of the family goal. It involves judicious activities and continuous improvement in all components of farming enterprises through appropriate management processes. Integrated Farming aims to combine the best of modern tools and technologies with traditional practices according to a given site and situation.

Integrated farming system

In practical words the Farming system is a resource management strategy to achieve economic and sustained agricultural production to meet diverse requirements of farm livelihood while preserving resource base and maintaining a high level of environment quality (Lal and Miller,1990). It interacts adequately with environment without perturbing the ecological and socioeconomic balance on one hand and attempts to meet the common goals on the other (Jayanthi *et al.*, 2002). IFS envisage harnessing the complementarities and synergies among different agricultural sub-systems/enterprises and augmenting the total productivity, sustainability and gainful employment. It is essentially cyclic (organic resources-livestock-land-crops). Therefore, activities related to one component affect the others. The farm wastes are better recycled for productive purposes in the IFS.

From the above description, it is evident that, IFS, is a complex and dynamic combination of different agricultural enterprises. The above all are dynamic component and are significantly affected by climatic deviations. The climatic elements have significant impact on individual component as well as on the dynamic inter relation of these enterprises.

Climate and climate change

In literal terms Climate is a measure of the average pattern of variation in temperature, humidity, atmospheric pressure, wind, precipitation, atmospheric particle count and other meteorological variables in a given region over long periods of time. Climate is different from weather, as weather only describes the short-term conditions of these variables in a given region. The climate of a region is generated by the climate system, which has five components: atmosphere, hydrosphere, cryosphere, lithosphere, and biosphere. The climate of a location is affected by its latitude, terrain, and altitude, as well as nearby water bodies and their currents. Whereas, climate change may refer to a change in average weather conditions, or in the time variation of weather around longer-term average conditions (i.e., more or fewer extreme weather events). Climate change is caused by factors such as biotic processes, variations in solar radiation received by Earth, plate tectonics, and volcanic eruptions.

The most general definition of climate change is a deviation in the statistical properties of the climate system when considered over long periods of time, regardless of cause. Important to mention that fluctuations over periods shorter than a few decades, such as El Niño, do not represent climate change. Global Warming is the increase of Earth's average surface temperature due to effect of greenhouse gases, such as carbon dioxide emissions from burning fossil fuels or from deforestation, which trap heat that would otherwise escape from Earth.

Climate of a place has a crucial role in growth and production of crop plants.

There is a global trend for increased frequency of drought, as well as heavy precipitation events over the most land areas. Agriculture and livestock sector is a source for the global climatic change by emitting green house gases (GHG) as well as a sink by providing ample scope to sequester carbon and reducing GHG emission by innovative management practices. Its negative impacts are more severely felt on agriculture and allied livestock sector for their livelihoods. The threat is much higher in semiarid regions due to the predicted increased temperature by 1ºC to 4ºC, CO_2 levels from normal to 540 ppm by 2020 and 770 ppm by 2050 and increases in rainfall variability.

The inter relationship of integrated farming system and climate change is having different important aspects of study such as crop weather relationship during germination, establishment, vegetative and reproductive growth stages, yield and yield attribute as well as during post harvest period. Hence, precise the assessment, prediction of wether parameters has become important in disaster management, disease management, as well as post harvest management.

The different crops have temperature requirement in a definite range (Table 1). For example, Sorghum requires warm weather for development and it cannot grow properly or set seed at temperatures below 18ºC - 21ºC. It cannot recover from freezing temperature except in the juvenile stage. It can adequately recover from drought and also withstand flooding. Sorghum has very efficient photosynthesis mechanism and shows little photorespiration.

Table 1. Temperature requirement of important crops

Crops	Minimum	Optimum	Maximum
Oat	4-5	25	30
Maize	8-10	32-35	40-44
Sorghum	8-10	32-35	40
Pearl millet	8-10	30-32	40
Berseem	3-4.0	25-27	32
Wheat	3-4.5	22	30
Rice	10-12	30-32	36-38

Impact of elevated CO_2, temperature & precipitation

Some predict positive impacts on agriculture from climate change like increased temperatures and higher carbon dioxide levels. Increased concentrations of CO_2 may boost crop productivity, only where moisture is not a constraint. Higher levels of CO_2 can stimulate photosynthesis in certain plants (30-100 per cent). Experimental observations confirm that when plants absorb more carbon grow bigger and more quickly. This is particularly true for C_3 plants (so called because the product of their first biochemical reactions during photosynthesis has three carbon atoms). Increased CO_2 tends to suppress photo-respiration in these plants, making them more water-efficient. But, the response of C_4 plants would not be as significant. C_3 plants correspond to crops like wheat, rice and soy bean whereas C_4 plants correspond to crops like maize, sorghum & sugarcane. The impact on yields of C_4 crops is more difficult to predict as yields may be reduced by 10-30 per cent due to increased summer dryness. The effects of an increase in carbon dioxide would be higher on C_3 crops than on C_4 crops, because the former is more susceptible to carbon dioxide shortage. Crop modeling studies on forage sorghum (C_4) and cowpea (C_3) showed increased dry matter biomass by 3% in sorghum but more prominent in cowpea by 46% under elevated CO_2 from 330 ppm to 770 ppm. The interaction study of enhanced CO_2 and temperature showed prominent negative impact on yields of both the crops (Rai *et al.*, 2014).

A 10-15% increase in monsoon precipitation in many regions, a simultaneous precipitation decline of 5-25% in drought-prone central India and a sharp decline in winter rainfall in northern India are also projected. This implies changes in output of winter wheat and mustard crops in north western India. A decrease in number of rainy days (5-15 days on an average) is expected over much of India, along with an increase in heavy rainfall days in the monsoon season. These changes are expected to increase the vulnerability of Indian agriculture. This is particularly important in India, where agriculture is highly sensitive to monsoon variability as 65% of the cropped area is rain-fed.

The agricultural sector is a driving force in the GHG emissions and land use effects. The three major causes of the increase in GHGs observed over the past 250 years have been fossil fuels, land use and agriculture. The agricultural processes (rice cultivation, enteric fermentation in cattle) comprise 54% of methane emissions, 80% of nitrous oxide emissions and major percentage of carbon dioxide.

Impact on livestock enterprise

Climate change poses formidable challenge to the development of livestock sector in India. The anticipated rise in temperature between 2.3 and 4.8°C over the entire country together with increased precipitation resulting from climate change is likely to aggravate the heat stress in dairy animals, adversely affecting their productive and reproductive performance (Sirohi and Michaelowa, 2007). The organ systems of animals respond to physical, chemical, biological and climatic stimuli from their surroundings and work in concert to perform the essential body functions. The performance (*e.g.*, growth, milk and wool production, reproduction), health and wellbeing of the livestock are strongly affected by climate both, directly and indirectly (Singh *et al*, 2012). Livestock are home other, which means, that they must regulate their body temperature within a relatively narrow range to remain healthy and productive. The ambient temperature below or above the thermo neutral range creates stress conditions in animals. The rearing and feeding systems under intensive livestock production systems need to be evaluated in the context of climate change. Adapting livestock production to the climate change demands adjustments through diversification, intensification and integration of crop-livestock husbandry to meet the exigencies, changing land use and stocking pattern in terms of number and species composition etc.

Climate affects animal production in four ways: (a) the impact of changes in livestock feed-grain availability, impacts on livestock pastures and forage crop production and quality; (c) changes in the distribution of livestock diseases and pests and the direct effects of weather and extreme events on animal health, growth and reproduction Smit *et al.*, 1996. The indirect effects of climate driven changes in animal performance result mainly from alterations in the nutritional environment. As regards to animal component, the direct effect on livestock could be small but the indirect effect brought about by the change in feed and fodder supply may severely alter the existing livestock systems. Research suggests that changes in climate would affect the quality and quantity of forage produced (Topp, 1996). In C_4 species, temperature increases up to 30-35°C may increase productivity of forage and pastures when nutrients and water are available in optimum quantity. In C_3 plants, temperature has similar effect but increase in CO_2 levels will have a positive impact on the productivity of these crops and livestock. In cereal crops, harvest indices will change and hence the quality of roughages and their availability will also be affected. Heat stress induces behavioral and metabolic changes, which include reduced feed intake and metabolic activity, and thereby a decline in their productivity.

A study was conducted under small holders' mixed farming system at village Kadesara kalan in Lalitpur district (semi-arid tract of Bundelkhand region) of Uttar Pradesh on eight lactating nondescript (local) buffaloes (average body weight 409 kg) with the objective of Feeding urea treated straw as measure to reduce the adverse effect of extreme temperature and humidity expressed in terms of temperature humidity index (THI) on milk production. The prevailing extreme climatic stress in terms of high temperature and reduced environmental humidity during summer severely affects the animal health as well as productivity. Upadhyay *et al.* (2007) stated that thermal stress on Indian livestock particularly cattle and buffaloes has been reported to decrease oestrus expression and conception rate. The anima fed with untreated or treated straw with 8% urea solution (*adlibitum*) offered 2 times a day. The buffaloes of treated group consumed 22.13% more straw as compared to untreated group. The DM intake as percent of body weight was 2.98 and 2.59%. Overall milk yield of the animals fed urea treated straw was 22.3 % more than the untreated animals.

Breeding varieties of fodder and cereal crops of shorter duration will help in meeting the demand for food and fodder. Promoting indigenous livestock breeds will lessen the effect of harsh climate on their production performance. Introduction of early warning system and other forecasting and crisis preparedness system in context of livestock is necessary. A better understanding of the causes and impact of climate change on livestock is the need of the hour.

Thus, looking at the above points it is clear that the farming system management requires suitable combinations of several factors like soil, water crop variety, livestock species, breed, trees etc. Each crop has its own weather optima for the expression of its full yield potential. So knowledge of the influence of weather parameters on growth and productivity of different components is essential for successful judicious appropriation. Thus, nowadays agrometeorological services are increasingly being demanded by the farming communities to cope more efficiently with climate variability and the increasing incidence of extreme events. The study of agricultural meteorology will guide in planning cropping systems/pattern, selection of sowing dates/cost effective ploughing, harrowing weeding etc., reducing losses of applied chemicals and fertilizers, judicious irrigation to crops, efficient harvesting of all crops, reducing or eliminating outbreak of pest and disease, animal rearing and feeding.

References

Jayanthi, C., Rangasamy, A., Mythili, S., Balusamy, M., Chinnusamy, C. and Sankaran, N. 2002. Sustainable productivity and profitability to integrated farming systems in low land farms. In: Extended summaries, (Eds: A.K. Singh, B. Gangwar, Pankaj and P.S. Pandey), National Symposium on Farming System Research on New Millennium, PDCSR, Modipuram, pp. 79-81.

Lal, R. and Miller, F.P. 1990. Sustainable farming for tropics. In: Singh, R.P. (Ed.) Sustainable agriculture: Issues and Prospective. Vol.1, pp. 69-89, *Indian Society of Agronomy*, IARI, New Delhi.

S. Rai, P. Ghosh, S. Kumar and J. Singh. 2014. "Research in Agrometeorolgy on Fodder Crops in Central India – An Overview, *Atmospheric and Climate Sciences*, **4** (1): 78-91

Singh S. K., Meena H. R., Kolekar D. V. and Singh Y. P. 2012. Climate Change Impacts on Livestock and Adaptation Strategies to Sustain Livestock Production *J. Vet. Adv.* **2**(7): 407-412.

Smita Sirohi and Axel Michaelowa. 2007. Sufferer and cause: Indian livestock and climate change. *Climatic Change,* **85**: 285–298.

Smit, B., Mc Nabb, and D., Smihers, J. 1996. Agricultural adaptation to climatic variation. *Climatic change* **33**: 7-29.

Topp, C.F.E., and Doyle, C.J. 1996. Simulating the impacts of global warming on milk and forage production in Scotland. 1. The effects on dry matter yield of grass and grass-white clover stands. *Agricultural Systems* **52**: 213-242.

Upadhyay R. C., Singh S. V., Kumar, A., Gupta, S. K. and Ashutosh, A. 2007. Impact of Climate change on Milk production of Murrah buffaloes. *Italian Journal of Animal Science* **6**(2): 1329-1332.

□□□

Chapter 30

Assessment and Mitigation of Greenhouse Gases in Integrated Farming Systems

R.S. YADAV, N.K. JAT, A. KUMAR AND T. RAM

ICAR - Indian Institute of Farming Systems Research, Modipuram-250110 (U.P.)

Atmospheric CO_2 level has increased to 391 ppm in 2011 (IPCC, 2013) from a pre-industrial concentration of around 280 ppm predicted to affect the human civilizations catastrophically on further increase in its concentration. Reduction and control of Green House Gases (GHG) emissions has become an important global issue. Unabated human activities in pursuit of comfort and development will further aggravate the GHG situation and hence inviting climate change. Agriculture, besides being serving as a sink of C from atmospheric CO_2, also contributes substantially to anthropogenic emission of GHG. Potential sources of GHGs emission on farms include the soil, animals, growing crops, farm operations, machinery, households and manure. Globally, agriculture contributes about 10 - 12 % of anthropogenic GHG emissions (FAOSTAT, 2013). The IPCC have indicated that the risk of severe climate change impacts will increase markedly with a temperature increase of 2 ^{0}C above pre-industrial level and to limit temperature by this level, CO_2 equivalent (CO_2-e) concentrations will have to be stabilized at levels of between 400 and 450 ppm CO_2-e. It is widely believed that it is possible to reduce GHG emission sufficiently to achieve the 2 ^{0}C target without destabilizing the global economy by rapidly implementing strong deliberate policy choices (Stern, 2006). Three elements of policy are required for an effective global response: i) the pricing of carbon (C) implemented through tax, trading or regulation, ii) the support of innovation and the deployment of low C technologies, and iii) the removal of barriers to energy efficiency and to inform, educate and persuade individuals about what they can do to respond to climate change (Stern, 2006). Under these responses, agriculture can be an important part of the solution to mitigate climate change by reducing the net GHG emission from both industrial and agriculture sectors, and by the sequestering of C in soil and biota (Lal, 2004). For these policies to be effective in general and agriculture in particular, assessment of CO2-e emissions from different activities of agricultural are essentially required.

The quantification of GHG emission from agriculture is fundamental to identifying mitigation solutions that are consistent with the goals of achieving greater resilience in production system, food security and rural welfare (Olander *et al.*, 2013). A C footprint is a measure of C emission in terms of the amount of GHG produced in units of CO_2-e (Lynas, 2007). Measurement of C footprint is an important tool in quantifying and benchmarking the C emissions arising from farm household activities and identifying the hotspots to reduce the C footprint of a given farming system/farm households. Identified hotspot activities for C reduction can be addressed by replacing with more efficient available technology options. There is a growing interest in reducing the carbon footprint of agricultural products (Williams and Wikstrom, 2011). As per International commitment, India is to provide the emission of GHG emission to UNFCC. It gives the GHG emission for the country as a whole by using activity from secondary data. However, for action at farm level, data on emission of GHG for mitigation of climate change is not available. Therefore, there is a pertinent and emergent need to study and quantify the C footprints of different farm productions globally in general and particularly in the country and design and develop such farm production systems, which are C neutral/minimum with a policy guideline on the subject to keep pace with world developments on the subject amidst global climate change scenario.

Integrated Farming System (IFS) is a participatory and comprehensive approach of developing location and situation specific farming systems harnessing the interactions among components of a farm for higher and sustained agricultural production for environmental, social, economic and nutritional security. It is a single window system encompassing all possible interventions for a farm in totality. Studying and developing sustainable IFS can be a way forward for mitigating the climate change. Integration of farming practices mitigating the climate change can only address the issues of global warming. Knowing the scale and source of GHG emission from different components of integrated farming system will help to find the ways to reduce them and pave the way in enabling to C labeling for farming products. Hence, this handout briefly explains the C accounting for assessment and mitigation of climate change through integrated farming system approach.

Carbon accounting

C accounting refers generally to processes undertaken to measure amounts of CO_2-e emitted by an entity. C accounting depends upon the purpose for which it is carried out. Broadly there are two approaches of C accounting: i) whole farm approach, and ii) life cycle approach (LCA). The whole farm approach is an approach of C accounting of whole farm activities *i.e.* from cradle to farm gate. It is a useful estimate for individual farm benchmarking purposes, which helps to find the hot spots or areas where GHG reductions might be made to mitigate climate change. LCA approach involves C accounting of whole farm activities *vis-à-vis* final disposal of farm product *i.e.* from cradle to grave. This approach is a product centric. The calculation shows GHG emissions per unit of product. Associated GHG emissions with a product can be used to highlight emissions within the life cycle of a product. Common resources available for standards and guidance for accounting of GHG emission are-

i) Intergovernmental Panel on Climate Change (IPCC) - IPCC guidelines provide methodologies for estimating national inventories of anthropogenic emissions by sources and removals by sinks of greenhouse gases. It categorized all anthropogenic emissions into four sectors *viz.* energy, agriculture, industrial process and product use, forestry, and other land use and waste. The 2006 IPCC guidelines generally provide advice on estimation methods at three levels of detail, from tier 1 (the default method) to tier 3 (the most detailed method). IPCC guidelines provide default emission factors for GHG emissions from all sectors, and also encourages the use of country-specific factors if the methods used to calculate them can be defended and withstand international peer-review.

ii) GHG protocol of World Resource Institute (WRI)/World Business Council for Sustainable Development (WBCSD) – Almost all GHG accounting guidelines, including ISO 14064 and PAS 2050 of BSI (2008) are based on this protocol. This protocol provides separate guidelines for GHG accounting and reporting during the life cycle of products and organizations.

iii) Publicly Available Specifications - 2050 (PAS 2050) of British Standard Institution - It specifies the requirements for assessing the life cycle GHG emissions of goods and services (BSI 2008).

C Accounting Tools

C accounting of a farm/enterprise/product can be carried out simply by collecting relevant data and subjected to calculations in MS Excel worksheet. However, looking into the importance of C accounting, many software tools have been developed as C calculator world over depending upon the clientele need. Common tools used for C accounting are as under-

a) The CLA CALM
b) Cplan
c) Cool Farm Tool
d) EasyPol Ex-ante

Steps in C accounting

One should select the methodology based on what one want to measure. Generally, maximum possible information should be collected for more accuracy and complete C accounting.

Selection of GHG

Selection of GHG for C accounting depends on the guidelines followed, the requirement of accounting, type of activity etc. For IFS, CH_4, CO_2 and N_2O may be significant. However, all standards and guidelines direct to include all six Kyoto gases.

Setting boundaries

A boundary in C accounting refers to selection of geographical area, length of accounting period and activities, which will be used in accounting of GHG emissions. Boundaries should be well defined to achieve the desired GHG emission goals and must be selected to meet standard requirements of GHG protocol. In an Integrated Farming Systems (IFS), boundaries for C accounting can be as under:

Livestock

Enteric fermentation

Animal Manure

Agricultural Soils

Direct emission

Indirect emission

Paddy cultivation

Farm Operations

Crop Residue Management

Household

C removal/sink and sequestration

Collection of GHG activity data

For accounting of GHG emissions from integrated farming systems, following activity data are recorded :

S.No.	Emission sources	Activity data	
1	Livestock	Enteric fermentation	No. of dairy and non dairy animals with age
		Animal Manure	Management of animal manure
2	Agriculture soils		Rate of application of N fertilizers, FYM, vermicompost
3	Farm Operations		Energy used in farm operation viz. harrowing, ploughing, sowing, harvesting etc.
4	Crop Residue Management		Incorporation or burning of crop residue, green manuring
5	Paddy cultivation		Area & moisture regime
6	Household		Energy use
7	C removal/sink and sequestration		Species & number of trees

Calculation of GHG

After defining the boundary, selecting the GHG and collection of activity data, assessment of emission of GHG is made either through using the applicable calculator or using the activity data and emission factor in spread sheet. The emissions at farm level are estimated by using the formula:

Emission = A * EF

Where,

Emission = Annual emissions in units of tones CO_2-e farm household^{-1} yr^{-1}

A = Activity data (kg of N used, litres of fuel used etc.)

EF = Default country specific or IPCC emission factors (CO_2 or CO_2-e per kg N, CO_2 or CO_2-e per litre fuel).

Accounting of GHG – A case study

A study to assess the emission of GHG from different components of farming systems was conducted in Meerut district of Uttar Pradesh. Four blocks viz. Sardhana, Hastinapur, Rohta and Rajpura blocks of Meerut district were selected. From each block, two villages and from each village, 10 farm households were selected for the study. The C accounting methodology followed the standard Intergovernmental Panel on Climate Change (IPCC) accounting and reporting guidelines with available default and country specific emission coefficients. The C accounting took into account the three Green House Gases (GHG) *viz.* CO_2, CH_4 and N_2O. CH_4 and N_2O emission was converted into CO_2-e by Global Warming Potential (GWP) of 1 for CO_2, 21 for CH_4 and 310 for N_2O. The study also took C sequestered by trees on the farms into account for net emission of GHGs. The operational boundary of the study consisted of agricultural soils, livestock production, paddy cultivation, farm operations, crop residue management and farm families.

Total mean emission of GHG from farming systems in the Meerut district was found to be 9.76 t farm household^{-1} yr^{-1} with net emission of 6.16 t farm household^{-1} yr^{-1} (Fig. 2). Among different categories of emission sources in the district (Fig. 1), livestock contributed 58.2 per cent of total CO_2-e emissions followed by agricultural soils (22.5 per cent) by N application in crops, household activities (11.4 per cent), farm operations (5.78 per cent), residue burning (1.35 per cent) and paddy cultivation (0.800 per cent). Amongst blocks, Hastinapur block contributed 37.5 per cent of the total emission of GHG (Fig. 2) followed by the Rohta block (25.1 per cent), Rajpura block (20.9 per cent) and

Sardhana block (16.5 per cent), while the net emission was maximum in Rajpura block (30.5 per cent) followed by Hastinapur block (29.1 per cent), Rohta (20.7 per cent) and Sardhana (19.7 per cent).

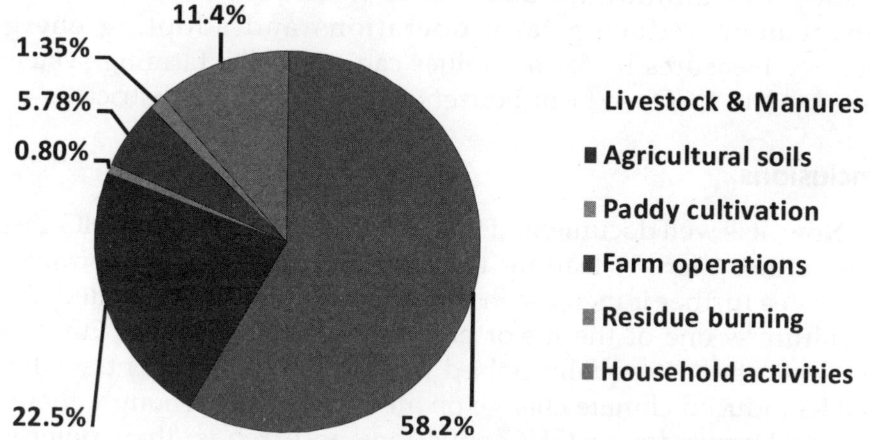

Fig. 1. Source wise emission of CO_2-e (kg) farm household^{-1} year^{-1} in Meerut district

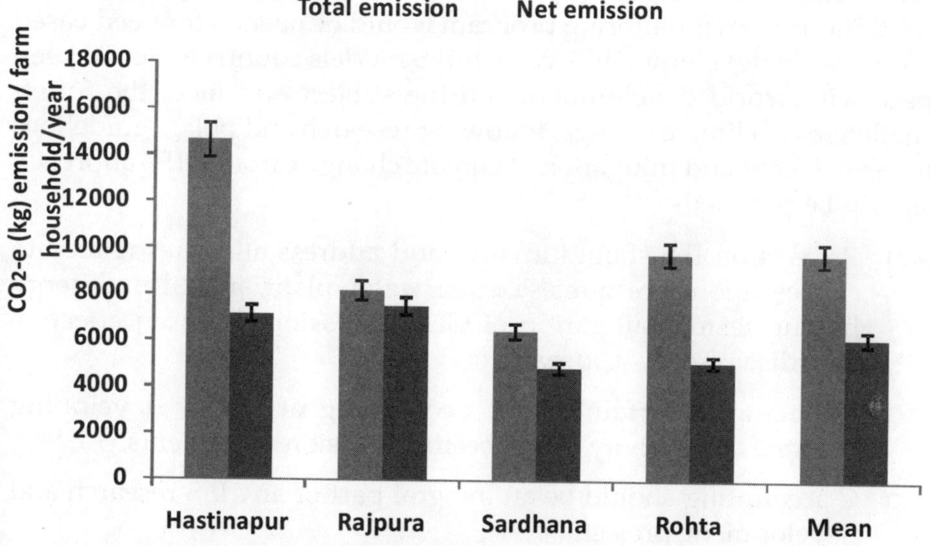

Fig. 2. Block-wise total and net emission of CO_2-e (kg) in the Meerut district

By adopting agroforestry with additional trees of 273 per farm household with an average size of holding of 2.07 ha makes the farming systems in the Meerut district C neutral. In other words, it is advocated that by planting 132 trees more ha^{-1} or having 209 trees ha^{-1} by each farm households in the Meerut district, farm households can be C neutral. At block level, planting 318, 334, 216 and 227 additional trees per farm household makes the farms C neutral in Hastinapur, Rajpura, Sardhana and Rohta block, respectively. Further, by adopting minimum balanced fertilization, better livestock management, reducing farm operations and adopting energy efficiency measures by farm families can make the farming systems C –ve by 1.18 t CO_2 –e farm household^{-1} year^{-1} in the district.

Conclusions

Now, it is well documented that continuous increase in GHG gases in the atmosphere is leading to climate change and this situation aggravates further if increase in these gases continues unabated. And agriculture is one of the major emitters of GHG globally. Future of next generations is equally poised in balance due to potential effects of GHG induced climate change on all walks of life. Though, there is growing knowledge of GHG emissions and trends, their potential effects, low carbon practices and technologies sectorally, but C accounting of existing and specific IFS with low carbon practices and continuous farm monitoring program would be needed for need based measures in developing low carbon IFS models countrywide. To keep pace with world developments on the subject and meet the future challenges of climate change, following research and policy guidelines for assessment and mitigation of climate change through IFS approach should be pursued-

- Projects on IFS should involve and address all issues related to conservation of resources *i.e.* soil, water, plant, animal and energy having tested mitigation of GHG emissions with an effective interdisciplinary approach.
- Reducing uncertainty in C accounting of IFS by developing country and activity data specific emission coefficients.
- C accounting should be an integral part of any IFS research and development project.
- Calibration of existing and development of C accounting models for a specific region, crop type, climate etc. in IFS domain.
- Generation of data base to predict response of GHG emissions to management changes at farm, regional and national scale to take advantage of C credit markets in future agriculture, if permissible.

- Participation of stakeholders/farmers in research and development of climate smart IFS and creating awareness about emissions of GHG on his/her farm.
- Sociopolitical research to understand and overcome the concerns of farmers/stakeholders in adoption of practices and policies to mitigate the climate change.
- Formulation of government policy incentives to promote low C practices as an attractive option for agriculture. Agricultural subsidies can be linked with reduction in GHG emissions of a farm.

References

BSI 2008. Publicly Available Specification 2050- Specifications for the assessment of the life cycle greenhouse gas emissions of goods and services. British Standards Institute

faostat.fao.org/ default.aspx

IPCC 2006. 2006 IPCC Guidelines for National Greenhouse Gas Inventories: Volume 4: Agriculture, Forestry and other Land Use. Intergovernmental Panel on Climate Change. Paris, France.

IPCC 2013. *Climate Change 2013: The Physical Science Basis. Contribution of Working Group I to the Fifth Assessment Report of the Intergovernmental Panel on Climate Change* [Stocker, T.F., D. Qin, G.-K. Plattner, M. Tignor, S.K. Allen, J. Boschung, A. Nauels, Y. Xia, V. Bex and P.M. Midgley (eds.)]. Cambridge University Press, Cambridge, United Kingdom and New York, NY, USA, pp 1535.

Lal R. 2004. Carbon emissions from farm operations. *Environment International* **30**: 981-990.

Lynas M. 2007. Carbon Counter, Harper Collins Publishers, UK, Glasgow, UK.

Olander L, Wollenberg E, Tubiello F and Herold M. 2013. Advancing agricultural greenhouse gas quantification. *Environmental Research Letter* **8**: Pp.07.

Stern N. 2006. Stern review: the economics of climate change. Cambridge University Press, Cambridge.

Williams H and Wikstrom F. 2011. Environmental impact of packaging and food losses in a lifecycle perspective: a comparative analysis of five food items. *Journal of Cleaner Production*, **19**: 43-48.

WRI/WBCSD 2004. The Greenhouse Gas Protocol: A corporate accounting and reporting standard (revised edition). World Business Council for Sustainable Development and World Resource Institute, Geneva.

❏❏❏

Chapter 31

Role of Plant Growth Regulators in Integrated Farming System

C.K. GUPTA, D. VIJAY, A. MAITY AND V.K. WASNIK

ICAR - Indian Grassland and Fodder Research Institute, Jhansi-284003 (UP)

Introduction

Crop production based farming has become less viable due to its poor return. Uncertainty in income and employment has led many farmers to abandon agriculture. Therefore, it is imperative to evolve suitable strategy for augmenting the income of a farm which remains sustainable and viable over longer period. Integration of various agricultural enterprises viz., cropping, livestock production, fishery, agro forestry etc. which is commonly known as integrated farming system (IFS) provide great opportunity for improvement of agricultural economy. These enterprises in integration not only supplement the income of the farmers but also help in increasing the family labour employment and resource use efficiency. The livestock based integrated farming system approach introduces a change in the farming techniques for maximum crop production and takes care of optimal utilization of natural resources. The farm wastes including livestock by-products are better recycled for productive purposes in the integrated system. A judicious mix of agricultural enterprises like dairy, poultry, piggery, fishery, sericulture etc. suited to the given agro-climatic conditions and socio-economic status of the farmers would be crucialbringing prosperity to the farming community.

Use of plant growth regulators (PGRs) in this integrated system may be one of the options to harness the maximum benefit by altering the crop growth patterns. They may be useful in regulation of various physiological processes related to growth and development of crops. It is obvious that changes in the level of endogenous hormones due to biotic and abiotic stress alter the crop growth and any sort of manipulation including exogenous application of growth substances would help for yield improvement or at least sustenance of the crop. PGRs have found immense use in agriculture like in preventing lodging in cereals, hastening maturity to decrease turnover time, preventing preharvest fruit drop, synchronizing maturity to facilitate mechanical harvest and reducing labour requirement.

Plant growth regulators (also known as growth regulators or plant hormones) are chemicals used to alter the growth of a plant or plant part. According to Devlin and Witham (1986), *"plant growth regulators* are organic compounds other than nutrients that in small amounts promote, inhibit, or otherwise modify a physiological process in plants. Plant hormones or phytohormones are regulators produced by plants, in which a low concentration regulates plant physiological processes." The growth regulators can either be natural or synthetic and are widely being used in the agricultural production. The plant growth regulators have various economic importances in agricultural field. Their importance has been a boon for the farmers and horticulturists. According to the American Society for Horticultural Science, plant growth regulators fall into six major classes. Table 1, below, lists these classes with the plant development function(s) that are controlled by the plant growth regulators. Table 1 also provides examples of the practical uses with which plant growth regulators are typically associated.

Table 1. PGRs classes, its functions and practical uses (ASHS, USA)

Class	Functions	Practical uses
Auxins	Shoot elongation	Thin tree fruit, increase rooting and flower formation
Gibberellins	Stimulate cell division and elongation	Increase stalk length, increase flower and fruit size
Cytokinins	Stimulate cell division	Prolong storage life of flowers and vegetables and stimulate bud initiation and root growth
Ethylene generators	Ripening	Induce uniform ripening in fruits and vegetables
Growth in hibitors	Stops growth	Promotes flower production by shortening of internodes
Growth retardants	Slows growth	Retard tobacco sucker growth

Till date many synthetic and natural compounds have been characterized for its use in the agriculture. The use of plant growth regulators in agricultural production within the United States and other developed countries began in the 1930s. The first discovery and use of plant growth regulators was with acetylene and ethylene, which enhanced flower production in pineapple. Subsequently, use of plant growth regulators has grown exponentially to become a major component of agricultural commodity production. Certain herbicides and insecticides that are not true plant growth regulators cause some plant-growth-regulating effects. For example, the widely used insecticide carbarylis used to thin apple fruit from trees and to aid in encouraging annualbearing.

Plant growth regulators/Plant hormones/Phytohormones

The word hormone is derived from Greek, meaning set in motion. Plant hormones affect gene expression and transcription levels, cellular division, and growth. They are naturally produced within plants, though very similar chemicals are produced by fungi and bacteria that can also affect plant growth. A large number of related chemical compounds are synthesized by the industry. They are used to regulate the growth of cultivated plants, weeds, and in vitro-grown plants and plant cells; these manmade compounds are called Plant Growth Regulators or PGRs.

Plant hormones are not nutrients, but chemicals that in small amounts promote and influence the growth,development, and differentiation of cells and tissues. The biosynthesis of plant hormones within plant tissues is often diffuse and not always localized. Not all plant cells respond to hormones, but those cells that do are programmed to respond at specific points in their growth cycle. The greatest effects occur at specific stages during the cell's life, with diminished effects occurring before or after this period. Plants need hormones at very specific times during plant growth and at specific locations. The production of hormones occurs very often at sites of active growth within the meristems, before cells have fully differentiated. After production, they are sometimes moved to other parts of the plant, where they cause an immediate effect; or they can be stored in cells to be released later. The concentration of hormones required for plant responses are very low (10"6 to 10"5 mol/L).

Traditionally five major classes of plant hormones are listed: auxins, cytokinins,gibberellins, abscisic acid and ethylene. However as research progresses, more active molecules are being found and new families of regulators are emerging; one example being polyamines such as putrescine or spermidine. This classification is based partially on the chemical structure and partially on the commonalities of plant physiological affects that certain substances exhibit. Members of one class may not relate from a structural point of view to another.

Auxins for instance include not only many indole 3-carboxylic acid derivatives but numerousphenyl acetic acids as well. Most cytokinins (such as zeatins) are derivatives from adenine but still differ widely in their chemical structure. Hence, the mechanism driving action may be different in each case and likewise each specific activity will differ also. This is demonstrated by the range of optimal concentrations required for different factors which spans many decimals (0.001-100 mg/L). Role of five classical plant hormones is listed below:

Auxins: Auxins are compounds that positively influence cell enlargement, bud formation and root initiation. They also promote the production of other hormones and in conjunction with cytokinins, they control the growth of stems, roots, and fruits, and convert stems into flowers.Auxins were the first class of growth regulators discovered.They affect cell elongation by altering cell wall plasticity. They stimulate cambium, a subtype of meristem cells, to divide and in stems cause secondary xylem to differentiate. Auxins act to inhibit the growth of buds lower down the stems (apical dominance), and also to promote lateral and adventitious root development and growth. Leaf abscission is initiated by the growing point of a plant ceasing to produce auxins. Auxins in seeds regulate specific protein synthesis,as they develop within the flower after pollination, causing the flower to develop a fruit to contain the developing seeds. Auxins are toxic to plants in large concentrations; they are most toxic to dicots and less so to monocots. Because of this property, synthetic auxin herbicides including 2,4-D and 2,4,5-T have been developed and used for weed control. Auxins, especially 1-Naphthaleneacetic acid (NAA) and Indole-3-butyric acid (IBA), are also commonly applied to stimulate root growth when taking cuttings of plants. The most common auxin found in plants is indole-3-acetic acid or IAA.

Role of Auxin

- Stimulates cell elongation

- Stimulates cell division in the cambium and, in combination with cytokinins in tissue culture
- Stimulates differentiation of phloem and xylem
- Stimulates root initiation on stem cuttings and lateral root development in tissue culture
- Mediates the tropistic response of bending in response to gravity and light
- The auxin supply from the apical bud suppresses growth of lateral buds
- Can induce fruit setting and growth in some plants
- Delays fruit ripening
- Promotes flowering
- Stimulates growth of flower parts
- Promotes femaleness in dioecious flowers
- Stimulates the production of ethylene at high concentrations.

Practical Application and its economic importance

1. **Germination:** IAA, IBA, NPA, 2, 4-D are widely used to soak seeds for germination.Auxins are widely used to break the dormancy of the seeds. The pre-treatment of seeds before sowing with IAA, IBA, 2, 4-D, etc found to have increased the length of the hypocotyls at low concentration and shortening at higher concentration.
2. **Rooting:** NAA and IBA are used for inducing the rooting of cuttings of woody plants.
3. **Flowering:** Minute quantity of auxins accelerate flowering in many crops. It is also reported that IAA promotes the formation of female flowers. The application of auxins also helps to increase the yield and flowering of wheat. NAA and 2, 4-Dinduce flowering in litchi and pineapple.
4. **Parthenocarpy:**It is produced mainly by NAA and IBA.Treatment of NAA in cucumber and melon fruits produces seedless fruits.
5. **Weeds control:** They are selective weeds killers and considerably used in the destruction of weds in crop fields, rail-roads, lawns and forests. Common weedicides of auxin nature includes 2, 4-D and 2, 4, 5-T, etc.

Gibberellins: Gibberellins, or GAs, include a large range of chemicals that are produced naturally within plants and by fungi. They were first discovered when Japanese researchers, including Eiichi Kurosawa, noticed a chemical produced by a fungus called *Gibberellafujikuroi* that produced abnormal growth in rice plants. Gibberellins are important in seed germination, affecting enzyme production that mobilizes food production used for growth of new cells. This is done by modulating chromosomal transcription. In grain (rice, wheat, corn, etc.) seeds, a layer of cells called the aleurone layer wraps around the endosperm tissue. Absorption of water by the seed causes production of GA. The GA is transported to the aleurone layer, which responds by producing enzymes that break down stored food reserves within the endosperm, which are utilized by the growing seedling. GAs produce bolting of rosette-forming plants, increasing internodal length. They promote flowering, cellular division, and in seeds growth after germination. Gibberellins also reverse the inhibition of shoot growth and dormancy induced by ABA.

Role of gibberellins

- Stimulate stem elongation by stimulating cell division and elongation.
- Stimulates bolting/flowering in response to long days.
- Breaks seed dormancy in some plants which require stratification or light to induce germination.
- Stimulates enzyme production -amylase in germinating cereal grains for mobilization of seed reserves.
- Induces maleness in dioecious flowers (sex expression).
- Can cause parthenocarpic fruit development.
- Can delay senescence in leaves and citrus fruits.

Practical Application and its economic importance

1. **Germination:** It has been widely used for germination experiments. GA is found to increase the length of hypocotyls and cotyledonary leaf. GA also has important role in seed germination of cereals.
2. **Flowering:** GA induces flowering in long day plants and in plants requiring inductive cold period. It also promotes the formation

of male flowers as it plays important role in development of male flower.

3. **Parthenocarpy:** GA induces the formation of parthenocarpic fruits in brinjal and guava. It is also used to induce the parthenocarpy in pome fruits and use to increase the fruit size and bunch length in grapes as well.

4. **Fruit Setting:** GA affects more positively on fruit setting. GA promotes increase in fruit setting in seedless grapes;an increased in fruit size, fruit set, total yield and quality in sweet lime have also been due to the treatment of GA.

GAs is used to increase the size of Thompson grapes berries. It is used by some breweries to increase the rate of malting by increasing the potential of starch digestion. Gibberellins are also use to increase sugar cane growth and sugar yields. Gas also have role in increasing stalk length and production, breaking of dormancy, inducing uniform crop emergence, producing staminate flowers, loosening of fruit clusters, increasing the fruit size, quickening of maturity, improving fruit quality, controlling cracking of fruits, reducing non-productive period, promoting vegetative growth and flower size, increasing sugar content in sugar cane, inducing flowering and so on.

Cytokinins: Cytokinins or CKs are a group of chemicals that influence cell division and shoot formation. They were called kinins in the past when the first cytokinins were isolated from yeast cells. They also help delay senescence of tissues, are responsible for mediating auxin transport throughout the plant, and affect internodal length and leaf growth. They have a highly synergistic effect in concert with auxins, and the ratios of these two groups of plant hormones affect most major growth periods during a plant's lifetime. Cytokinins counter the apical dominance induced by auxins; they in conjunction with ethylene promote abscission of leaves, flower parts, and fruits.

Example: Kinetin, Benzyl adenine.

Role of Cytokinins

- Stimulates cell division.
- Stimulates morphogenesis -shoot initiation/bud formation in tissue culture.

- Stimulates the growth of lateral buds-release of apical dominance.
- Stimulates leaf expansion resulting from cell enlargement.
- May enhance stomata opening in some species.

The major use for cytokinins derives from its ability to delay senescence and maintain greenness. The artificial, highly active cytokinin,benzyladenine, is the main compound used. The treatment of holly for festive decorations enables its harvest many weeks prior to use. Post-harvest sprays are now available to prolong the storage life of green vegetables such as asparagus, broccoli, and celery.

These hormones also have been used to increase the shelf life of fruits, quickening of root induction and production of efficient root system. It is also used to increase the yield and oil contents in ground nuts, to break thé dormancy, delay senescence, causes cell division and so on.

Abscissic Acid (ABA)

It is regarded as one of the principal accelerators of abscission and also bud dormancy. It is natural growth inhibitors, and acts as a stress hormone that helps to cope with adverse environmental conditions.

Role of Abscissic Acid

- Stimulates the closure of stomata-water stress brings about an increase in ABA synthesis.
- Inhibits shoot growth but will not have as much affect on roots or may even promote growth of roots.
- Induces seeds to synthesize storage proteins.
- Inhibits the affect of gibberellins on stimulating.
- Has some effect on induction and maintenance of dormancy.
- Induces gene transcription especially for proteinase inhibitors in response to wounding which may explain an apparent role in pathogen defense.

Ethylene: Ethylene is a gas that forms through the breakdown of methionine, which is in all cells. Ethylene has very limited solubility in water and does not accumulate within the cell but diffuses out of the cell and escapes out of the plant. Its effectiveness as a plant hormone is dependent on its rate of production versus its rate of escaping into the atmosphere. Ethylene is produced at a faster rate in rapidly growing and dividing cells, especially in darkness. New growth and newly germinated seedlings produce more ethylene than can escape the plant, which leads to elevated amounts of ethylene, inhibiting leaf expansion. It is commonly known as ripening hormone as it is highly effective in fruit ripening. It is actually the natural product of ripening fruits. It is a gas at temperature under which a plant can survive. It is a powerful inhibitor of root and stem elongation.

Role of Ehtylene

- Stimulates the release of dormancy.
- Stimulates shoot and root growth and differentiation (triple response)
- May have a role in adventitious root formation.
- Stimulates leaf and fruit abscission.
- Stimulates Bromiliad flower induction.
- Induction of femaleness in dioecious flowers.
- Stimulates flower opening.
- Stimulates flower and leaf senescence.
- Stimulates fruit ripening.

Ethylene gashas a wide variety of uses, but its gaseous nature precludes its applicationsin non-enclosed spaces. Ethylene itself can be used to enhance the ripening of fruits such as bananas in storage following their shipment in an unripe condition. An ethylene-producing liquid chemical, 2-chloroethylphosphonic acid- commercially called Ethrel or Ethephonhas been introduced and it is sprayed onto the plant at a slightly acid pH.When it enters the cells and encounters the cytoplasm at about neutral pH,it breaks down to release gaseous ethylene which is commercially used for fruit ripening. These hormones have a potential area of use in a wide variety of fruits such as grapes, cherries, and citrus. Promotion of female flower production in cucurbits -cucumber, squash, and melon so as to increase the number of fruits produced per plant. It is also used for the promotion of flower initiation and controlled ripening in pineapples.

Other known hormones

- **Brassinosteroids** - are a class of polyhydroxysteroids, a group of plant growth regulators. Brassinosteroids have been recognized as a sixth class of plant hormones, which stimulate cell elongation and division, gravitropism, resistance to stress, and xylem differentiation. They inhibit root growth and leaf abscission. Brassinolide was the first identified brassinosteroid and was isolated from extracts of rapeseed (*Brassica napus*) pollen in 1979.
- **Salicylic acid** - activates genes in some plants that produce chemicals that aid in the defense against pathogenic invaders.
- **Jasmonates** - are produced from fatty acids and seem to promote the production of defense proteins that are used to fend off invading organisms. They are believed to also have a role in seed germination, and affect the storage of protein in seeds, and seem to affect root growth.
- **Plant peptide hormones** - encompasses all small secreted peptides that are involved in cell-to-cell signaling. These small peptide hormones play crucial roles in plant growth and development, including defense mechanisms, the control of cell division and expansion, and pollen self-incompatibility.
- **Polyamines** - are strongly basic molecules with low molecular weight that have been found in all organisms studied thus far. They are essential for plant growth and development and affect the process of mitosis and meiosis.
- **Nitric oxide (NO)** - serves as signal in hormonal and defense responses (e.g. stomatal closure, root development, germination, nitrogen fixation, cell death, stress response). NO can be produced by a yet undefined NO synthase, a special type of nitrite reductase, nitrate reductase, mitochondrial cytochrome c oxidase or non-enzymatic processes and regulate plant cell organelle functions (e.g. ATP synthesis in chloroplasts and mitochondria).
- **Strigolactones** - implicated in the inhibition of shoot branching.
- **Karrikins** - a group of plant growth regulators found in the smoke of burning plant material that have the ability to stimulate the germination of seeds.

Plant growth regulators and their use in agriculture

PGRs in field crops

Crop	Chemical/ Dosage	Response
Rice	Kinetin, GA_3 and Triacontanol (1000 ppm)	Grain filling and partitioning. Delayed senescence
Cotton	IAA, NAA (30 ppm), CCC (cycocel)	Increase grain yield. Decrease boll shedding. Increase photosynthetic rate and yield. Increase number of bolls, boll weight and lint yield.
Sunflower	Benzyl adenine (BA) (250-ppm) GA+BA(150ppm)	Increase yield and Achenes weight and number.
Groundnut	Mepiquat chloride (125-ppm)(2,3,4-D-chloroph ynoxy triethyl amine)	Increased grain yield and chlorophyll synthesis and decrease chlorophyllase activity. Stimulation of assimilate transport in germination.
Sugarcane	Ethephon, Glyphosine	Reduced growth rate and regulate ripening.
Tapioca	Ethrel (250 ppm)CCC (1000 ppm)	Early tuberisation. Increase the weight of storage roots.
Pigeon pea	Ethrel (40 ppm) and GA_3 (20 ppm)CCC(0.64 mM)	Increase grain yieldRespond well for RWC, chlorophyll and stomatal conductance.
Carrot and Potato	GA_3 (50 ppm)	Induction of flowering in long day and seed setting.

Plant growth regulators in ripening of fruits

Crop	Chemical	Dose	Response
Mango	Ethephon	1000 ppm	Accelerated fruit ripening and improves surface colour
Citrus	Ethrel	1000 ppm	Induce yellow colourwith in seven days
Banana	Ethrel	1000 ppm	Accelerated ripening by two days
Papaya	Ethephon + NaOH	2000 ppm	Ripening within 24 hours
Sapota	Ethephon + NaOH	500 ppm	Ripening within two days

Plant growth regulators in flowering and fruit set

Crop	Chemical	Dose	Time of spray and number of spray	Response
Coconut	2,4-D	30 ppm	One month after opening of spathe, through micro sprayer	Fruit setting percentage increased to 32.5% against 23% in control
Banana	2,4-D	25 ppm	Within a week after opening of last bud	Prevents seediness in poovan
Mandarin orange	2,4-D or NAA	20 ppm, 100 ppm	Spray at flowering.	Increase fruit set
Thompson seedless	GA3	25 ppm	Dip cluster at calyptra falling stage	Increase fruit set
Pine – apple	Planofix 10 ppm +2% urea 0.04% sodium carbonate +20 ppm ethrel	50 ml/ plant applied in to the crown200 to 300 ppm	At 35-40 leaf stageSprayed during fruiting	Induced uniform flowering to increase the fruit size
Snake gourd, Bitter gourd, Bottle gourd	Ethrel	100ppm	4 times 10 –15 days after sowing at weekly intervals	Increase in yield
Ribbed gourd, Pumpkin	Ethrel	250ppm	4 times 10-15 days after sowing at weekly intervals	Increase in yield
Sweet potato	Ethrel	250ppm	At fortnight interval from 15 days after planting	Increased tuber yield

Plant Growth Regulators and Food Safety

From the regulatory control perspective, plant growth regulators are classified under "pesticides". The same classification applies both internationally, under the Codex systems and competent authorities of developed countries, as well as locally in the Pesticides Ordinance. As a well accepted principle, all pesticides, including plant growth regulators, have to be registered with the competent authority before application in agriculture. Their safety and efficacy will be thoroughly assessed during the registration process. Proper use of these registered pesticides including plant growth regulators in accordance with good agricultural practices (GAP) will result in minimal residue in food of insignificant food safety risk. GAP is a set of nationally authorised conditions to use the pesticide safely (e.g. in relation to public health and environmental safety concerns) for effective and reliable pest control. Various conditions, such as the type of commodities authorised for using the pesticide, the recommended application rates, frequencies, and amount as well as the duration between the last application of the pesticide and harvest, are prescribed in the GAP.

Conclusion

In conclusion, plant growth regulators are a group of chemicals for controlling and enhancing the natural plant growth processes to better meet the requirements of food supply in general. It can be effectively utilized in various IFS modules to regulate plant growth pattern dependi9ng on need of the farmer. It gives an opportunity to shorten or prolong the crop growth cycle in integrated system. As far as biosafety of PGRs use in agriculture is concerned mechanisms are in place under the Codex system to oversee residues of pesticides (including plant growth regulators) in food for setting standards and public health protection.

References

Devlin, R.M. & Witham, F.H. (1986). Plant Physiology (1st ed.) Delhi: CBS Publishers & Distributors.

Lincoln Taiz and Eduardo Zeiger (2010) Plant Physiology (Fifth Edition): Sinauer Associates.

Nickell L.G. (1982) Plant growth regulators, agricultural uses. Springer-Verlag. Berlin, Heidelberg, New York.

Pandey, S.D. & Sinha, B.K. (2006). Plant Physiology (4th ed.) Noida, UP: Vikas Publishing House Pvt Ltd.

Srivastava, L.M. (2002). Plant growth and development: hormones and environment. Academic Press. p. 140. ISBN 0-12-660570-X.

□□□

Chapter 32

Participatory Rural Appraisal (PRA) for Socio-economic Analysis

PURUSHOTTAM SHARMA

ICAR - Indian Grassland and Fodder Research Institute, Jhansi-284003 (UP)

Introduction

Participatory Rural Appraisal (PRA) is a growing family of approaches to enable local people to share, enhance and analyze their knowledge of life and condition to plan, to act, to monitor and evaluate. PRA combines the traditions and methods of participatory research, applied anthropology, and field research on farming systems, agroecosystem analysis and rapid rural appraisal (RRA). Participatory Rural Appraisal is an intensive, systematic but semi-structured learning experience carried out in a community by a multidisciplinary team which includes community members (Chambers, 1994). Participation is critical components for success and is associated with increased mobilization of stakeholder's ownership of policies and projects empowerment, cost reduction and transparency etc (Pretty 1995). The four major sectors of PRA applications are natural resources management, agriculture, poverty and social programmes, health and food security (Chambers, 1994).

Basics of PRA

According to Chambers, 1994, the three basics of PRA are

1. Participatory methods to facilitate analysis by rural people.
2. Behavior and attitudes of outsiders. Passive role for outsiders (hand over the stick), and active and confident participation by local people to express their own knowledge, to conduct their own analysis, and to assert their own priorities. DIY (do-it yourself) is important principle of PRA for local people.

- The third basic is sharing (knowledge and experience). For practitioners, trainers and local people share among themselves, local people with outsiders, outsiders themselves with each other and with local people. Philosophy of PRA emphasizes open access to information and avoiding professional possessiveness.

PRA is a means to people's participation

- To understand the social structure of the village like income groups, size of land holding, size of cattle population, livelihood, predominant occupations, dependency on other organizations
- To develop common understanding on the resource both natural and man-made i.e. types of soils, types of lands and their extent, water sources, land use pattern etc.
- To understand farmers decision making process
- Farmers need assessment
- To identify problems and prioritize, intervention points and their appropriate technologies.
- Direct observation at farmer's field in existing situation.
- To plan for management and utilization of common property resources
- Inclusive participation of resource poor, weaker sections and women farmers.

PRA may be exploratory or topical and can be used for many purposes i.e. farmer oriented planning, need assessment, feasibility analysis, prioritization, monitoring and evaluation. For better usefulness of PRA, it requires attitude which encourages participation of farmers, respect for community members, curiosity to know their existing system, patience, not rushing, not interrupting, listening not lecturing and methods which encourages the community to express, share, enhance and analyze their knowledge.

Principles of PRA

- Active role for community members
- Openness in sharing knowledge, experience etc.
- Optimal Ignorance of the issues which are not relevant
- Appropriate imprecision

- Triangulation of facts with other resources
- Interdisciplinary team
- Handing over stick to the community members
- Methods and Techniques.

Types of Participation

- Passive participation
- Participation in information giving
- Participation by consultation
- Participation for materials incentives
- Functional participation
- Interactive participation
- Self mobilization.

Key Elements of Participatory Application

- Bottom up approach
- To solve production problems
- Build upon indigenous technical knowledge (ITKs) and adoption
- Provide options, not solutions
- Group action
- Institutional building
- Capacity building
- Flexibility
- Equity
- Sustainability
- Facilitative/handing over the stick.

Table 1. Difference in Conventional v/s Participatory Approach

S.No.		Conventional approach	Participatory approach
1	Purpose	to collect information	to empower local people
2	Goals	pre-determined	evolving
3	Approach	uniform/ blue print	flexible, diverted
4	Modes of operation	Extractive	participatory
5	Focus of decision making	external centralized	local people
6	Methods	highly structured précised measurement	open ended visual
7	Role of researcher	controller expert	catalyst, facilitator
8	Role of lock people	passive reactive	active creating
9	Ownership of results	by outsiders	by local people
10	Out put	reports, publications	enhanced local action, capacity building

(Cited from: Manage, Hyderabad)

Methods and Techniques

The applications can be grouped in four major types of process (Chambers, 1994) are:

- participatory appraisal and planning;
- participatory implementation, monitoring and evaluation of programs;
- topic investigations;
- training and orientation for outsiders and villagers.

The information about the agro-eco systems and socio-economic conditions in the villages are obtained through suitable Participatory Rural Appraisal (PRA) tools and semi structured interview (SSI) schedule by a multi-disciplinary team of scientists. The production problems according to the micro-farming situations are identified, prioritized and problem cause analysis is done for identification of intervention points through farmer scientist interface meetings.

The possible solutions for each intervention point are called out and the appropriate ones are selected for technology intervention. Then they are detailed including the economic parameters like the total cost of intervention, return per rupee spent along with criteria for technical feasibility and farmers' perception as well as refined suggestions for revival. The programme is developed in multi-disciplinary and institutional mode in the transparent manner through farmer participatory bottom-up approach (Table 2).

Table 2. Methods and Techniques

S.N.	Steps involved	PRA tools used
1.	Characterization of production Systems/ micro farming situations analysis. Learning through Participatory Approach	Agro ecosystem analysis (AES), space analysis / mapping/Modeling/ transact.
2.	Inventorying technologies being followed by the farmers and identifying technology gaps through Farmers practice	Technology mapping
3.	Understanding variations with respect to time in various properties of agro-ecosystem analysis	Time analysis
4.	Identifying the problems of different production systems and prioritization	Matrix/problem ranking
5.	Problem cause analysis	Flow analysis/Problem tree
6.	Locating the technology alternatives available at SAUs/ICAR and deciding the alternatives for testing through discussion with target groups	
7.	Preparation of action plan	Decision analysis
8.	Implementation of the project by	Conducting OFT/VT
9.	Designing suitable modules for the Intervention	
10.	Assessment of technological interventions as per the objectives of the project	Agronomic/statistical/ economical and farmers' perspectives
11.	Evaluation of technologies and identifying the needs for refinement	Rural peoples' knowledge
12.	Repetition of steps from 7-10	
13.	Evaluation of technologies for stability, sustainability productivity and equitability in different production system	
14.	Impact analysis of refined technologies- By using interview schedule	
15.	Extrapolation of technologies to sustain production systems	

List of PRA techniques (Chambers, 1994 and Cavestro, 2003)

- Direct observation
- Do it yourself
- Participatory mapping and modeling
- Transect walks and guided field walks
- Seasonal calendars
- Daily-activity profiles
- Semi structured interviewing
- Types, sequencing, and chain interviews
- Permanent-group interviews
- Time lines
- Local histories
- Local researchers and village analysts.
- Venn diagrams
- Participatory diagramming
- Wealth and well-being rankings
- Direct-matrix pair-wise ranking and scoring
- Matrices
- Traditional management systems and local-resource collections
- Portraits, profiles, case studies, and stories
- Key probes
- Folklore, songs, poetry, and dance
- Futures possible
- Diagrams exhibition
- Shared presentations and analysis
- Night halts
- Short questionnaires
- Field report writing
- Survey of villagers' attitudes toward PRA
- Intriguing (Indigenous) practices and beliefs.

Organization of PRA

1. Selection of PRA team members
2. Objectives
3. Formation of sub-topics
4. Selection of methods, designs and respondents
5. Interview
6. Sub-team meeting
7. Whole team meeting
8. Report writing.

Semi-structured interview (SSI)

Semi-structured interviews are guided conversations, relatively informal, relaxed discussion based around a predetermined topic through a series of guide questions (not structured questionnaire) relevant to the villagers with individuals, key informants, interest groups or other small groups of villagers i.e. women's groups (Cavestro, 2003).

Basic information of the District/village

As per the selection criteria of the site, that is representative of AES/PSR in district, demonstration value of site, diverse production system representing varying situation, allow implementation of varying farming situations at sector - the core team of scientists select villages and their basic information is collected both from villages and secondary data.

Village transect walk

Transect walk taken along with able and willing community members/farmers helps to observe the facts directly on the field. The Land type, soil type, water source, crop/tree/livestock/fish enterprises, problems, (include all possible variations) opportunities to mend them and the potentialities to further develop the resources can be understood by farmers and the PRA practitioners (Table 3).

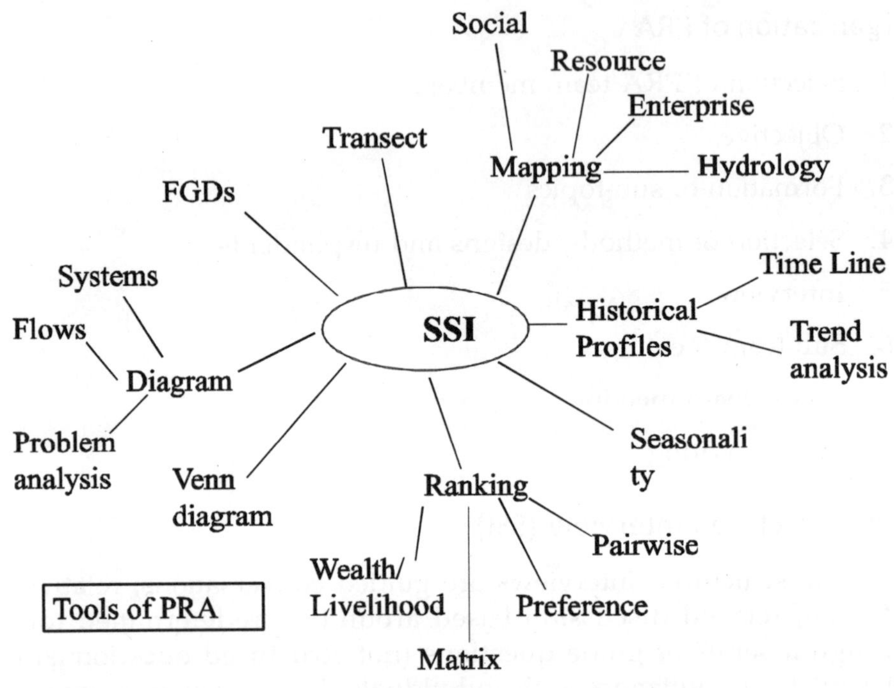

Classification Table 3 Example of transect

	Cultivated Land	Land cultivated at slopes	South to North cultivated land	Uncultivated
Slopes	1 - 1.5%	1.5 - 2 %	1 - 3 %	3 - 5 %
Soil type	Kaber & Red	Red & Rakar	Kaber & Red	Rakar & Red
Land Use	Groundnut, Soybean, Maize, urd, moong, til tomato and chillies Wheat, Gram, peas, mustard and cucurbites	Groundnut, urd, moong, Gram, mustard	Groundnut, soybean, maize, urd, moong, til, chillies, Wheat, gram, peas, mustard and cucurbites and local grasses	Urd, moong and fallow
Source of irrigation	Pond, wells	Diesel pump	Pond, diesel pump	-
Soil erosion	Low	Moderate	Moderate	High

Vegetation	Neem, subabool, babool, lime, guava and weeds of different types	Neem, babool, imli, banian tree, sirus and mango	Mango, imli, subabool, lime and guava	Neem, reonja, karondi and wild bushes
Insects/ pests	Rats, termites and aphids	Rats and termites	Rats, termites and aphids	Rats and termites
Diseases	Tikka of Groundnut, wilt of gram, rust and smut of wheat	Tikka of Groundnut and wilt of gram	Tikka of Groundnut, wilt of gram, rust and smut of wheat	-
Problems	Menance of Rats, termites and problem of stray cattle	Menace of Rats, termites and problem of stray cattle	Menace of Rats, termites and problem of stray cattle	Menace of Rats, termites and problem of stray cattle

Social map

Social map of the village shows families living in the village and their population data, no. of farm families, social structure of the village and social institutions, Interdependence of families on each other, informal/local leaders and their position, all the primary necessities of the village, i.e. temple, schools, hospital hospitals etc. The map helps team to understand the basic situation of the village and social dynamics which further help in forming self-help groups and thrift and saving groups.

Crop- pattern map

The map shows main cropping pattern followed in the village and help to understand the shift in the pattern.

Resource map

The map include Enterprise map, Hydrology map, Soil map and other Resource map.

Seasonality analysis

Seasonal diagrams helps to understand seasonal variations in amount of rainfall, availability of residual moisture, fodder, employment opportunities, migration pattern, expenditure pattern, income flows and availability of minor forest products etc.

Seasonal analysis (Activities)

The main activities, problems and opportunities of the village are identified by using seasonal calendar (Table 4).

Table 4. Seasonal analysis (Activities)

Particulars	Jul	Aug	Sep	Oct	Nov	Dec	Jan	Feb	Mar	Apr	May	Jun
Crop Husbandry												
Kharif												
Rabi												
Animal Husbandry												

Seasonal analysis (Problems)

The seasonal analysis of the problems related to the diseases and pests affecting crops and livestock are carried out. In this method, the pests and diseases affecting various crops are documented pertaining to their months of infestation. Other information like unavailability of labor and farm machinery are also recorded by involving the farmers. The information collected is depicted in the tabular form (Table 5) or by chart.

Table 5. Seasonal analysis (Problems)

Particulars	Jul	Aug	Sep	Oct	Nov	Dec	Jan	Feb	*Mar*	Apr	May	Jun
Crop Husbandry												
Kharif												
Rabi												
Animal Husbandry												

Timeline Exercise / Trend Analysis

Over a period of time what changes are taking place in the status of land water, vegetation and their effect on Socio-Economic pattern of the villages can be known by using this technique. The following are some of the aspects that can be understood:

- Trends in the amount of top soil and their loses
- Trends in land utilization pattern

- Forest cover
- Amount of rainfall
- Water bodies, their size and location
- Animal population
- Vegetation
- Usage of timber
- Transport facilities
- Migration pattern
- Influence of outside forces
- Cattle production
- Changes in predominant occupations.

Matrix Ranking

By employing this technique, we can understand villagers / user groups preferences for various items and what factors influence and contribute in their decision making process. Some of the examples where matrix ranking can be used are listed below:

- Selection of crop in a particular piece of land
- Variety preference of a crop
- Trees ranking for social forestry in common lands
- Trees ranking for horticultural crops either in private or common lands.

Venn Diagram

The Venn Diagram helps us to understand the network of relationships of various institutions working for the village. The Venn diagram also depicts the role played by various agencies, firms or persons for agricultural development in village with key informants. Degree of influence and relationship between them are identified and the relationship between them is depicted using circles. The size of circles indicates the importance of respective organizations and their future scope.

Resource flow map

Resource flow map helps to understand the in and out supply position of different agro inputs and products. The diagram is made on the basis of information obtained from representative farmer and through survey of the village. It was constructed considering the existing common practices as well as possible proposed way for better utilization of bio resources.

Mobility map

The mobility map depicts the most frequently visited place for most of the purpose due to its vicinity and well connectivity with the village (Table 6). It depicts the interactions within the community and with other community in the village and helps to know the social, agricultural and economic growth of a given society.

Table 6. Mobillity analysis

S.No.	To	Mode of transport	Distance (Km)	Fare (Rs)	Purpose
1					
2					
3					
4					

Daily Routine Diagram

Daily routine refers to the activities carried out by the villagers during the day till they retire to bed at night including various agricultural and domestic activities performed by individuals. Based on daily routine project activities can be planned.

Livelihood Analysis

The livelihood analysis reveals the main source of income and main aspect of expenditure of different categories of farmers thus their system of managing the family and farm unit.

Wealth Ranking

In order to identify different economic classes, the farmers are encouraged to wealth rank all the farmers in the village. The farmers consider various criteria i.e., diverse sources of income, ownership of different agricultural machinery and implements and a number of school going children in each family for wealth ranking. Despite diverse criteria for wealth ranking it appeared that wealth status of an individual villager is determined by the size of his land holding in the village

Indigenous Technical Knowledge (ITK)

The traditional, and technical knowledge as well as their innovations are documented and if needed can be validated.

Gender Analysis

Gender analysis of the village shows that role played by the male and females of family in the household activities, livestock's and agricultural activities as well as market, social and leisure activities in all categories of people (Table 7).

Table 7. Gender Analysis

Aspects	Resource Rich Farmers			Resource Poor Farmers		
	Male	Female	Both	Male	Female	Both
A) Crop Husbandry						
B) Animals husbandry						
C) Horticultural Crops						
D) Fish Farming						
E) House hold activities						

Focused Group Discussions (FGD)

FGDs help in understanding an issue thoroughly those who are either positively or negatively affected by an action.

Problem causal and solution

Participatory planning as a PRA tool is used to identify various problems in the village. It is a process where in farmers is facilitated to identify the problems related to crop production, livestock production, besides problems related to overall development of the village. Once the problems are identified, the farmers were encouraged to prioritise them. The information is then shifted to possible solution. The farmers are encouraged to analyse the problems themselves and seek solution as well. The important problems are identified in the village e.g.

1. Low milk yield of cows
2. Low milk yield of buffaloes
3. Lack of credit facilities
4. Lack of education facilities
5. Lack of medical facilities
6. Low productivity of groundnut
7. Poor yield of soybean
8. Low productivity of wheat
9. Poor yield and income from fisheries
10. Problem of drudgery faced by farm women
11. Low Productivity and poor quality of wool from sheep
12. Problem of soil erosion.

Action plan

Through suitable Participatory Rural Appraisal (PRA) techniques, different categories of farming situation with different areas are identified. Accordingly their needs and resources also varied. Different technological modules encompassing the areas of arable farming, horticulture, and alternate land use system and off farm activities are formulated in consultation with the farmers.

The developed modules included introduction of crop varieties, crop and pest management practices, livestock management practices and activities for income augmentation as per the farmers' need, their problems, resources and willingness to act upon. Based on these modules, an action plan is prepared. Natural resource conservation is given due importance keeping in view of achieving stability of production and sustainability of production system.

Brief description of problems

The figures illustrated in problem caused diagram of various problems and their possible causes indicate the point of interventions.

Problem (e.g.) Low productivity of cattle and buffaloes :

There are large number of cattle and buffaloes Cows are mostly of non-descript type yielding 1-2 kg milk per day. The Murrah buffaloes are also comparatively poor yielder because of various constraints. The total milk production of village is around 300 kg per day, which reduces considerably during lean period due to limited availability of green fodder. Marketing of which is done in Jhansi market.

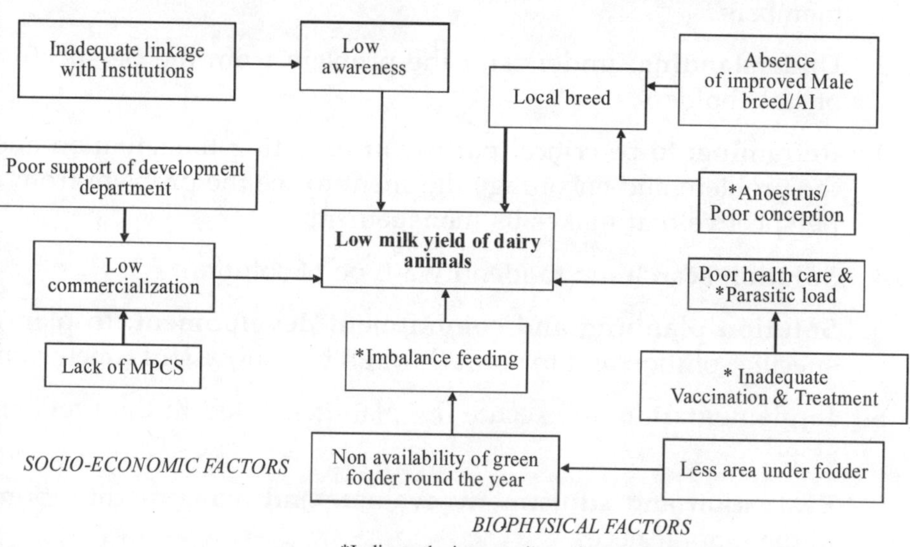

Fig.1. Problem Cause Digram For Low Milk Yield of Dairy Animals

Intervention points

- Balance nutrition
- Strategic supplementation of dry fodder.
- Disease control
- Poor potential of breed

Technological Interventions

- Introduction of fodder like maize, pearl millet guar and perennial grasses.
- Supplementation with mineral
- Vaccination of HS and FMD
- Treatment of repeat breeding and anoestrus

The production system wise problems, theme areas and suitable technology are identified. The interventions can be in the demonstration/trial form.

The eight stages in problem solving with PRA (Cavestro, 2003) are as follows :

1. **Rapport formation:** form a relationship with community members.
2. **Understanding:** understand the problem from the perspective of stakeholders.
3. **Reframing:** to be critical partner in reflecting the situation and the problem and encourage the client to see the problem from a perspective that makes its management
4. **Solution searching:** to identify a type of solution
5. **Solution planning and commitment development:** to plan a specific solution and to see it through to a successful conclusion.
6. **Implementation:** to execute the plan generated in the previous stage
7. **Evaluation and adjustment**: evaluate and make modifications in the plan, if necessary
8. **Ending and consolidation:** to help the partner consolidate the problem solving skills they have learned or the solutions they have adopted.

Utility of PRA in farming system research

Norman, while studying Farming Systems suggested that new technologies should, wherever possible, increase farmers' flexibility to adapt their production and broader livelihood systems to stochastic shocks and to the constantly changing economic environment (Chambers, 1991), more emphasis need to be placed on technologies that reduce risk and new technologies should complement, and not conflict with, the complex livelihood systems of poor households. The sustainable livelihood (SL) field work usually are very location specific; and the full application of the SL concept requires well trained and experienced interdisciplinary teams that are skilled in participatory methods. Some highlights of importance of participatory mode are :

- Encourage greater participation by farmers through addressing their specific needs rather than simply trying to "fit" technologies to specific enterprises that had been preselected by researchers/ mandates.
- Help in identifying separate area-based farming system.
- Increased sensitivity of technical scientists to the complexity and variability of farmers' production environments.
- Opportunity for technical and social scientists to cooperate in the diagnosis of farmers' situations and in the design, testing, and evaluation of new technologies, beyond the disciplinary boundaries.
- Shows importance of complementary policy/support systems (e.g., input distribution systems and product markets) in determining the appropriateness of new technologies.
- Farmers' participation reduce scepticism about validity of system or technologies.
- Develop linkages between farm and household in influencing decision making and flows of resources and benefits.
- Interdisciplinary linkages with crop and livestock scientists.
- Better data collection methodology highlighting new dimensions e.g., multiple decision makers, differentiation in terms of distribution of benefits).
- The types of technologies that could be developed/evaluated are in context of the policy/support system.
- The development of participatory rural appraisal (PRA) techniques incorporated concerns i.e. farmers interpreted their production situations, articulation of their constraints and needs to researchers; participation in design and evaluation of new technologies. Basically, PRA techniques improved the potential usefulness of farmers' participation to both farmers and researchers. (Biggs, 1989).
- Household analysis increases sensitivity to gender issues.
- Help in developing more appropriate methods to analyze the results of on-farm research and make recommendations based on them (PRA techniques - especially matrix ranking and scoring).

- Improved farmer participation in all phases of the research process, from problem identification to technology design, development, and evaluation.
- Empowerment of farmers and improved farmer-to-farmer interaction / farmer-scientist interaction.
- Better utilisation of bioresource
- Help to adopt eco-regional sustainable approach.

Rao and Hallb (2003) highlighted the importance of farmer participatory evaluation of fodder traits in the development of improved dual-purpose varieties for future genetic enhancement and technology transfer strategies. They found that for farmers in semi-arid, tropical India dual-purpose crops (coarse cereal and leguminous crops) are by far the most important feed source contributing about 50–60% of total feed is obtained from crop residues. There is large scope of farmer knowledge on these issues, as well as its complexity and contextual nature. For the effective genetic enhancement of crop residues to improve animal production, ways of dealing with this complexity and diversity need to be devised in the research and technology delivery systems in India.

References

Biggs, S.D., 1989. Resource-poor Farmer Participation in Research: a Synthesis of Experiences from Nine National Agricultural Research Systems. The Hague, Netherlands: International Service for National Agricultural Research.

Chambers, R. 1991. "Complexity, Diversity and Competence; Toward Sustainable Livelihood from Farming Systems in the 21st Century." *Journal of the Asian Farming Systems Association,* **1**(1): 79-89.

Chambers Robert .1994. Participatory Rural Appraisal (PRA): Challenges, Potentials and Paradigm. *World Development (Elsevier),* **22** (10):1437-1454.

Chambers Robert .1994. The origin and practice of Participatory Rural Appraisal. *World Development (Elsevier),* **22**(7): 953-969.

Cavestro Luigi. 2003. P.R.A. - Participatory Rural Appraisal concepts methodologies and techniques. Universita' degli Studi di padova, Facolta' di Agraria, Dipartimento Territorio e Sistemi Agro-forestali, Master in Cooperazione allo Sviluppo Nelle aree Rurali.

Norman, D.W. The Farming Systems approach: A historical Perspective. *Conference.ifas.ufl.edu/ifsa/papers/invite/norman.doc*

Pretty N Jules .1995. Participatory learning for sustainable agriculture. *World Development (Elsevier),* **23** (8): 1247-1263.

Rao P Parthasarathy , Hallb A.J.2003. Approaches to improve the utilization of food-feed crops. *Field Crops Research (Elsevier)*, **84**(1-2): 189–198.

❑❑❑

Chapter 33

Gender Issues in Integrated Farming Systems Development

MANJU SUMAN

ICAR - Indian Grassland and Fodder Research Institute, Jhansi-284003 (UP)

Introduction

Gender play very important role in integrated farming development. Gender ' is a ociological construct whereas sex is a biological one. What men and women do and how they relate socially is an important aspect of gender. Gender is a dynamic concept. As culture is dynamic and socio –economic condition change over time , so gender patterns change with them. Two persistent aspects of gender are division of labour' and inequality. The importance of gender as a basis of the social division of labour is that men and women do different types of task and have different jobs within the family, in house hold production and other. An estimated 52-75% of Indian women engaged in agriculture are illiterate, an education barrier that prevents women from participating in more skilled labor sectors. The integrated farming system approach introduces a change in the farming techniques for maximum production in the cropping pattern and takes care of optimal utilization of resources. Participation of Female engaged more in Dairying related activities. While, male in crop production activities. Participation of female was found more in dairy related activities in comparison to male, and also do more work/day. So gender issues should have prime important in achieving a more integrated farming system development.

The world "gender " was first used by Ann Oakley and others in the 1970s to describe those characteristics of men and women which are socially determined, in contrast to those which are biologically determined. The concept of gender' refers to the cultural interpretation of biological differences between men and women (Stolenand Vaa,

1991). So we can say that gender' is about the relationship between men and women in the societies in which they live , and different roles that they play in the societies. Hence 'gender ' is a sociological construct whereas sex is a biological one. What men and women do and how they relate socially is an important aspect of gender. The social nature of gender is evident in the variation between attribute s and activities considered male and female when comparison is made across cultures, between classes and ethnic groups in the same country, or in change over time.

Gender is a dynamic concept. As culture is dynamic and socio – economic condition change over time , so gender patterns change with them. Two persistent aspects of gender are division of labour' and inequality. The importance of gender as a basis of the social division of labour is that men and women do different types of task and have different jobs within the family , in house hold production and other. 'inequality emphasis the subordinate position of women to men with respect to resources, rewards and power. The division of labour along the lines of gender is a key concept in attempts to understand and address inequalities. Gender is a valuable concept in that it encourages the identification of social structure, practices and ideologies that perpetuate and reinforce the unequal position of women and men, and the unequal relationship between them. This is a necessary basis for the development of strategies to redress those inequaliaties.

Looking at gender ' from a development perspective

The main reason for considering gender in development efforts is thet women and men differ in their activities and in the social and economic resources they can command. The neds and wants vary between women and men due to their position and status in the family , community and nation. There will thus be difference in the impact of development trends or polices. Without identifying such differences it is not possible to devices policies that meet the specific needs of women and men and address existing inequalities. Whitehead indicates 'gender' as more relevant to development along the following lines:

No study of women and development can start from the viewpoint that the problem is women, but rather men and women, and more specially the relation s between them (Ostergaard, 1991:6).

The issues of women are the essence of growing importance of gender because of their relative subordinate position to men. However, 'gender 'as a substitute for women is also used to suggest that information about women is necessary information about men. That one implies the study of the other. This usage insists that the "world of women is a part of the world of men, created in and by it" (Scott. 1988:32). Looking at gender differences thus entails identifying the distribution of tasks, activities and rewards associated with the gender division of labour as well as the relative position of women and men. A gender better under standing of the gender division of labour and type and extent of women's participation in economic development provide a stronger base for economic planning leading to more realistic polices. In addition, identification of the different tasks, opportunities, constraints and incentives allows for more accurate identification of target groups and more appropriate support polices and efficient use of resources, which in trun allows for more effective polices. Furthermore, more equitable polices can be gained through attention to differences in social and economic position of men and women, enabling the distribution of benefits and costs of particular policies to achieve an equitable impact. It would also allow identification of gap and weaknesses in policy from the point of view of particular groups of women and men (Goldey, 1995).

The need to incorporate gender awareness into development effort was recognized some three decades ago, when planner's began to realize that expecting a country to develop towards modernization with the female half of its population unable to take full part in the process was like asking someone to work with one arm and leg tied behind their back (Ostergarrd, 1991"xii).

Since the early 1970s, however, the issues of women in agricultural development has gained momentum. It is now a priority for development organizations and planners. There is increasing recognition of the need to inter women into mainstream agricultural development in order to stimulate the entire agricultural labour force and to maximize output. Women represent a major force to rural change, a largely untapped resources that could boost rural economics and lead to higher growth rate and increased food production (FAO, 1990). Attempts made at international and national forums to address the gender inequalities as a nation of equitable sharing of responsibilities which lead to nation building activities have been focused on following different alternative development perspectives and policy approaches, viz, Women in Development (WID), Women and Development (WAD), Gender and Development (GAD).

However, within the development context, effectiveness of approaches wil depend on the good will of governments, mostly represented by men and their willingness to allow far reaching improvement on gender inequalities, in their countries.

Gender labour

In India, the typical work of the female agricultural laborer or cultivator is limited to less skilled jobs, such as sowing, transplanting, weeding and harvesting, that often fit well within the framework of domestic life and child-rearing. Many women also participate in agricultural work as unpaid subsistence labor. According to United Nations Human Development Report only 32.8% of Indian women formally participate in the labor force, a rate that has remained steady since 2009 statistics. By comparison men constitute 81.1%.

Literacy

An estimated 52-75% of Indian women engaged in agriculture are illiterate, an education barrier that prevents women from participating in more skilled labor sectors. In all activities there is an average gender wage disparity, with women earning only 70 percent of men's wage. Additionally, many women participate in agricultural work as unpaid subsistence labor.

Role of women's in Agriculture

Women play in various role in agriculture : Land preparation, sowing, weeding, thinning crop harvesting, winnowing, storage, livestock management and household activities. Women in agriculture families perform many farm related activities, both within and outside the house held in most part of the country. They constitute a large part of the total work force in agriculture. Although the pattern of division of labor between men and women varies greatly from region to region, women are involved in most of operations in agriculture, including subsidiary enterprise like: diary, poultry, beekeeping mushroom cultivations sericulture, fish culture, social forestry etc. So for as crop husbandry is concerned, women participate in almost all activities, right from preparatory tillage to harvest.

Traditionally seed cleaning, seed grading, sowing, planting, transplanting, weeding, thinning, gap filling, inter culturing, harvesting, threshing, shelling, hulling, winnowing, storing, regarding livestock related roles: cleaning of animal shed, feeding cattle and collecting fodder from field, chaffing the fodder, milking, shelling of milk, bathing of animal, preparation of milk product, Dahi, Ghee, Paneer, etc. Looking other milk animals and poultry birds are the main job for women. Activities such as processing and storage at home are performed exclusively by women. It may, however be mentored that regardless of the economic status, higher cast women usually avoid doing field work. There activities are confined within and at best around the homestead. By and large, the main role of women in high income group is supervisory.

Women multiple Roles

Women's multiple roles and responsibilities being farmer's waver, mothers and traders are after over looking development agencies / projects. They are also after excluded from cooperatives and farmers organization. Their lack of access to resources including advice and credit, and exclusion farm the benefits of innovation and income constitute an important negative influence on the rate Agriculture development. It is important to recognize that in equity stems not only form lack of access to resources but also from the quality of human resources. Whatever the reasons. Many of the poorest and most disadvantage rural women lack appropriate skills and education. Without these measures to improve access to resources and to confer more power will not be effective.

Women's key contribution to food security at the house hold regional and national level is similarly ignored. Many new technologies have simply increased the labor Borden of women farmers without increasing their share of farm review.

Gender time distribution in agriculture

Table 1. Time spent (Hrs) by farmers in different agricultural activities in selected area under study.

S. No.	Particulars	Male		Female	
		Irrigated Zone	Rainfed Zone	Irrigated Zone	Rainfed Zone
1	Ploughing	5:20 (80.2)	6:49 (68.8)	-	-
2	Root Collection	2:43 (52.1)	3:56 (55.2)	1:55 (24.0)	1:58 (26.0)
3	Leveling	3:46 (83.3)	4:47 (81.3)	-	-
4	Ridge & Furrow	3:19 (77.1)	5:18 (74.0)	-	-
5	Sowing	4:5 (79.2)	5:2 (79.2)	3:33 (10.4)	1:56 (16.7)
6	Sapling	2:6 (21.9)	1:52 (22.9)	2:50 (9.4)	2:0 (4.2)
7	Irrigation	5:47 (82.3)	3:51 (21.9)	-	1:59 (14.6)
8	FYM+ NPK	5:49 (84.4)	6:41 (81.3)	-	4:17 (1.0)
9	Plant Protection	1:14 (64.6)	7:34 (77.1)	-	5:54 (91.7)
10	Harvesting	5:26 (69.8)	7:7 (86.5)	5:4 (77.1)	6:27 (93.8)
11	Collection of harvested materials	3:53 (66.7)	5:7 (70.8)	2:38 (69.8)	4:50 (84.4)
12	Threshing	3:42 (82.3)	5:5 (77.1)	-	-
13	Sila Collection	-	-	3:47 (52.1)	5:12 (74.0)
14	Storage	3:12 (80.2)	4:4 (68.8)	4:23 (75.0)	4:55 (72.9)

Gender time distribution in livestock management

Table 2. Time spent (Hrs) by farmers in different livestock management l activities in selected

Area under study

S. No.	Particulars	Male		Female	
		Irrigated Zone	Rainfed Zone	Irrigated Zone	Rainfed Zone
1	Feeding	1:39 (100.0)	2:25 (91.7)	2:28 (100.0)	3:6 (95.8)
2	Milking	0:24 (46.9)	0:18 (40.6)	0:46 (99.0)	0:36 (95.8)
3	Housing	0:30 (14.6)	0:16 (12.5)	0:29 (97.9)	0:27 (89.6)
4	Health care	1:4 (97.9)	1:0 (89.6)	1:7 (97.9)	1:3 (91.7)
5	Milk Utilization	0:28 (56.3)	0:31 (41.7)	0:44 (93.8)	0:30 (83.30
6	Preparation of Dung cake	-	-	0:46 (97.9)	0:34 (89.60
7	Dairying	3:13 (100.0)	3:38 (95.8)	6:14 9100.0)	5:57 (97.9)

- Study revealed that the (table 1) participation of female was found maximum for crop harvesting in both the zones (77.1% with average 5:4 h/day in IR and 93.8% with average 6:27 h/day in RF). But for male it was 69.8% with average 5:26 h/day in IR and 86.5% with average 7:7h/day in RF.
- Study revealed that (table 2) Gender wise : Female contributed more towards dairying in both the zones (6:14 h in IR, 5:57 h in RF) in comparison to male (3:13 h in IR, 3:38 h in RF), and, Zone wise: Female contributed more in IR in comparison to RF, while, reverse trend in case of male (Suman, M, 2011)
- Almost all the farmers was agree that dairy farming is beneficial for economic upliftment in both the zones (97.9 % in IR and 100% in RF).
- In IR 80.2 % farmers agree that feeding of green fodder to animals reduce the cost of milk, while, in RF it was 93.8%.

Developing Technologies for Farm Women

In order to develop women-oriented technologies, the women specific jobs and their perspectives have to be identified following Samanta (1995), the principles for developing technologies for farm women are :

Improving farm women's productivity/work efficiency.

1. Increasing their income generating capabilities.
2. Increasing their employment opportunities.
3. Reducing the drudgery and health hazards in working for farm and home.

There is need to incorporate knowledge and skill of women into the development to modern farm technologies by the scientists. The blending of indigenous wisdom of farm women with the modern technologies is also development and generation process id of at most importance. Because of their involvement in the farming as well as in the house hold activities, the farm women after are hard pressed for time. While developing the technology, point much be kept in mind by the scientist, So that particular orbicular takes less time which is performed by the women, giving them ample time for other work and rest, A 'women angel' should be built into all technologies, which are relevant to women.

Women roles in livestock production

Gender involvement in farming system activities

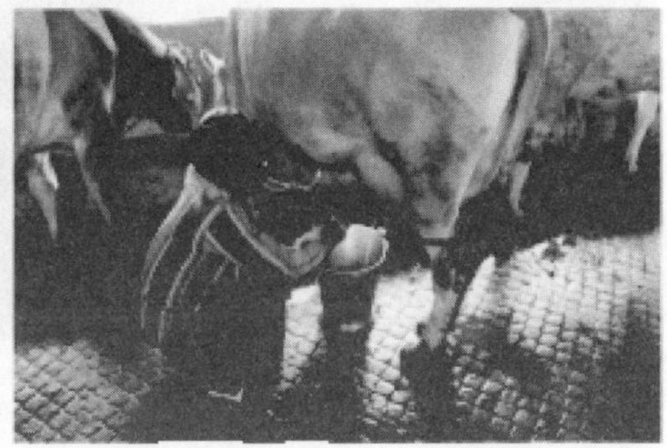

Women role in Agriculture

Conclusion

- It is concluded that Participation of Female engaged more in Dairying related activities in both the zones, while, male in crop production activities in both the zones.
- Participation of female was found more in dairy related activities in comparison to male, and also do more work/day.
- For dairying, per day work by females in IR is found more than RF as they have more work opportunity in IR, may be due to, Number of Bovines in IR was more .
- Women in India constituting about half of the total population , are engaged in some kind agricultural productive work. But their significant role as producers in agro-rural sector is hardly recognized, nor is adequately recorded in the national data of cences. This lead to neglect of women by extension agencies. Hence gender inequality is likely to be considered a serious obstacle in agricultural productivity and other socio-economic

aspects of human life. As all development efforts ultimate ly concentrate on the benefits of human life and both men and women are active actors in the development process addressing the gender issues should have prime important in achieving a more integrated farming system development.

Suggestions

- Literacy amongst women should be increased.
- Most needed areas of training by the women were found as plant protection aspects in relation to fodder production and animal health care.
- Some drudgery reducing implements for use for women should also be purchased by farmers.
- Accordingly gender issues must be considered in the design or modification of extension projects and programmes.

Reference

Stolen, K.A.and Vaa, M.1991. Gender and changes in developing countries, Dramnca, Norwegian University Press, 312.

Ostergaard, L. (ed). 1992. Gender and Development :A practical guide, Routledge, London and New york. 220.

Scott, J.W. 1988, Gender and the politics of history. Columbia University Press. New Yark, 242.

FAO. 1990. Women in Agricultural : FAO's Plan of Action Rome, 41.

Suman, m, Singh, M.Mallyya and Kumar, A. 2011. Gender issues in livestock production with emphasis on fodder production and livestock management, presentation of RPF–III in IRC, IGFRI, Jhansi.

❑❑❑

Chapter 34

Indigenous Technical Knowledge (ITK) and Environment-friendly Technologies for IFS

R.P. DWIVEDI

ICAR - Central Agroforestry Research Institute, Jhansi-284003 (UP)

Introduction

Indigenous Technical Knowledge (ITK) has immense potential for innovation, especially at the grassroots level. India is a country populated by a number of indigenous communities, most of which have their own set of unique traditional knowledge and technology base. Many of these knowledge and technologies are at par with the modern knowledge and technology system and have been provided the indigenous communities with comfort and self-sufficiency. These traditional knowledge and technologies have played a significant role in the overall socio-economic development of the communities. India needs environment-friendly technologies for integrated farming system because Indian agriculture or farming system is facing major environmental issues from global climate change i.e. emission of GHGs, water, soil and air pollution emanating from industry, transport, fertilizer application and persistent organic pollutants and loss of biodiversity. The solution of these environment problems lies in ITKs which were practiced in agriculture since the dawn of civilization in India. The advent of the concept of sustainable agriculture in late eighties in Indian agricultural scenario has evoked interest on ITK that has the element of use of natural product to solve the problems of agriculture and allied activities.

Indigenous Technical Knowledges (ITKs) are treasure troves of ancient wisdom, beliefs and traditional knowledge passed on from generation to generation for preservation, effective utilisation and conservation of natural resources, soil, plant and other organisms. It is mainly passed down generations through folklore, myths, customs, folk songs, proverbs, puppetry and other traditional methods (Swathi and Dinesh Babu, 2009). According to World Bank (1998), ITKs are local, tacit knowledge, transmitted orally or through imitation and demonstration, experimental rather than theoretical, learned through repetitions, and are constantly changing.

India is home to hundreds of ITKs thanks to the ancient civilisation and the localised way of cultivating crops and rearing of fish as well as livestock. These ITKs play significant role in agricultural production, as Das *et al.* (2002) observed that 'the enhancement of the quality of life of the Indians who in great majority live in and depend on agricultural production systems would be impossible by keeping this rich tradition of ITK aside'. Das *et al.* (2003) compiled the ITKs practiced in agriculture in India. They found 1998 ITKs being practiced by the farmers in the area of veterinary and animal sciences, pest and disease management in crops, grain/ seed storage, horticultural crops, and cropping systems and fisheries.

What is Indigenous Knowledge

Traditional knowledge (TK), also known as indigenous knowledge (IK) or local knowledge (LK) generally refers to the matured long-standing traditions and practices of certain regional, indigenous, or local communities. It also encompasses the wisdom, knowledge, and teachings of these communities.

(Source: http://en.wikipedia.org/wiki/Indigenous_knowledge)

Indigenous knowledge (IK) has also been referred to as the unique, traditional, local knowledge existing within and developed around the specific conditions of women and men indigenous to a particular geographic area (Louise Grenier, 1998).

Indigenous knowledge is the local knowledge –knowledge that is unique to a given culture or society. IK contrasts with international knowledge system generated by universities, research institution and private firms. It is the basis for Local-Level decision in agriculture, health care, food preparation, education, natural resource management, and a host of their activities in rural communities (Warren,1991). Indigenous knowledge is the information base for a society, which facilitates communication and decision-making. Indigenous information system are dynamic, and are continually influenced by internal creativity and experimentation as well as by contact with external system.

Indigenous knowledge (IK)-is the participants' knowledge of their temporal and social space. Indigenous knowledge as such refers not only to knowledge of indigenous people, but also to that of any other defined community.

Indigenous knowledge system (IKS) - delineates a cognitive structure in which theories and perception of nature and culture are conceptualized. Thus it includes definition, classifications and concepts of the physical, natural, social, and economic environment. The dynamics of IKS take place on two different levels, the cognitive and the empirical. On the empirical level, IKS are visible in institutions, artifacts and technologies.

Indigenous technical knowledge (ITK) - is specifically concerned with actual application of the thinking of the local people in various operations of agriculture and allied areas.

Different Kinds of Knowledge

No matter what importance western knowledge system assumes in the industrialized countries, it is only one of the knowledge systems. It exists along with many non-formal indigenous (local, traditional or indigenous) knowledge systems in a lot of countries.

Formal knowledge: It refers to knowledge systems developed within the predominantly western-based education systems. It is called formal because it tends to be supported by written documents, rules and regulations, and technological infrastructure.

Indigenous or local knowledge: In contrast, the concept of "indigenous or local knowledge" (IK) refers to the knowledge, know-how, and practices developed and maintained by people with long histories of close interaction with their natural environment. *(http://www.scidev.net/dossiers/index.)*

Indigenous knowledge is fundamental to local decision-making regarding daily activities like hunting and gathering, fishing, agriculture, animal husbandry, water conservation, health, etc. Moreover, unlike formal scientific knowledge, indigenous knowledge is generally transferred as oral wisdom from one generation to the other, and is seldom, if ever, documented. Combining all forms of knowledge other than the formal ones as 'indigenous knowledge' would lead to its generalization and oversimplification, and may negate the outstanding contribution local knowledge can make to sustainable development.

Importance of indigenous knowledge

In the recent years, the role of IK in a range of sectors is being talked about. It includes intercropping techniques, pest control, crop diversity, and seed varieties in agriculture; plant varieties, and fish breeding techniques in biology; traditional medicine in human healthcare; soil conservation, irrigation, and water conservation in natural resource management; and oral traditions and local languages in education. The realization of IK's contribution to these sectors has led to an increasing interest in it by academicians, and policymakers alike.

Many government and non-governmental organizations, as well as international organizations such as the World Bank, International Labor Office, UNESCO and FAO are now appreciating the role IK can play in achieving sustainable development in a country. This interest is also apparent in the policies and programmes of various countries.

The erosion of IK systems

Although some IK is lost naturally as practices get modified or are left unused for long time periods, the current rate of loss can be attributed to modernization and cultural homogenization, the current educational systems that believe macro-level problems can only be addressed through the global knowledge pool, and the slow growth of institution supporting grassroots innovations.

"We are reminded of the global and historical tendency of complex technologies associated with economic powers to squash smaller, local technologies . . . We are urged to identify the valuable elements of smaller technologies and to create a place for them in the new century."(Kroma, 1996)

With rapid population growth, in-migration, and government relocation schemes (in case of large development projects), standards of living is often deteriorated. As poverty augments, short-term economical gains are chosen over environmental friendly local practices. The introduction of mono cropping patterns in agriculture and forestry results in a loss in biodiversity, thus leading to a decline in IK. Deforestation leads to the disappearance of several precious yet unknown medicinal plants, and as a result the knowledge associated with those plants also declines.

An elderly woman in northern India was selecting seeds for storage while being interviewed by a researcher about the impacts of modern agriculture. She commented, "It takes a sharp eye, a sensitive hand, and a lot of patience to tell the difference between these seeds. These are not the things that are honoured any more."(Zweifel, 1997).

A lot of traditional knowledge is also being lost because of the communication gap, since neither children nor adults spend as much time in their communities. Because IK is generally transmitted orally, it is susceptible to change, particularly when people move to new regions, or when people's lifestyles tend to be different from that of their ancestors.

Recognizing indigenous knowledge

Sharing indigenous knowledge on a global scale by development agencies such as the world bank's indigenous knowledge for development program (World Bank, 2002) and UNESCO's Best Practices of Indigenous Knowledge (UNESCO, 2002) database. Meanwhile, national and multinational corporations have taken indigenous knowledge as a valuable commodity and are sharing the knowledge in the commercial world for profit.

Indigenous knowledge resource centers

There are 37 indigenous knowledge resource centres across the world with 16 in Africa, 8 in Asia, 5 in Europe, 4 in latin America, 1 in Middle East and 3 in North America. In India Centre for Advanced Research on Indigenous Knowledge Systems, Mysore; Centre for Indigenous Knowledge on Indian Bioresource, Lucknow and Centre for Indigenous Knowledge of Indian Herbal Resources (Central Institute of Medicinal and Aromatic Plants), Lucknow are the prominent ones.

Sources of Indigenous technical knowledge (ITK)

Documentation of ITK has to be systematic so that the information collected can be properly documented, giving the name and addresses of the users of the ITK, location, details of the practices of the ITKs etc.

The sources of ITKs are as follows :

1. Farmers and the community members especially elders are the best source of ITK.

2. Folk-lore, songs, poetry and theatre can reveal a great deal about people's values, history and practices. These are often not written down and need to be recorded.
3. Community records: Although ITK is mostly transmitted by word of mouth, some indigenous forms of record keeping do exist. These include writings, paintings, carvings and many other forms.
4. People working with communities such as extensionists can be a valuable source of ITK. Other resource persons are local school headmaster, credit co-operative society officials, village milk co-operative members, men and women labourers and village panchyat members and sarpanch.
5. Secondary sources include published and unpublished documents, databases, videos, photographs, museums and exhibits.

Why is Indigenous Knowledge Important

In the emerging global knowledge economy a country's ability to build and mobilize knowledge capital, is equally essential for sustainable development as the availability of physical and financial capital. (World Bank, 1997) The basic component of any country's knowledge system is its indigenous knowledge. It encompasses the skills, experiences and insights of people, applied to maintain or improve their livelihood.

Significant contributions to global knowledge have originated from indigenous people, for instance in medicine and veterinary medicine with their intimate understanding of their environments. Indigenous knowledge is developed and adapted continuously to gradually changing environments and passed down from generation to generation and closely interwoven with people's cultural values. Indigenous knowledge is also the social capital of the poor, their main asset to invest in the struggle for survival, to produce food, to provide for shelter or to achieve control of their own lives.

The Indian Council of Agricultural Research (ICAR) has brought CD inventory of ITK in agriculture which is listed below (Table 1).

Table 1. CD of Inventory of Indigenous Technical Knowledge (ITK) in Agriculture

S. No.	Name of CD	Price (Rupee)
1.	Inventory of Indigenous Technical Knowledge in Agriculture (Document 1)	Rs.300/-
2.	Inventory of Indigenous Technical Knowledge in Agriculture (Document 2)	Rs.300/-
3.	Inventory of Indigenous Technical Knowledge in Agriculture (Document 2 Supplement 1)	Rs.300/-
4.	Inventory of Indigenous Technical Knowledge in Agriculture (Document 2 Supplement 2)	Rs.300/-
5.	Validation of Technical Knowledge in Agriculture (Document 3)	Rs.300/-
6.	Cross Sectoral Validation of Indigenous Technical Knowledge in Agriculture (Document 4)	Rs.300/-
7.	Indigenous Technical Knowledge in Agriculture-Geographical Indication of Plant Species (Document 5)	Rs.300/-

(*Source:* www.icar.org.in)

Sundaram, et al (2004) listed the 1460 ITKs under the National Agriculture Technology Project (NATP) of ICAR (Table 2).

Table 2. ITKs under the National Agriculture Technology Project (NATP) of ICAR

S.No.	Indigenous Technical Knowledge (ITK) Class	Numbers collected
1.	Rainwater management	41
2.	Soil and water management	10
3.	Tillage and interculture	10
4.	Crops and cropping systems	117
5.	Pest and disease management	237
6.	Soil fertility management	53
7.	Farm implements	131
8.	Post harvest technology	34
9.	Grain and seed storage	67
10.	Horticulture	112
11.	Veterinary science and animal husbandary	595
12.	Fisheries	2
13.	Method of weather forcasting	22
14.	Food product development	19
15.	Ethno botany and agrobiodiversity	3
16.	Agro-animal based yarns, natural dyes and weaves	6
17.	Low cost housing material	1

Some popular ITKs are

Cow urine: control of smut in sorghum, fungal and bacterial diseases, termite control in whear, seed treatment of chickpea to prevent termite, control of storage pests of pulses, improving the fruit quality of pomegranate, curing injuries in animals, curing cough promlems, as antiseptic for wounds.

Cow dung: Used for grain storage, for storage of pumpkin, used as crop-protective chemical, protecting grapes from stem0gridler, dipping solanaceous seedlings in slury for better growth, well-decomposed cow dung is spread on the field of late sown wheat for proper temperature maintenance, used for nutrient management, termite control and wilt control in ginger.

Neem: Neem cake is used for termite control, control of early stem borer in rice, Neem leaves in paddy nursery to control pests and diseases.

Use of plant material in animal-health care

Different plant-origin materials are being used in indigenous practices for treatment of the animals for various ailments (Table 3).

Table 3. Use of different plant parts for health care of animals

Ailment	Name of plant	Part used
Foot-and mouth disease	Kidamari (Aristolochiara bracte-olae), Neem (Azadirachta indica)	Leaf, New growing branch
Mastitis	Nagod (Vitex negundo)	Root
Diahhorea	Kantas (Strychnas potatorum)	Leaf
Eye infection or injury	Angoria (Bakabits aegyptica)	Bark
Flatulence	Kakdi (Cucumis sativus)	Seed
Tympany	Garmalo (Casia fistula)	Pod

(*Source:* Handbook of Agriculture, ICAR, 2012)

Keeping in view the need for development of appropriate technology by the scientists and its diffusion by the change agents, the knowledge on various aspects of the of the plant species incliding their habitats, characteristics and acxtive ingredients is of paramount importance for its further use.

Integration of ITK in agricultural education and research

The indigenous knowledge that is in use over generations has not found appropriate place in development of agricultural technology. Neither it has focused in the syllabus meant for UG/PG students in agricultural and allied subjects. It should be in included in UG/PG course curriculum. Asking RAWE students to collect ITKs during their camps and work experience programmes in the villages. Agricultural research should address the evaluation of efficacy of ITKs by conducting trials under controlled, field and on-farm conditions including identification of active ingredients present in materials which are used in ITKs.

ITK in agricultural extension

The following steps are suggested to popularize ITKs through agricultural extension mechanism :

- Human resource development by conducting long-duration training programmes on collection, documentation and validation of ITKs for in-service personnel.
- Mobilization resources for collection and documentation of ITKs
- Dissemination of information on ITKs through print and electronic media.
- Inclusion of topic on ITKs in foundation course for the agricultural scientists of National Agricultural Research System.

Initiative for protection of indigenous knowledge

Legal scholars, anthropologists and native scholars are now proposing new legal regims designed to defend indigenous cultures by radically expanding the norm of copyright. Centre for Indigenous Knowledge for Agriculture and Rural Development (CIKARD) has been closely involved with a growing network of indigenous-knowledge resource centre-these centers provide the mechanisms for protecting indigenous knowledge.

Today, many indigenous knowledge systems are at risk of becoming extinct because of rapidly changing natural environments and fast pacing economic, political, and cultural changes on a global scale. Practices vanish, as they become inappropriate for new challenges or because they adapt too slowly. However, many practices disappear only because of the intrusion of foreign technologies or development concepts that promise short-term gains or solutions to problems without being capable of sustaining them. The tragedy of the impending disappearance of indigenous knowledge is most obvious to those who have developed it and make a living through it. But the implication for others can be detrimental as well, when skills, technologies, artifacts, problem solving strategies and expertise are lost.Indigenous knowledge is part of the lives of the rural poor; their livelihood depends almost entirely on specific skills and knowledge essential for their survival. Accordingly, for the development process, indigenous knowledge is of particular relevance for the following sectors and strategies: Agriculture, Animal husbandry and ethnic veterinary medicine,. Use and management of natural resources, Primary health care (PHC), preventive medicine and psychosocial care, Saving and lending, Community development and Poverty alleviation.Indigenous knowledge is not yet fully utilized in the development process. Conventional approaches imply that development processes always require technology transfers from locations that are perceived as more advanced. This has led often to overlooking the potential in local experiences and practices. Introduced varieties and commercially marketed seeds are replacing local varieties - along with them, the concomitant local knowledge disappears. For many years, the international community is establishing - with considerable effort - gene banks to preserve the genetic information of local varieties or indigenous species. However, the seeds and clones do not carry the instructions how to grow them. This knowledge needs to be captured, preserved and transferred as well.

Indigenous knowledge is relevant on three levels for the development process

- It is, obviously, most important for the **local community** in which the bearers of such knowledge live and produce.
- Development agents (Governments, NGOs, donors, local leaders, and private sector initiatives) need to **recognize** it, **value** it and **appreciate** it in their interaction with the local communities. Before incorporating it in their approaches, they need to understand it - and critically validate it against the usefulness for their intended objectives.
- Lastly, indigenous knowledge forms part of the **global knowledge**. In this context, it has a value and relevance in itself.

Indigenous knowledge can be preserved, transferred, or adopted and adapted elsewhere.

The **development process interacts with indigenous knowledge.** When designing or implementing development programs or projects, three scenarios can be observed:

The development strategy either

- Relies entirely or substantially on indigenous knowledge,
- Overrides indigenous knowledge or,
- Incorporates indigenous knowledge.

Planners and implementers need to decide which path to follow. Rational conclusions are based on determining whether indigenous knowledge would contribute to solve existing problems and achieving the intended objectives. In most cases, a careful amalgamation of indigenous and foreign knowledge would be most promising, leaving the choice, the rate and the degree of adoption and adaptation to the clients. Foreign knowledge does not necessarily mean modern technology, it includes also indigenous practices developed and applied under similar conditions elsewhere. These techniques are then likely to be adopted faster and applied more successfully. To foster such a transfer a sound understanding of indigenous knowledge is needed. This requires means for the capture and validation, as well as for the eventual exchange, transfer and dissemination of indigenous knowledge.

Indigenous Technical Knowledge (ITK) in Agriculture is especially concerned with actual application of thinking of the local people in various operations of agriculture and allied areas. The term Indigenous Knowledge (IK) refers to the unique, traditional, local knowledge existing within and developed around the specific conditions of women and men indigenous to a particular geographical area (Grenier, L. 1998).Integrated Farming system is a whole farm management system which aims to deliver more sustainable agriculture. It is a dynamic approach which can be applied to any farming system around the world. It involves attention to detail and continuous improvement in all areas of a farming business through informed management processes. Integrated Farming combines the best of modern tools and technologies with traditional practices according to a given site and situation.

The International Organisation of Biological Control (IOBC) describes Integrated Farming as a farming system where high quality food, feed, fiber and renewable energy are produced by using resources such as soil, water, air and nature as well as regulating factors to farm sustainably and with as little polluting inputs as possible. Particular emphasis is placed on a holistic management approach looking at the whole farm as cross-linked unit, on the fundamental role and function of agro-ecosystems, on nutrient cycles which are balanced and adapted to the demand of the crops, and on health and welfare of all livestock on the farm. Preserving and enhancing soil fertility, maintaining and improving a diverse environment and the adherence to ethical and social criteria are indispensable basic elements. Crop protection takes into account all biological, technical and chemical methods which then are balanced carefully and with the objective to protect the environment, to maintain profitability of the business and fulfil social requirements.Integrated Farming Framework, which provides additional explanations on key aspects of Integrated Farming. These include: Organisation & Planning, Human & Social Capital, Energy Efficiency, Water Use & Protection, Climate Change & Air Quality, Soil Management, Crop Nutrition, Crop Health & Protection, Animal Husbandry, Health & Welfare, Landscape & Nature Conservation and Waste Management Pollution Control.

The Food and Agriculture Organization (FAO) of the United Nations promotes Integrated Pest Management (IPM) as the preferred approach to crop protection and regards it as a pillar of both sustainable intensification of crop production and pesticide risk reduction". IPM thus is one indispensable element of Integrated Crop Management which in turn is one essential part of the holistic Integrated Farming approach towards sustainable agriculture.The concept of sustainable rural livelihoods is central to the debate about rural development, poverty reduction and environmental management; where the emphasis is on the goals of rural poverty reduction, empowerment of people and the promotion of increased security of livelihoods of rural people. A livelihood comprises the capabilities, assets and activities required for a means of living. Livelihoods are secure when households have secure ownership of or access to resources (both tangible and intangible) and income earning activities to offset risks, ease shocks and meet contingencies. Understanding the livelihoods with an economic, ecological and social perspective is imperative to get information for strategic planning to improve the livelihoods of rural people. Small and marginal farmers constitute 78 per cent of the farming community in India (The Hindu,

March 27, 2006). These farmers can be spared of the drudgery they are suffering from generations if science and technology helps them in increasing crop productivity. In addition, unemployment has emerged as a major problem plaguing youth in our country, both in urban and rural areas. But for rural youth, the problem assumes greater proportion since land- man ratio has gradually declined over the past years due to ever increasing population and rapid industrialisation. Continuous fragmentation due to ownership multiplication and division of joint families into nuclear families has also resulted in low per capita availability of land. Agricultural labourers constitute one of the vital inputs in the agricultural production process. Farm women and youth agricultural labourers perform most of the farm operations and it remained as a cause if drudgery faced by them.

Environment-friendly agroforestry

Agroforestry is a production system in which crop, livestock, fodder, fisheries, bee keeping and trees are grown together in a unit of land. Due to modernization and urbanization, the agricultural land is decreasing in a fast rate. The negative consequence of deforestation is the reduction of tree cover in the country. As per the Forest Policy of Govt. of India, 1988 there should be 66 % and 33 % forest cover in the hills and plains respectively. The shrinking natural resources has resulted in the poor supply of fuel wood, fodder, fruit and food to the majority of Indian farmers.

It is important to examine the social dimensions particularly with reference to gender and social equity and ethics very carefully, while designing agroforestry research and development programmes. Only such an integration can help us to impart a pro-nature, pro-poor and pro-women orientation to technology development and dissemination. Agroforestry being a multi-disciplinary and multi-institutional subject, synergies are required at various levels among scientists, planners, development agencies and all other stakeholders to harness maximum benefits of inputs. Linkage between research, education, extension and farmers have been established but their interactions need to be strengthened for appropriate technology generation and its effective transfer.

Forests and Agroforestry in Livelihood support

The major products for which the country depends overwhelmingly on forest lands are fuel wood and green fodder. A recent survey has indicated the dependence of people in fuel wood to meet the energy requirement to the tune of 78% in rural areas and 26% in urban settlements. It is an accepted fact that most of the fuelwood is in the form of dead or dry wood and is extracted from forests on head loads by the villagers living near forests. This extraction of fuel wood goes unrecorded. But even after allowing for this unrecorded production, the total production from these forests, within silviculturally permissible limits would not exceed 40 million cubic meter or 25 million tonnes. Fuel wood production from various sources was around 169 million tonnes in 1990, which increased to 201 million tonnes by 2001. Forests account for over 50% of this supply, the rest comes obviously from the farmlands and community lands. About 10 million rural people are estimated to be engaged in collection of fuel wood. Besides this many more are engaged in the marketing of fuel wood in rural and urban areas.

It is estimated that the demand for timber (wood other than fuel wood) is likely to grow from 82 million cubic meters in 2006 to 116 million cubic meters in 2025. The supply of wood is projected to increase from 65 million cubic meters (12 from forests and 53 from plantation & agroforests) in 2006 to 116 million cubic meters (12 from forests and 104 from plantation & agroforests) in 2025. The deficit of 17 million cubic meters (in 2006) could be met by promoting agroforestry. The productivity of timber in India is only 0.7 cu. m/ha/year whereas the world average is 2.1 cu. m/ha/year. Increased production of timber and forest produce and their downstream processing could bring about accelerated employment and socio-economic development in rural areas.

The agroforestry system yields almost everything which people need. Sustainable land management, conservation and increase in productivity are important aims of agroforestry. Through agroforestry we can continue to improve productivity and profitability of small farms on an environmentally and economically sustainable basis. The Research and Development (R&D) efforts made so far has clearly indicated that agroforestry is sustainable, environmentally safe and economically attractive land use option which is already making major contributions towards the supply of timber to support local value addition by industries, meet the local demands of fuelwood, fodder and bio-energy; conserving biodiversity; environmental ameliorations; diversification away from agriculture and large employment generation specially for the socially disadvantaged sections of the population and poverty alleviation in rural areas. Besides, promotion of agroforestry is only viable option to achieve the National Forest Policy target of 33 % tree cover.

With the promotion of agroforestry in rural areas, small holders have emerged as the timber suppliers of the 21st century. In 2006, out of a total demand of 82 million m^3 of timber and other wood, 53 million m^3 i.e. 65% was supplied from plantations out side forests and agroforestry alone. Agroforestry is also meeting almost half of the total demand of 201 million ton of fuelwood in the country. The contribution of agroforestry will increase substantially as more and more improved planting material is made available to farmers.

Extension Strategies for ITKs

The extension in the ITK has five main strategies:

- Raise the awareness of the public on the significance of the ITK,
- Increase the understanding and participation willingness of farmers,
- Establish cooperation with those who influence the participation of traditional farmers,
- Include the farmers into implementation process of the ITK and giving credit to them,
- Dissemination of ITKs.

Round the year livelihood opportunities

ITK provides livelihood opportunities round the year to the small and marginal farming community (Table 4). The operations involving labour input are: July-October: Management of nursery, seedling plantation, budding in Aonla and Ber, management of budded trees etc.; November-March: Intercultural activities of agroforestry systems, weeding, spray of bio-insecticide and bio-pesticides on trees and crops, harvesting of fruits, packaging and marketing of agroforestry produce; April-June: Digging of pits, layout of plantations, pruning in Ber, preparation for nursery, putting farmyard manure and compost in the pits and nursery sites, collection of fuelwood, fodder and tree fodder, bark, gum, resin, honey bee, seeds of multipurpose tree species, drying of seeds and tree leaves for medicinal purpose and making tree leaves' *Dona-Pattal* (Paper-cup and plate).

Forestry, agroforestry and related industries are a major source of employment in the primary, secondary and tertiary sectors all over the country. Forestry activities alone generate employment of approximately 240 million person days per day. At least 10 percent of the present tribal population of 50 millions derive their livelihood directly from forests. Agroforestry intervention in Himalayan region can provide employment opportunity to almost 6 million humandays which will have grate bearing on the migration of landless and marginalised communities from rural to urban settlements. Silvipasture generates 120 humandays/ha/yr employment directly. In southern Tamil Nadu, *Prosopis juliflora* provides employment to the tune of 6.34 million mandays and 7.03 million womendays per annum for charcoal making and bioenergy production. Biofuel plantations are likely to create employment for 0.44 million humandays only for seed collection. Bamboo sector generates employment for 48-60 million humandays for harvesting and 60-70 to million humandays for loading, unloading, handling etc., annually. These are only few examples which indicate the vast scope of employment generation and livelihood security from these sectors.

Table 4. Round the year management of ITK work in agroforestry

Sl. No.	Months	Management of ITK work in agroforestry
1.	April to June	Digging of pits, making rows for tree plantation, pruning in indigenous Ber trees, nursery preparation, application of Farm Yard Manure (FYM) and compost in pits and nursery, collection of fuel wood, gum, bark, tree leaves for medicinal purpose, collection, drying and packaging of multipurpose tree seeds, etc.
2.	July to October	Care of nursery, transplanting, budding and gootee in fruit trees, removal of unwanted tillers from budded branches, etc.
3.	November to March	Inter-cultural activities in trees, weeding, spraying of insecticide and pesticides, plucking of fruits, etc.

(*Source:* Solanki and Dwivedi 1999)

Conservation of Traditional Knowledge

International attention has in the past years has shifted to intellectual property laws to preserve, and promote traditional knowledge. The Convention on Biological Diversity of 1992 acknowledged the contribution traditional knowledge can make in protecting species, ecosystems and landscapes, and therefore included laws pertaining to its access and use.

Indigenous people and local communities have since long resisted the use of art and crafts, traditional symbols and designs; the use or modification of traditional songs; the patenting of traditional uses of medicinal plants; and the copyrighting and distribution of traditional stories. Certain communities have also sought to ensure that their traditional knowledge is used equitably - according to restrictions set by their traditions, or requiring benefit sharing for its use.

Experience has proved that developmental activities that tend to ignore local technologies; local knowledge systems, and those which fail to formulate policies without a concern for local environment generally fail to achieve the desired outcomes.

Conservation of Traditional breeds

Union Agriculture Minister Shri Radha Mohan Singh while addressing a programme at National Institute of Veterinary Epidemiology and Disease Informatics (NIVEDI) in Bengaluru on 9th January, 2015, urged upon ICAR to have major programs to improve the indigenous breeds as the products from the indigenous cattle also have health attributes to humans. He said that empirical evidence suggests that our indigenous breeds of cattle are relatively less prone to diseases as compared to exotic and cross-bred. He further said that the efforts should be made to improve indigenous stock in terms of production and productivity. He mentioned that the best indigenous breeds like Gir, Tharparkar, Rathi, Sahiwal etc. have better heat tolerance, feed conversion efficiency and disease resistance. (PIB, Govt. of India, 2015).

A step towards conserving indigenous knowledge systems and promoting grassroots innovations: an example from National Innovation Foundation (NIF)

NIF was launched in 2000 by the Government of India as India's national Initiative to strengthen the Grassroots technological innovations and Traditional Knowledge. For the last seventeen years, the Honeybee Network and Society for Research and Initiatives for

Sustainable Technologies and Institutions (SRISTI) have been scouting innovations by farmers, artisans, women, etc. at the grassroots level. The Gujarat Grassroots Innovations Augmentation Network (GIAN) scales up innovations, from the Honey Bee database of innovations, through value additions in innovations to sustain creativity and ethics of experimentation. Following are the steps taken by these organizations in order to ensure a unique platform for grassroots innovations:

- Scouting and documentation of innovations: This requires extensive fieldwork, travel in rural and urban areas, search for 'odd balls' - the experimenters - and local community and knowledge experts in the society.
- Knowledge dissemination: The entire effort of scouting and documentation leads to the development of a database of innovations and traditional knowledge aimed at building the electronic National Register. The process involves using various information technology and database applications for horizontal networking amongst innovators and traditional knowledge experts as well as other stakeholders.
- Value addition and R&D: Most of the innovators and/or traditional knowledge experts need optimization in design and/or product formulation through blending with modern science and technology inputs. Market prospects for many innovations will be very low without proper value addition. Efficiency gains can be made by creating technology networks.
- Business development and micro venture.
- Learning : Cross regional, peer-to-peer learning exchanges aimed at enhancing local capacity to identify and apply indigenous/traditional knowledge and practices (http://nifindia.org/activity.htm).

Conclusions

The most important challenge for the future extension managers would be the Management of ITKs. The success of a farmer in the years to come is going to be primarily dependent upon his level and sharing of indigenous knowledge. The permanent solution for farming system lies in sustainable land management.

There is an instant need to document and preserve the Indigenous Technical Knowledge (ITK) of different communities, many of which are at the brink of extinction. There is a lack of proper alliance between the practice of indigenous and modern knowledge. There are serious issues related to intellectual property rights. An appropriate association between the traditional and modern knowledge and technology systems has immense potential to benefit the society. As ITK is a relatively new area of study, capacity building to understand ITKs has been considered to be the first job to be accomplished, followed by collection and documentation of ITKs, conducting research to validate the wisdom of ITKs, supported by data and finally patenting of the relevant ITK is the need of the hour.

Validation of ITK is a logical step to qualify and quantify effectiveness of the practices. Suitable modifications of the local practices, through R&D will help to develop appropriate and acceptable methodologies that are more suited to our farming situations. The farmers practicing ITKs may also collaborate with research scientists in refining the knowledge for evolving appropriate technologies. Local ITK practitioners need to be honoured and recognised and publications of local innovation in regional magazines including names and photographs of the innovators can motivate traditional practitioners to disseminate their knowledge.

References

Das, P., Das, S. K., Arya, H. P. S., Subba Reddy, G. and Mishra, A. 2002. Inventory of ITK in Agriculture. *NATP Mission mode project on collection, documentation and validation of Indigenous Technical Knowledge.* Indian Council of Agricultural Research, Document I, p. 5.

Das, P., Das, S. K., Arya, H. P.S., Singh, R. P., Mishra, A., Bujarbaruah, K. M., Bujarbaruah, G., Subba Reddy,L. R.,Verma, M., Rani, G., Gupta, H. S., Satapathy, C. and Kavia, Z. D. 2003. *Inventory of ITK in agriculture. NATP Mission mode project on collection, documentation and validation of Indigenous technical knowledge.* Agricultural Research, Document 2, p. 2.

Grenier, L. 1998. Working with indigenous knowledge: a guide for researchers. http://www.idrc.ca/en/ev-28703-201-1-DO_TOPIC.html

Handbook of Agriculture, ICAR, New Delhi 2012.

Kroma S. 1996. The science of Pacific Island peoples. Indigenous Knowledge and Development Monitor, 4(2). http://www.nufficcs.nl/ciran/ikdm/.

Press Information Bureau. 2015. Govt. of India 9th January. Print release.

Solanki K.R. and Dwivedi R.P. 1999. Krishivaniki me gramin yuvaon ki bhagidari. Unnat Krishi, **38**(5): 45-47.

State of Forest Report. 2003. Forest Survey of India. Dehradun.

Sundaram KV, Moni M ans Mrityunjay M Jha 2004. Natural resources management and livelihood security: survival strategies and sustainable policies. Concept Publishing Company, New Delhi.

Swathi L. and DineshBabu, P. S. 2009. Indigenous technical knowledge and ancient proverbs of the coastal fisherfolk of Kerala and their implications. *Indian J. Traditional Knowl.*, **8**(2): 296-297.

THE HINDU. 2006. India's national newspaper, March 27, 2006.

Warren, D. M. 1991."The Role of Indigenous Knowledge in Facilitating the Agricultural Extension Process". Paper presented at International Workshop on Agricultural Knowledge Systems and the Role of Extension. Bad Boll, Germany, May 21-24, 1991.

World Bank, (1997) "Knowledge and Skills for the Information Age, The First Meeting of the Mediterranean Development Forum"; Mediterranean Development Forum, URL:http://www.worldbank.org/html/fpd/technet/mdf/objectiv.htm

www.icar.org.in/en/node/4622 CD of inventory of ITK in agriculture

http://www.scidev.net/

http://en.wikipedia.org.

www.nifindia.org.

www.sristi.org.

www.gian.org.

Zweifel H. 1997. Biodiversity and the appropriation of women's knowledge. Indigenous Knowledge and Development Monitor, 5(1). http://www.nufficcs.nl/ciran/ikdm/.

□□□

Chapter 35

ICT Readiness of Stakeholders for Sustainable Integrated Farming Systems

SATYAPRIYA

ICAR - Indian Agricultural Research Institute, New Delhi - 110 012

Introduction

The marginal and small farmers constitute 78.2 per cent of the farming community in India. The unique Indian situation of small fragmented holdings and lack of capital investments is not suitable for single commodity farming being practiced in developed countries. So, the integrated farming system appears to be a viable solution to the Indian agriculture for increasing productivity and income of the small and marginal farmers with constrained resources.

Farming systems approach is location specific. Hence, farming systems models adoptability or feasibility research has to be undertaken in all agro climatic zones under all possible farming situations. Presently this is being done in a limited area. In order to make the integrated farming systems management more effective, the lead farmers are to be trained in the Agricultural University farms. The findings of the individual studies at the micro level will serve as a technological base for guiding the farming under different identical micro environments.

Under the above circumstances, aś strategic approach on "Sustainable integrated farming systems for different Agro climatic zones of India this lecture is designed.

Scenario of Indian Agriculture

India and Agriculture/fodder/allied sector

Total population in India	–	1029 million
Rural Population	–	743 million (72%)
Pop. Engaged in Agriculture	–	234 m

Cultivators	-	127 m
Agricultural Labour	-	107 m
% Contribution to GDP (2000-01)	-	22.3%

*Data Source – Agricultural Statistics at a Glance- 2004, DES, DAC, MOA, GOI

Above table and figure shows that ICT is the only tool which can spread the technology/information to such large no. of beneficiaries with minimum time and cost.

India and Connectivity

- 107 Million Telephones
- 15 Million PCs (Proposed 75 Million by 2010)
- 5 Million Internet Connections (45 M. by 2010)
- Connectivity is highly skewed
- Internet connectivity more skewed
- Technology for Rural Connectivity available
- Efforts required to promote Rural Connectivity

Agricultural/Fodder Extension: Where we are today: Challenges:

We can understand the trend of Agricultural/Forestry Extension since 1950 in India by going through the following point:

1. 1950s–1970s: Community Development, National Extension Services, Multipurpose Agriculture Development Programmes;
2. 1970s-1990s: Specialized Extension Systems: Training & Visit System – Fortnightly Workshops, Subject Matter Specialists at district, Sub-District levels;
3. 1998-2002: Broad basing of Extension – Bottom-up Planning, Participatory Mode, Integrated delivery of Extension Services (Agriculture, Animal Husbandry, Horticulture, Sericulture, Fisheries etc.), involvement of University Scientists and Farmers' Groups, using ICTs for effective communication; and
4. 2003 onwards: Extension Reforms in most Asian Countries, Public Private Partnerships, increased role of NGOs, Farmers' Organizations.

Extension Communication: The Most Critical Gap

- Lack of feedback mechanism in the Research Extension Linkages identified as one of the weaknesses in existing Extension Systems (by many studies, evaluation reports)
- Communication is the most critical element in Research-Extension-Farmer-Marketing loop
- Media and ICTs offer strong potential to improve this mechanism

Still the Indian farmers depend on input dealers or their progressive counterparts for the agril. information, in such scenario ICT can be effective to knock down the door of every farmers to increase their profitability to strengthen the nation.

New ICTs

The new ICT includes digital devices such as computes, E-mail, internet, multimedia, video conference, mobile phones etc., which have the potential of providing vast amount of relevant information to rural populations in timely, comprehensive and cost effective manner.

Following are most frequently used kinds of information technologies.

1. Electronic Database Access and Search Systems

Common labels used in practices are:-

- Database
- Teletext
- Videotext

2. Feedback Systems

3. Advisory Systems

Advisory system can include

- Decision Support System
- Expert Systems
- Knowledge Systems

4. Networks

- The Internet
- Electronic mail
- Computer conferencing – Computer conferencing is simply and extension of E-mail

5. Multimedia

Databases

Databases may contain information such as traits of animal's breeds, plant and animal diseases and possible control methods, data and formulae to be used to calculate a feeding ration, weather forecasts, libarary catalogues and documentation's systems.

Teletext

Teletext is a system that links computer to television by which textural and graphical information can be transmitted on a one way to the farmers.

Videotext

Videotext is a computerized information storage and retrieval system. Videotext plays an immense role in extension education.

Feedback Systems

Common label used in practice is Management Information Systems. For a long time farmers have used accounts, milk records etc., as a feedback systems. Modern technology makes possible much faster and more effective feedback. For example sensors can measure how much milk per cow is produced today. Computer models can predict how much could be expected when taking into account the amount, which was produced yesterday.

Expert Systems

Export systems are an important development in information technology. These advise the farmers which alternative to choose from a wide range of possible alternative by processing data from a large number of variables according to certain decision rules. These systems can apply the decisions rules more effectively than the farmers can himself. Expert systems and other simulation Technology requires a high level of agriculture and animal husbandry research with a capacity to develop relevant data and models to process these data.

Networks

Farmers in many countries, especially under rain-fed condition do not live close to each other but still like to know what problems other farmers face and how they are solving them in other districts. Modern information technology to some extent allows widely separated farmers to maintain contact, irrespective of distance. For example, some Australian farmers groups, which participate in the national land care programme, have joined land care net, a "Commercial Electronic Network". Farmer groups as well as staff at research institutes and other governamental agencies can place information on the network, ask the question of the other groups, send electronic mail to each other and gain access to international data bases through the internet.

In our country the flow of information pertaining to crop/agril. information is given below:

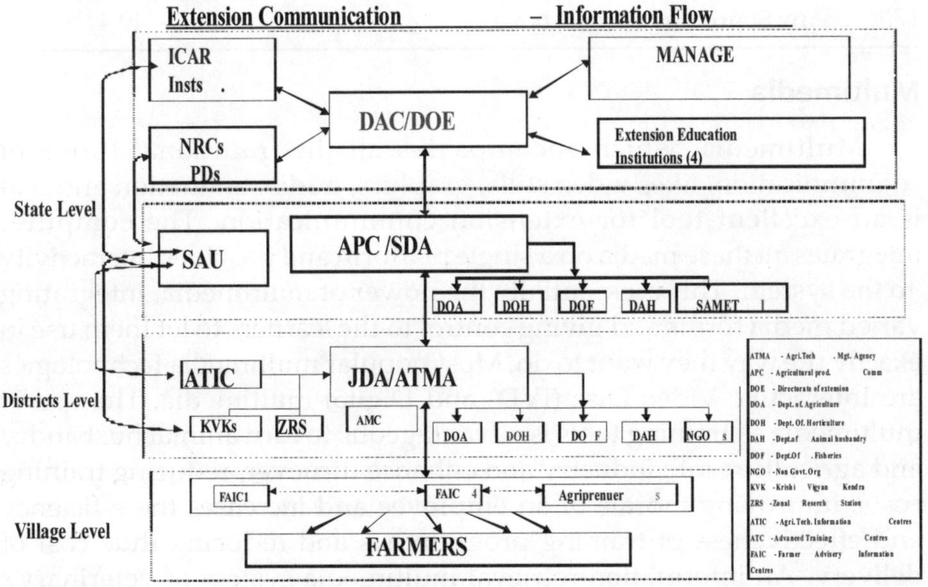

Accessing Modern Agricultural Technology

S.No.	Source	% of hrs
1.	Participation in Training	0.9
2.	Krishi Vigyan Kendra (KVK)	0.7
3.	Extension worker	5.7
4.	Television	9.3
5.	Radio	13.0
6.	Newspaper	7.0
7.	Village fair	2.0
8.	Government demonstration	2.0
9.	Input dealer	**13.1**
10.	Other progressive farmers	**16.7**
11.	Farmers' study tour	0.2
12.	Para-technician / private agency / NGO	0.6
13.	Primary cooperative society	3.6
14.	Output buyers / food processor	2.3
15.	Credit agency	1.8
16.	Others	1.7
17.	Any Source (all of the Above)	40.4

Multimedia

Multimedia, which encompasses all the traditional forms of communication, likes video stills, graphics, audio, animation and text is an excellent tool for extension communication. The computer integrates all these media on a single platform and provides interactivity to the system. This is essentially the power of multimedia, integrating varied media forms and giving control to the learners to let them use in exactly the way they want to do. Most popular multimedia technologies are Interactive Video Disc (IVD) and Digital multimedia. The use of multimedia as training tools is advantageous to turn animal husbandry and agriculture into industry and enhance turnover, reducing training costs, improving morale of an employee and increases the efficiency and effectiveness of training programmes and reducing thde cost of delivery. An information retrieval multimedia system of veterinary/agricultural extension can be used as primary medium for information delivery.

Features of a multimedia system

Audio, Video (recorder, Camera), Animation (cartoons), Text, Image (Pictures, photographs), Video Conference (talking & seeing on a computer), Virtual Reality (VR) (graphic technology, movement, sound & tough etc.)

Role of Information and Communication Technology

In this respect some potential areas in which ICTs play a catalytic role in developing the rural sectors are as under:-

Decision-making process

Sound decision-making depends upon availability of comprehensive, timely and up to date information. Information also needed to facilitate the development and implementation of development policies. E-mail and Internet can be used to transmit information to and from rural inaccessible areas.

Advisory services

The on line services offered by the ICT provide information on education and training, monitoring and consultation, diagnosis and monitoring and finally transaction and processing. It also helps in providing information services to the farmers on package of practices, disease/pest early warning system, weathre forecasting, input supply, credit availability, crop insurance, and post harvest technology at the earliest possible time.

Market outlook

It is cheaper and faster to trade on-line than on paper based medium, telephone or fax. E-commerce, enables the entrepreneurs to access global market information at their door step through websites and open up new regional and global market to promote their products that fetch better price and increase farmer's earning.

Empowering rural community

With the launching of new ICTs, rural communities can able to raise their voice and become motivated through training and dialogue with others, to express their views and opinions and become part of the decision making process in the execution of different rural development programme for their benefit. This approach should be participatory and lead to improved policy formation and execution.

Multi-sectoral service

New ICTs have the potential to penetrate under serviced areas, facilitate development with relevant local contents and faster delivery of information on technical assistance and basic human needs such as food. Agricultue, health and social sectors. The Internet enables the remotest village to access regular and reliable information from a global libray (websites). At the same time also provide opportunity to disseminate information about their community and the best out puts to the whole world.

1 Targetting mariginal groups :

   ICTs can benefit all stakeholders including the civil society, in particular youth and women of the rural society, and other disadvantaged groups that can be targeted for any developmental projects.

2. Creating employment opportunities :

   Through the establishment of rural information centers, ICTs can create employment opportunities in rural areas by providing them engagements in ICT related different jobs.Even the centres can also provide training and those trained may become small scale entrepreneurs.

3. Maintenance of land holding records :

   ICT facilitates the maintenance of land holding records and provides on line land registration services to the people in the rural sector thus avoiding the cumbersome process involved in it. Land records can also be made available on line which helps in creating transparency.

4. Triggering knowledge revolution :

   Introduction of ICT encourages in establishing computer aided and Internet connected Rural Knowledge Centres which helps in triggering a knowledge revolution in agriculture and allied sectors and will lead to an efficient and eco-sensitive precision farming movement.

Detail of Some ICT enabled services in India

1. The unique **"Warna Wired Village"** pilot project, launched under the aegis of National Information Technology Task Force as the Warna Nagar Cooperative Complex in Kolhapur and Sangli district, Maharashtra, provided Internet access to all the cooperative societies in seventy villages, this successfully demonstrating the acceptability and usage of modern information and communication technologies (ICTs) on agriculture, helath, eduction and GIS (Geographical Information System) on agriculture, health, education and GIS (Geographical Information System) in the fore front of development agenda thereby promoting sustainable agriculture and rural development.
2. The "**Information Village**" (1998) Project of M.S. Swaminathan Research Foundtion established a rural information network along a hub and spoke model aimed at bringing the benefits of high end information and communication technologies to the rural families in Veerampattinam, Pondichery. A Valua Addition Centre, which is the hub of the information network has been establihed in the project headquarter in Villianur village and is connected with ten villages through its ICT network. Information on agriculture, health, government policies, educational opportunities, the weather and many more information relevant and useful to the rural people of the region are designed put into a format the makes it more accessible to rural communities with low level of literacy, translated into local language and transmitted to the communities - owned and operated info-ships known as Village Knowledge Center Via the network. The critical issues involved with the "Info Village" Project are the participation of ultra poor in rural areas in managing the use of ICTs and secondly the involvement of rural women in the development process and greater emphasis on training and evaluation aspects.
3. The **"Gyandoot" Project** is an intranet based Government to Citizen (G1C) service started in January 2000 in Dhar district of Madhya Pradesh with an aim to create a cost effective, replicable, economically sefl-reliant and financially viable model for passing the Information and Communication Technologies (ICTs) to the rural masses. The main aim of this project is to access the services offered by the government departments through Gyandoot Kiosk (Soochanalay) by any citizen on payment of a nominal transaction

fee. The Gyandoot Project became very successful by ensuring equal access to emerging technologies and the services for the oppressed and exploited segments of the society through community owned, technologically innovative and sustainable Kiosks in a poverty striken, tribal dominated district of Dhar, Madhya Pradesh.

4. The **"SARI"** (Sustainable Access in Rural India) Project demonstrates the creation and deployment of Madurai district of Tamil Nadu. Teh project provides information over a network of 1000 internet connections encompassing 350 villages of the district which could be accessed by the villagers through net, leading to improvement in health, empowerment, learning and economic development amongst the poorest and most disadvantaged communities.

5. **"Drishtee"** an organizational platform, facilitates the creation of approximately 50,000 information Kiosks all over India within a span of six years and provides Information and Communication Technology services to vast rural and semi urban population of the country through the usage of state of the art software through access to government programmes, market information and private information exchanges.

6. **"e-Chaupal"** the unique web based initiative of ITC's International Business Division, offers information and communication technologies related to latest local and global information on weather, scientific farming practices and market prices through the web portal all in Hindi. This project facilitates easy access of these information by the farmers at their door step. Taking the literacy and infrastructure constraints at village level, e-Chaupal Sanchalak i.e. a lead farmer is acted as the interface between computer terminal and the user framers.

7. **"SEWA" (Self Employed Women's Association)** Technology Information Centers started in January, 2001 in eleven districts of Gujarat. SEWA is the laragest primary women's trade union in India serving over 250,000 poor self-employed women members in Gujarat, which provides training to the women organizers to build up their capacity and strengthen the micro enterprises of their members through information and communication technology services. Further, through Internet access system the various cooperatives established by the members are linked and

new programmes are being developed to support economic empowerment of their members.

8. **"Intelligent Community"** It is a joint venture project launched by Government of Andhra Pradesh and US based information Technology Company, Hewlett - Packard to create the world's first community in five mandals of Kuppan constituency of Andhara Pradesh. This project provides browser based services to the rural population to access information on government run programmes, agricultdsure, educational facilities, and health services on test run basis. Besides that with the assistance of a global voluntary agency on community leadership training i.e. World Crops, the Chief Minister's scheme for Unemployed entrepreneurs, has set up a chain of communication Information Centres (CISs) i.e. local cyber dhabas, which serve as copier shop, telephone booths and dissemination points for a variety of government welfare schemes. In addition to these ICT Projects, in National Dairy Development Board, ICT is being used at milk collection centers and in cooperatiaves to measure butter fat content of milk, test the quality of the milk and promptly make the payment to farmers. It has resulted in the removal of incentives to those who adulterate milk, reduced the time for payments from 10 days to less than 5 minutes and instilled the confidence in farmers on cooperative setup. All these factors have helped the milk market to expand to grater dimension. The Central Institute for Research on Goats (CIRG) has developed E-mail Conference System for Goat Outreach on its goat-nic.in server unsing free softwar called 'majordoma' which is available on www.greatcircle.com on a free Linux operating system. Three e-mail conferencing systems, viz., goat-net@cirg.nic.in, livestock net@cirg.nic.in and fishnet@cirg.nic.in. have been launched by the institue to help information inflow among technologists, farmers, development officers and planners. Under Animal Health Project funded by Department for International Development (DFID), Rajiv Gandhi College of Veterinary and Animal Sciences, Pondicherry in collaboration with University of Reading, UK, has installed a interactive touch screen information Kiosk in one of the peri- urban region of Pondicherry on a trial basis. It has information on important cattle diseases in addition to management of cattle and methods of acquiring information with the touch of the screen which had text and pictures with sound back-up.

9. **Expert System**: An expert system is a computer based advisor. An. expert system manipulate knowledge rather than numbers . It can serve as an in house expert ii) a specific field. It can combine book information with the knowledge, skills and experience of human experts. micro computers can be used for data organization, input calculations, record management, :Port production and for establishing computerized networks.

Expert System on Fodder and Grasses by IGFRI, Jhansi

Technical information and scientific recommendations of IGFRI are being passed on to the farmers through on-campus and off- campus training, frontline demonstrations (FLDs), publications etc. These methods are rigorous, time consuming and resource intensive. In the era of Information Technology (IT), the institute has developed

E-learning content for faster, accurate and effective dissemination of technologies.

Efforts on the development of "Expert systems on various fodder and grasses" has been initiated in English and Hindi language for the use of farmers on a mouse click.

Other effective tools of ICT for livestock based IFS system

1. Farm School On All India Radio :

   Farm School on All India Radio was started in 1974 as a joint venture of TNAU and All India Radio stations in Tamil Nadu and Pondicherry. Latest technologies are taught to farmers by offering a series of 13 lessons broadcast over 3 months at the rate of one lesson per week. At the end, one day contact programme is organised to provide an opportunity for the participants to discuss with the scientists who offered the lessons.

2. Audio and Video Cassette Lessons :

   One of the means to reach farmer's home is Audio and Video cassettes. Farmers can hear and see at their will and leisure and also repeatedly. Audio and Video cassette lessons in agriculture and allied sciences are being duplicated and supplied to the farming community at a nominal cost.

3. Agricultural Information Services :

   The Agricultural Information Services is responsible for disseminating the latest technologies on agriculture and allied subjects through the publications.

4. Agri Portals :

   Agri portals have been established as an electronic network in collaboration with input agencies. The information on the technology package at village levels to the farmers are made possible through these portals.

5. Gyan Vani Radio Network :

   Gyan vani radio network is yet another milestone activity of the Tamil Nadu Agricultural University under transfer of technology efforts. This collaborative venture between Tamil Nadu Agricultural University and Indira Gandhi National Open University, New Delhi has been started during January, 2002 to produce technical programmes through FM broadcast for the benefit of all kinds of listeners.

6. Narrowcasting of Agricultural Programmes Through Lpt :

   The TNAU in collaboration with Doordarshan Kendra, Chennai has recently started the "Narrow casting of agricultural programmes through LPT" for the benefit of farmers and general public in and around Coimbatore district.

7. Kisan Call Centre :

   With a view to bridge the gap between the farmers and scientists, the Kisan Call Centre established through out the country. Farmers uses the TOLL FREE Number. 1551 and contacts the scientists for suggestions and clarifications to solve their problems.

ICT in Agriculture/Forestry: Current Status

- Central Government Initiatives: DAC-NATP, DACNET, AGMARKNET, KCC DARE-ARIS, IVLP,NATP, ATIC, Ministry of CIT(NIC)- CICs, Warna Project.
- State Government Supported: Rajasthan- Jan Mitra, Gujarat-GAU-SAT-KRU, M.P.- Gyandoot, Maharashtra-Maha-Agrinet, A.P.-Rural e-Seva, Karnataka-Rhythu Sampark Kendras and Bhoomi, Kerala- Kissan Kerala, Akshaya Kendras.

- Corporate Sector Initiatives: e-Chaupals of ITC-IBD, Tata Kisan Kendras of Tata Chemicals, Parry's Corners of EID Parry and Chirag Kendras of n-Logue.
- NGOs and other private Sector: IVRP-MSSRF, Drishti.com, Agriwatch.com, TARA Kendras.

Information Kiosks: Current Status

S.No.	Major Project	No. of Kiosks	Services
1.	E-Chaupal (ITC-IBD)	5200	Market Prices, Procurement support, other info Service
2.	MKCL	2500	Educational Services, Extension Services
3.	DISK (NDDB)	2500	Dairy Information Services
4.	Chirag-Kendras (n-Logue)	2100	General Information Services
5.	Akshaya Kendras (Kerala)	2100	Agricultural Information, market informtion
	Total	26373	

Swaminathan using IT for rural development

M S Swaminathan, the father of the Green Revolution has now started using IT to rural development. M S Swaminathan Research Foundation (MSSRF). They are talking about the Info Village project, which started in 1998 in Pondicherry. It connects 10 villages through a hybrid wired-and-wireless network consisting of PCs, very high frequency (VHF) duplex radio devices, and related equipment to facilitate both voice and data transfer. The project creates content for villagers to obtain the information they could use to make improvements in their living conditions.

Information Kiosks: Major Projects in Pipline

- Ministry of Panchayati Raj to establish Information Kiosks in 2,37,000 Panchayats (Ministers' Speech, 9th August 2005: Mission 2007).
- Department of Information Technology (DIT) plans to initiate over 1,00,000 Common Service Centres (CSCs) under the National eGovernance Plan (NeGP).

- ITC is adding 7 new eChaupals a day and plans to scale up to 20,000 eChaupals by 2010 covering 100,000 villages in 15 states. (Sivakumar CEO, ITC-IBD).
- Governments of Uttaranchal, Rajasthan, A.P., Kerala have also announced Major Rural Connectivity Projects.
- ICTS for Agricultural Extension : Opportunities.

Strategy and Options

- Conscious Effort at Policy level to integrate ICTs enabled Extension as a part of Agricultural Extension/fodder extension;
- Necessity of a nodal planning and implementation agency at Country/ Province level;
- Continuous hunt and integration of new Communication technologies in Cyber Extension Strategy;
- Use of Media –Radio and T.V. in participatory programming and interactive mode;
- Use of Call Centers for Fodder Extension;
- Continuous monitoring and updating of ALL Extension systems and Databases;
- Mechanism of Capturing lessons and sharing –horizontally, vertically (within nations, across the nations).

The Road Ahead ...

- Use of ICTs provides Excellent opportunity to fodder/agricultural Extension as a tool to enhance the profitability and sustainability of IFS.
- We must grab this opportunity with both hands and redefine our critical importance to our ultimate clients : The Farming Community (Farmers and Farm Families).

References

http://www.agmarket.nic.in xv

http://www.agriwatch.com xvii

http://www.itcportal.com

http://www.ikisan.com

http://www.indiaagriline.com

http://www.nddb.org

http://www.isapindia.org

http://www.tatatkk.com

http://www.icar.org.in

http://www.igfri.res.in

□□□

Chapter **36**

Role of NGOs/PPP in Farming System Research and Development

S.N. PANDEY

Programme Director, Development Alternatives, TARA gram, Pahuj, Jhansi-284003 (UP)

Introduction

Agriculture development is critical for securing nutrition, improving employment opportunities and income in the rural sector. Indirectly, agriculture development also contributes to environmental management, as the sector add to the critical needs such as absorbing carbon, managing watersheds, and preserving biodiversity. Agriculture development is also very important for the world, particularly the developing economies, in view of the spikes in the world food prices in the recent past.

According to a publication by FAO, the world may be in need of an agricultural system that can produce 50 percent more food to feed the world's estimated 9 billion people by 2050. However, this would be a challenge if we do not have a coordinated approach to develop the agriculture sector in the world. The growing demand for food and nutrition can be managed with a good policy framework and sustained investments.

At present, the farmers concentrate mainly on crop production which is subjected to a high degree of uncertainty in income and employment to the farmers. In this contest, it is imperative to evolve suitable strategy for augmenting the income of a farm.

Integration of various agricultural enterprises *viz.*, cropping, animal husbandry, fishery, forestry etc. have great potentialities in the agricultural economy. These enterprises not only supplement the income of the farmers but also help in increasing the family labour employment. The integrated farming system approach introduces a change in the farming techniques for maximum production in the cropping pattern and takes care of optimal utilization of resources. The farm wastes are better recycled for productive purposes in the integrated system. A judicious mix of agricultural enterprises like dairy, poultry, piggery, fishery, sericulture etc. suited to the given agro-climatic conditions and socio-economic status of the farmers would bring prosperity in the farming.

Public–private partnerships (PPPs) in agricultural research and development particularly in farming system context are increasingly viewed as an effective means of conducting advanced research, developing new technologies, and deploying new products for the benefit of small-scale, resource-poor farmers and other marginalized groups in developing countries. Partnerships between private investors, citizens and states are a topical theme in debates about farming system governance in developing countries.

For years, Public- Private Partnerships between private companies and states have formed a common means of providing environmental infrastructure or services where state funds or expertise are lacking. More recently, however, the concept of 'partnerships' has diversified, and now may include broader collaborations between states, companies and civil organizations, and for a wider range of policy initiatives. The NGOs and the private sector have started to play a greater role in extension in the last two decades. There is an increasing realization that public extension by itself cannot meet the specific needs of various regions and different classes of people, farming communities. These include non-governmental organizations (NGOs) at the international, regional and national levels, as well as groups concerned with the empowerment of marginalized sections of society, such as women, indigenous peoples, and youth groups, and other community based organizations. Such organizations have worked to foster grassroots based approaches to the protection and preservation of the region's environment. Major public interest groups are increasingly contributing efforts towards sustainable development through participation, advocacy, demonstration projects, monitoring and research, as well as cooperation and networking with other NGOs and government departments.

NGO Roles and Activities

Traditionally, community based organizations played an important role in the management of common property resources such as forests, wasteland, water management and community resources, Community institution building, Gender development, technology generation and transfer for the people oriented development. The encouragement of public participation in agriculture development and management through legislation in recent years has also enhanced the role of People-Public-Private Participation as major groups.

The conviction that NGOs and government agencies can and should work together in a complementary relationship has become stronger and the credibility that NGOs have acquired from successful campaigns and implementation has created strong and growing public support for their new advocacies.

Active regional and international NGOs working on sustainable development concerns include: the Asian NGO Coalition for Agrarian Reform and Rural Development (ANGOC); and the Asian Alliance of Appropriate Technology Practitioners (APPROTECH ASIA), TERI, BAIF, Development Alternatives (DA), MYRADA, DHAN, PRADAN, CEE etc.

A Public–Private Partnership (PPP) is a government service or private business venture which is funded and operated through a partnership of government and one or more stakeholders (Private, voluntary organization, people institution, Panchayat, etc.). These schemes are sometimes referred to as PPP, P3 or P^3.

There are many voluntary organizations (NGO) engaged in promotion of integrated farming system which is viewed as natural resources management at the grassroot level. These organizations are primarily involved in motivation of farmers, creating awareness, establishing linkages with information centers and marketing outlets. Some of these agencies are also involved in providing micro-finance, promoting micro-enterprises and establishing linkages with various Government schemes for development.

At the community level, they are in the front line in providing assistance in the acquisition of basic needs and amenities, in identifying issues, raising awareness, and providing information to grassroots communities, articulating the communities' problems and needs and bringing these to the attention of those who can affect change by involving the other stakeholders. The basic function of these institutions in the managing natural resources in the following ways:

A) Awareness-Raising, Campaigning and Advocacy

- Offer a forum to interact with external agencies interested in supporting the development intervention in the subject area,
- use of promotional material and brand building of the massage on concern & issues related to the *natural* resource management
- sharing information for backward and forward linkages for multiplication and replication of the NRM modal/development initiatives involving other stakeholders
- organizing and participation of the topic -specific workshop, meetings, exposure visits/ orientation programme etc.
- Organizing campaign for awareness building, resolving conflict, opinion, ideas and alternatives at regional and national levels.

However numerous means have been used by the community based organization through Public–Private Partnership (PPP) to attract the planner and policy makers for recognizing the efforts from the community levels.

The Development Alternatives (DA)a national non for profit organization is regarded as one of the best-informed and organized grassroots movement in India with the help of NABARD and government for the promotion of WADI (Small orchard) and watershed Development Programme in Bundelkhand region of Central India. The modal has been attracted by many national and international planers to replicate in the other part of the region.

B) Institutionalization of Public–Private Partnership (PPP)

The participation of major public interest groups is also being institutionalized through formal representation in decision making and management bodies. There is equity in the decision-making between government, private sector and NGO (civil society) representatives in the programme formulation, management, and implementation at the local regional and national level. The mandate for civil society participation has also been provided through the development of programmes that facilitate direct community-based management of natural resources particularly to promote people's participatory institutions like Panchayati Raj institutions, co-operatives and self-help groups. The involvement of the NGO as a operational partner help to share and resolve conflicts locally in democratic fashion. It also helps to mobilize direct and indirect contribution (Money, labour, material etc.) from as a community fund and same for maintaining the community assets created during the programme. The local institutions important for mobilizing resources and regulating their uses with the view to maintain a long term base for productive activity by the involving government agency as a partner. The government and private sector mobilize the financial and intellectual resources to replicate and multiply the resource in effective institutionalization process by engaging different stake holders.

C) Provision and Validation of Information, Knowledge, and Communication channel

The provision and validation of information, knowledge, and communication channel forms basis for successful collaboration between governments, NGOs, community representations and other stakeholders, including project proponents, the private sector, the scientific community and the press and broadcast media. In circumstances where NGO are formally involved in inter-agency bodies and in public consultation processes that allows ongoing verification and clarification between parties, thereby facilitating progress and reducing the need for conflict resolution. The involvement of the NGOs in the natural resource management programme perform multiplicity of functions such as

- Available resources can be put to their most efficient and sustainable use with location specific knowledge, which is best generated and interpreted locally.
- Monitoring changes can be quick and less costly. Adaptive changes made in resource use gets accelerated by institutionalized local decision making.
- Community norms and consensus conditions people behavior. Institutional practices, which are environ mentally sound, may require more than just individual incentives and persuasion. Community wisdom is more practical and sustainable.
- In the absence of local institution, all conflicts are dealt with at higher levels, yielding slower and often less appropriate out come. Local institution helps in resolution of most of the local conflict.

D) Provision of Funding Support

Many governments and private actors in region provide or facilitate, funding to strengthen NGO activities in natural resource conservation. Governments, donors and NGOs have also joined together to form endowment funds to support environmental projects. Examples are in Japan's Fund for Global Environment, based on contributions from the government and the private sector, was established to provide assistance for the global environmental conservation activities of NGOs. The Global Environmental Facility (GEF) recognizes that civil society has become an important force in implementing Agenda (GEF 1998). About 20 per cent of the funds expended by the GEF involve NGOs in the design, planning and/or implementation of these projects. The GEF also supports a Small Grants Programme (SGP), implemented through the UNDP, and has started a Medium-Sized Grants Programme where NGOs can access up to US$1 million for biodiversity conservation projects.

Operationalising partnerships

One of the main difficulties lies in identifying appropriate roles and responsibilities for the various stakeholders in the natural resource programme. The problems of finding the appropriate institutional 'mix' has been noted for donor programmes in general and for natural resource management programme in particular with the community participation. The appropriate balance of power and responsibility will depend on preconditions such as the extent of conflict endemic within communities, the existence of entrepreneurial traditions and the effectiveness of government line department. For joint action approaches to be successful there must be agreement on the respective roles of different partners and a shared vision, where partners agree on and share common goals and objectives.

Partnerships between the concerned agencies need to be structured with each playing a role according to its comparative advantage.

Governments and research organization for example can provide technical assistance and guidance, economic incentives (especially where resource use generates externalities or users are the poor), an enabling legal framework with clear territorial rights, formal conflict resolution mechanisms and financial and technical support for decentralized monitoring.

Communities contribute local ecological information, knowledge of economic and social conditions which enables them to devise well adapted rules and procedures, low cost customary conflict resolution mechanisms and self-monitoring mechanisms.

The role of NGO as a service provider of the different technology transfer mechanisms separately with the help of research organization. However, in practice pathways are often composed of complementary links among the actors (government, private sector, NGOs).

Umbrella Programme on Natural Resource Management (UPNRM) GOI, NABARD and GIZ

Exploitation and degradation of productive resources like land, water and forest negatively affect the livelihood of India's 450 million rural poor, who are mostly dependent on agriculture. The conventional response to the degradation had been large scale, state sponsored grant programmes. With the realization that government money alone will not be able to solve the problem; the challenge is to link rural livelihood development with environmental sustainability in a way that allows economic growth through sustainable management, rather than exploitative usage, of precious resources. UPNRM's objective is simple yet challenging; reduce poverty by creating livelihood opportunities, increasing farm income, strengthening the agriculture value chain and conserving the natural resources..

The Programme's tailored loans and partnership facilitation supports the Government of India's thrive to incentivize and augment private sector investments like :

1. "Credit Plus" Facility - Better services than conventional financial products:
2. Public Private Partnerships (PPP) – Partnering for inclusive and equitable rural investments, UPNRM supports partnerships between public sector agencies and corporates that link rural businesses (farmer, rural business organisation, etc.) to existing value chains.

One of the agro-forestry projects, being implemented by Patneswari Agriculture Cooperative Limited (PACL) in Koraput district, Orissa, has been registered as a Clean Development Mechanism (CDM) project by the United Nations Framework Convention on Climate Change (UNFCCC) and is expected to mitigate 33400 tons of carbon dioxide annually.

Engaging the private sector: Five projects have been sanctioned in public private partnership mode, a model that has not been so far tested for rural development, e.g., the provision of clean water in remote villages in partnership with Waterlife India Pvt. Ltd.

The Way Forward - Fostering Business Viability and Environmental Sustainability

The main stakeholders are in natural resource management programme development process includes government agencies, NGOs, the private and banking sectors, political institutions and communities. Government not only provides services related to renewable natural resources but also much of the fabric necessary for the functioning of civil society (in the form of legal and administrative systems).

As local communities gain awareness of their new rights and responsibilities in resource management, demands for legal, financial, technical and logistical support from the public sector increase. These increasing demands on the state are ironic given that one of the founding reasons for the decentralization of natural resource management has been the inability of government departments to manage common pool resources due to insufficient human resources.

During the last decade, the Indian government explicitly stated that attempts would be made to involve NGOs and corporate as collaboration partners in various development programmes particularly in natural resource management related.

The NGO sector is however characterized by great disparity both in terms of size, professional qualifications, geographical and sectoral areas of work and also by diversity in approaches, scales of financial support, networks and linkages.

Reference

McKean, M. 2002. Common property: what is it good for and what makes it work?

MoRAE 2005. Guidelines for watershed development. Ministry of Rural Areas and Employment. Government of India.

3.ODA 2006. ODA's review of participatory forest management. Synthesis of findings, ODA, London.

Uphoff, N. 1992. Local institutions and participation for sustainable development. Gatekeeper series 31: IIED, London, UK.

Annual Report, 2010. Development Alternatives.

Sustainable development success story 2000. Integrated Planning and Management of Land Resources, Agriculture and Forests FAO.

□□□

Chapter 37

Application of Nanotechnology for Conservation of Resources in Agriculture

A. MAITY, R. SRINIVASAN, D. VIJAY, M. SRIVASTAVA, K.K. SINGH, C.K. GUPTA, M. CHAUDHARY AND A. RADHAKRISHNA

ICAR - Indian Grassland and Fodder Research Institute, Jhansi-284003 (UP)

All over the world, farmers work hard but do not make money, especially small farmers because there is very little left after they pay for all inputs (seeds, livestock breeds, fertilizers, pesticides, energy, feed, labour, etc.) with tighter profit margins and increasing environmental constraints, strategic planning of Integrated farming systems (IFS) is becoming both more important and more difficult. IFS (or integrated agriculture) is a commonly and broadly used word to explain a more integrated approach to farming as compared to monoculture approaches. It refers to agricultural systems that integrate livestock and crop production or integrate fish and livestock and may sometimes be known as Integrated Biosystems. In this system an inter-related set of enterprises used so that the "waste" from one component becomes an input for another part of the system, which reduces cost and improves production and/or income. The difficulty in its implementation is especially true for dairy production system in India. Livestock production is complex with a number of interacting processes that include crop and pasture production, crop harvest, feed storage, grazing, feeding, and manure handling. It becomes a challenge to the stakeholders to make the whole system profitable. When the principles of conservation agriculture merge with utilisation pattern of different components of IFS in a view of entrepreneurship, resource conservation becomes an integral part of it. Resource conservation doesn't always mean that crop will be supplemented with less amount of food, but will be in judicious way. But think of a technology which can help the crop stand to utilise whole amount of nutrients supplied to it without any leaching or runoff loss; a technology which can help to predict presence of any chemical regarding storage, preservation or disease prevalence even when our nose can't sense it; a technology that helps to supply a drug or supplement precisely to the target and many more advantages that are beyond the range by traditional technologies. Nanotechnology, though we have just begun to work with this technology and used in a tactical and safe way, can solve a number of problems that are not possible by existing technologies.

We know that agriculture is the backbone of the developing countries, with more than 60% of their population relying for their livelihood on this sector. The intense problems that jeopardise the agricultural productivity, such as several abiotic and biotic stress factors, require more precise and effective answers; and products with higher worth are further required to mitigate the stress. Nanotechnology can improve our perception and also deliver improved products with minimum use of resources. It can contribute to the development of superior systems for monitoring environmental conditions and delivering need-based nutrients or apposite pesticides at target and thus potentially enhance yields or nutritional values by minimising the utilisation of resources. Thus, nanotechnology can be of very high relevance and importance in the future agriculture, food systems and industry when resources are shrinking day by day.

Nanotechnology is one of very recently advocated techniques in Indian as well as world agriculture in that direction. It uses the knowledge of life science, material science and information technology and it involves manipulation of matter on an atomic or a molecular scale. Nanoparticle literally ranges from 1 – 100 nanometre (nm). NPs have lots of advanced or altered physical, chemical and electrical properties due to its microscopic size as compared to their bulk materials. From time immemorial humans have used this technology in making paintings, steel, vulcanizing rubber etc. In present days, our scientists are constantly trying to exploit the benefits of those altered properties that are resultants of the tiny size of NPs. Altered physical properties include increased surface area, enhanced electrical conductivity, improved charge capacity, reduced melting point, rearrangement of molecules leading to ultra hard structure, changed magnetism, self-purification of materials and many more in the list. Thus by using structure at nanoscale, we can greatly expand the range of performance of existing chemicals and materials with respect to those altered properties. As a result, for some examples, carbon in naosize becomes hundred times stronger than steel, aluminum turns highly explosive; gold melts at room temperature, ceramics and metals become extra strong and ductile; paints turn exceedingly shiny and water repellent, antireflective, self-cleaning, resistant to ultraviolet or infrared light, antifog, antimicrobial, scratch-resistant, or electrically conductive; computer storage and operation capacity get extraordinarily higher, desired chemicals can be loaded in nanoscale carrier to deliver in target site etc. Important changes in behavior are caused not only by diminishing size, but also by the emergence of totally new phenomena such as quantum confinement, a typical example of which is that the color of light emitting from semiconductor nanoparticles depends on their sizes. Altered chemical properties due to reduced size of NPs include enhanced reactivity and catalytic property that lead to development of entirely new and accurate biological sensors. It facilitates early detection of any chemicals present even in negligible quantity with respect to human, animal or plant diseases; seed and food preservation, defense sector etc. This precise biosensing ability aids in targeted drug delivery at defined site (used in cell, tissue or organ specific medication).

This unprecedented growth, expansion and inception of this technology by different fields have lured the agricultural scientists to conduct different basic study to merge it with already existing technologies to get reed off stagnant agricultural productivity. The door has been opened wide enough for every sectors of agriculture like nanofertilizers, nanosensors, nanocarrier based chemical delivery system; seed invigoration with nanoparticles, nanofilm coating on agriproducts etc. to exploit its benefits at maximum. Despite the fact that nanotechnology may serve as a prime tool to develop several products in material science, medicine and defence, we have just begun to scuff the surface in the field of agriculture.

Application of nanotechnology in resource conservation agriculture

Application of nanotechnology revolves around the word 'resource-conservation' by means of the basic nature of this technology. By the property of enhanced activity of nanoparticles due to its lower size and higher surface area, lesser amount of it can deliver the same degree of result or improved efficiency as compared to its bulk material of higher quantity. There are several areas of its application.

Maintenance of soil health

Management of soil health is a key issue in sustainable crop production because of nutrient ions fixation in the broken edges of clay particles and restriction of the nutrients availability. Nanoparticles can be adsorbed on to the clay lattice thereby preventing fixation and releasing plant utilizable plant nutrients into the soil. This process improves soil health and nutrient use efficiency by crops.

Nano-fertilizer for sustained use efficiency

In the past few decades, use efficiencies of N, P and K fertilizers have remained constant as 30-35%, 18-20% and 35-40%, respectively, leaving a major portion of added fertilizers in the soil and their subsequent entry into aquatic system causing eutrophication. Nano-fertilizer, an innovative strategy can be used to improve fertilizer use efficiency of crops. Fertilizer particles can be coated with nano-membranes that facilitate slow and steady release of nutrients in the soil. This process helps to reduce loss of nutrients while improving fertilizer use efficiency of crops. Manganese carbonate or zeolite based nano- fertilizer products are capable of releasing nutrients slowly and steadily that commensurate with crop demand. Nano-composites with all essential nutrients that ensure balanced fertilization can also be synthesized to nourish the plants.

Nanoherbicides for weed management

In rainfed areas, herbicide usage is far from satisfactory due to uncertainty in moisture availability. Nano-encapsulated herbicide molecules are capable of releasing herbicide molecules only when moisture is available in soils thereby circumvent weed seeds from germinating. This method would save time and costs of weeding. Multi-species approach with single herbicide in the cropped environment resulted in poor control and herbicide resistance. Continuous exposure of plant community having mild susceptibility to herbicide in one season and different herbicide in other season develops resistance in due course and become uncontrollable through chemicals. Foliar applied herbicide in perennial weeds such as *Cynodan dactylon*, *Cyperus* spp. and *Solanum elaeagnifolium* fail to kill them completely. The target domains of the present day herbicide in a plant cell are destruction of structure and function of the plant-specific chloroplast, inhibition of lipid biosynthesis, and interference with cell-division by disrupting the mitotic sequence or inhibiting the mitotic entry, inhibition of cellulose biosynthesis and deregulation of auxin-induced cell growth. Compared to foliar absorption, root absorption is a simpler process. Roots do not have cuticles like leaves; although, mature roots are covered by a suberized layer. This means that there are few barriers to herbicide absorption by plant roots. Since roots are essentially lipophilic, lipophilic herbicides will be readily absorbed. Although molecular mechanisms of action are not yet completely understood even for some commercially available herbicides, about 60% of conventional herbicides interfere with the Photo Electron Transport system of the chloroplast. At present, no commercial herbicides exist that interfere with CO_2-fixation and sugar production. Molecular characterization of underground plant parts for a new target domain and developing a receptor based herbicide molecule having specific binding property with nanoherbicide molecules like carbon nanotubes capable of killing the viable and dormant underground propagules of weed seeds.

Nanosensors for precise detection

Nanosensors can be developed to determine nutrient, moisture and physiological status of plants, which is very much essential to achieve the mission of precision agriculture. These sensors are helpful in taking up appropriate and timely corrective measures. Nano-particles are mini laboratories having the potential to precisely monitor temporal and seasonal changes in the soil-plant system. Nanotechnology may be developed and deployed for real time monitoring of the crop growth and field conditions including moisture level, soil fertility, temperature, crop nutrient status, insects, plant diseases, weeds, etc. Networks of wireless nanosensors positioned across cultivated fields provide essential data leading to best agronomic intelligence processes with the aim to minimize resource inputs and maximizing output and yield. Such information and signals include the optimal times for planting and harvesting crops and the time and level of water, fertilizers, pesticides, herbicides, and other treatments that need to be administered given specific plant physiology, pathology, and environmental conditions.

Targeted delivery of chemicals through nano-carrier

Nanoscale devices can be developed to detect and treat diseases in crops long before symptoms are visually seen and cause any yield loss. Smart delivery systems can deliver the desired chemical at exact location and monitor the effects of delivery of bioactive molecules. Implanting nano-particles in the plants could determine the health status of plants. Many nanoscale carriers, including encapsulation and entrapment, polymers and dendrimers, surface ionic and weak bond attachments and other mechanisms may be used to store, protect, deliver, and release by control of intended payloads in crop production processes. One of the advantages of nanoscale delivery vehicles in agronomic applications is its improved stability of the payloads against degradation in the environment, thereby increasing its effectiveness while reducing the amount applied. This reduction helps address agricultural chemicals run-off and alleviate the environmental consequence. The nanoscale delivery vehicles may be designed to "anchor" to plant roots or the surrounding soil structures and organic matter if molecular or conformational affinity between the delivery nanoscale structure and targeted structures and matters in soil could be utilized. Controlled release mechanisms allow the active ingredients to be slowly taken up, hence, avoiding temporal overdose, reducing the amount of agricultural chemicals used, and minimizing the input and waste. Environmental consideration including precision farming can reduce pollution to a minimum.

Seed invigouration

In Indian farming, farmers rarely test the seeds for its quality before sowing. Usually, farmers store the seeds of previous season for a few months and rush to take up sowing hurriedly without testing its germinability on receipt of rainfall. This results in poor crop stand that eventually causes low yield. Seeds registering normal germination percentage at the testing centers rarely reflect in the field conditions. This necessitates introducing nano-paricles in seeds in order to quench the reactive oxygen species generated as a consequence of storage. Nano- particles also suppress the damage caused by lipid peroxidation. Recent studies have shown that ZnO, Ag, CuO, Al, Fe and TiO_2 nano-particles could accomplish the task of catalyzing anti-oxidant activities.

Nanobiotechnology

The nanobiotechnology is yet another appealing field of science that has a potential to manipulate materials at molecular levels. New ways to study molecules, DNA and cells for food, nutraceutical and pharmaceutical applications are coming out every day. Advanced instruments and research methods for DNA and protein identification and manipulation are overcrowding the market. Nanobiotechnology can provide novel laboratory-on-a chip proteomics technology for assessment of metabolic pathways in important biocontrol agents. It also can provide rapid and reliable DNA methods for detection of phytotoxins and pathogens in digested and composted animal wastes to determine their subsequent safe use in agriculture.

Agricultural Engineering Issues

Nanotechnology has many applications in the field of agricultural machinery. These cover: application in machine structure and agricultural tools to increase their resistance against wear and corrosion and ultraviolet rays; producing strong mechanical components with use of nano-coating and use of bio-sensors in smart machines for mechanical-chemical weed control; production of Nano-cover for bearings to reduce friction. The use of nanotechnology in production of alternative fuels and reduction of environmental pollution are also worth mentioning.

Application of Nanotechnology in Animal Sciences

Nanotechnology has the ability to provide appropriate solutions for addressing the issues of food items, veterinary care and prescription medicines as well as vaccines for domesticated animals. Taking certain medications at nano level such as antibiotics, vaccines, and probiotics, would be effective in treating the infections, nutrition and metabolic disorders. Appropriate timing for the release of drug and self-regulatory capabilities are the main advantages of the use of nanotechnology in the application of drugs. The C-60 carbon particle (bucky ball) is spherical molecule having nearly 1 nm diameter. It is non-toxic to the live cells and biocompatible in nature. It can be used as a carrier to deliver the water soluble peptides and drugs. The nano particles can penetrate the skin through minor abrasions; these are reported to be used as sensor to detect the altered cell behaviour. The nano-magnets can be used as drug delivery system specially to treat the cancerous growth without any harm to the surrounding tissues. Different types of proteins like albumin, gelatin, gliadin and legumin can be used to prepare nanoparticle-based drug delivery system. Inert nanobeads are used to neutralize the antigen causing osteoarthritis in racing horses. Nanoparticle based nutrient supplementation has beneficial effects on growth performance and body composition. Nanotechnology is used to produce the chicken/ goat meat in the laboratory in large quantities, maintaining the same nutritive value, taste, texture without any hazard (vegetarian meat). Use of nanotechnology in designer egg production is well-known. It can produce eggs with low cholesterol, less yolk content, more nutrients, and desired antibodies. Animal waste is a serious concern in the animal production industry. Stricter environmental policies will prevent irresponsible discharge of animalwaste. The unpleasant odors that emanate from intensive animal production facilities adversely affect air quality, and in turn, living conditions and the real estate value of the adjacent area. However, bioconversion of animal waste into energy and electricity can result in new revenue, renewable energy, high quality organic fertilizer, and improved environmental quality while value added. Nanotechnology-enabled catalysts will play a critical role in efficient and cost effective bioconversion and fuel cell for electricity production as well as enabling efficient energy storage which will greatly facilitate and benefit the development of distributed energy supplies, especially in rural communities where infrastructure is lacking. Such an approach may result in the elimination of the need for system wide electricity grids, hence accelerate the rural development and improve productivity, and business and living environment and will be especially beneficial to developing countries.

Nano technology application in fisheries and aquaculture

Nanotechnology has tremendous potential to revolutionize fisheries and aquaculture sector. Nanotechnology tools like nano-materials, nano-sensor, DNA nano-vaccines, gene delivery and smart drug delivery have the potential of solving many puzzles related to fisheries nutrition and health production, reproduction, prevention and treatment of disease. Nanotechnology will help fish processing industry for producing quality products by detecting bacteria in packaging, producing strong flavour, colour quality and safety.

Nano-Food Industry

Today, nanotechnology is no longer an empty buzzword, but an indispensable reality in the food industry. The impact of nanotechnology is huge, ranging from basic food to food processing, from nutrition delivery to intelligent packaging. It is estimated that the nanotechnology and nano-bio-info convergence will influence over 40% of the food industries up to 2015. Microencapsulation of foods to preserve its flavour and functional qualities such as vitamins and volatiles, nanofilm technology for efficient packaging to extend the shelf life of food stuffs are well known facts.

Tracking of products for commercial protection

Identity Preservation (IP) is a system that creates increased value by providing customers with information about practices and activities associated with a particular crop or other agricultural products. Certifying inspectors can take advantage of IP as a more way of recording, verifying, and certifying agricultural practices. Quality assurance of agricultural products safety and security could be significantly improved through IP at the nano-scale by continuous tracking and recording of the history. The IP system is highly useful to discriminate organic versus conventional agricultural products.

Nanobar codes have been applied in multiplexed bioassays and general encoding because of their possibility of formation of a large number of combinations that render them attractive for this purpose. The particles to be utilized in nanobarcodes should be easily encodeable, machinereadable, durable, sub-micron sized particles. Nanobarcodes have been used as ID tags for multiplexed analysis of gene expression and intracellular histopathology, authentication or tracking in agricultural food and husbandry products. With the enhanced importance of traceability in food trade, such technologies will be helpful in promoting biosafe international food trade.

Quantum dots (QDs) for staining bacteria

QDs are better than conventional organic fluorophores (dyes) to stain bacteria due to their more efficient luminescence compared to the organic dyes, narrow emission spectra, excellent photostability, symmetry and tenability according to the particles sizes and material composition. By a single excitation light source, they can be excited to all colors of the QDs due to their broad absorption spectra.

Nanotechnology for environmental safety

Nanotechnology can help in the clean-up of wide range of pollutants that are non-biodegradable in nature. Living organisms clean the environment using molecular machinery to break down toxic materials. Nano products function very well in an ecosystem where toxicants have destroyed a group of beneficial microorganisms. Nanotechology, offers the potential of novel nanomaterials for the treatment of surface water, groundwater and wastewater contaminated by toxic metal ions, organic and inorganic solutes and microorganisms. Due to their unique activity towards recalcitrant contaminants, many nanomaterials are under research and development for use for water purification. Water filtration may be improved with the use of nanofiber membranes and the use of nanobiocides, which appear promisingly effective. One of the processes using nanoparticles is photocatalysis. This process can be used for the decomposition of many toxic compounds such as pesticides, which take a long time to degrade under normal conditions. The removal of toxins from wastewater is an emerging issue due to its effects on living organisms. Many strategies have been applied for wastewater treatment with little success. Photocatalysis can be used for purification, decontamination and deodorization of air.

Nanotechnology in irrigation water filtration

The emerging technologies that will benefit farmers all over world, especially in developing countries include several nanomaterials which are considered economically effective in purification of irrigation water. Nano-enabled water treatment techniques based on membranes filters derived from carbon nanotubes, nanoporous ceramics, and magnetic nanoparticles inspite using chemicals and UV light are common in traditional water treatment. Filters made from carbon nanotube could be employed in removing contaminants and toxicants from potable water. Carbon nanotube fused mesh that can remove water-borne pathogens, heavy metals like lead, uranium and arsenic has been suggested by researchers. Employing nanoceram filter with positive charge can trap bacteria and viruses with negative charge. This sophisticated filtering machine removes microbial endotoxins, genetic materials, pathogenic viruses, and micro-sized particles (Argonide, 2005).

Nanotoxicity

The ultra-small sizes that make the nanoparticles of immense usefulness, unfortunately the same characteristic also causes several adverse effects and may cause significant hazards to environment, animals, human beings and plants when used non-judiciously. Animals may inhale nanoparticles resulting into various ill effects and disorders. The particles may enter the blood stream. Nano-pesticides may reduce environmental contamination through the reduction in pesticide application rates, but they may also create new kinds of contamination of soils and waterways due to enhanced transport, longer persistence and higher toxicity. Air-borne nanoparticles present some specific hazards for human health; they may enter the body through the respiratory system. Due to the entry of nanoparticles into lungs and blood stream, there is possibility of inflammation, protein fibrillation and induction of genotoxicity. Nanoparticles as pesticides, fertilizers or in other formulations, when air-borne, may deposit on above ground parts of plants. They may plug stomata and create a fine physical and toxic barrier layer on stigma preventing pollen tube penetration. They may also enter the vascular tissue and impair translocation of water, minerals and photosynthate. Because of these risks, the use of nanotechnology in fertilizer and pesticide formulations has to be addressed very cautiously, and this warrants mandatory need to critically analyse and examine the risks involved with the nano-formulations.

Nanotechology and societal stigma

The effects of exposure to engineered nanoparticles may be different from the effects caused by naturally occurring nanoparticles. Engineered nanoparticles may be better to evade the body's defences because of their size or protective coatings. Moreover, the health and environmental risks raised due to the exposure to engineered nanoparticles need further study. Up-coming nanotechnologies in the agricultural field seem quite interesting and promising. However, the probable risks in using nanoparticles in agriculture are no diverse than those in any other business. Through the rapid distribution of nanoparticles to food products, whether it is in the food itself or part of the packaging, nanoparticles will virtually come in direct or indirect contact with everyone. The probability could be that "the merger of nanotech and biotech may cast unknown consequences on soil, health, biodiversity and the environment. Since there is no standardization for the use and testing of nanotechnology, products incorpo-rating the nanomaterials are being produced without check. The ability for these materials to infiltrate the human body is well known, but there is really no information on the effects that they may have. While there is no evidence of harm to people or the environ-ment at this stage, nanotechnology is a new and evolving area of study that could cause a great deal of harm due to its still ambiguous chemical properties. With the current application and advancements soon to come, nanotechnology will have a great impact on the direction that agriculture will take. Scientists are blazing a trail for a new technology and looking at every possible avenue to improve upon current methods in every possible field. In the field of agriculture, there are still many possibilities to explore and a great deal of potential with up-coming products and techniques. Therefore, extensive studies are required to understand the mechanism for nanoparticles materials toxicity and their impacts on natural environment.

Conclusion

Nanotechnology has great potential in agriculture as it can enhance the quality of life through its applications in fields like sustainable and quality agriculture and the improved and rich food for community. The data base of research initiatives on "Nano-Agriculture" vividly exhibits its potentials and prospects in Indian farming. The Knowledge gained will pave ways and means to explore this exciting field of science to achieve the goals enshrined in the millennium development. This frontier area of technology has the potential to revolutionize the agricultural and food industry with new tools for the molecular treatment of diseases, rapid disease detection, enhancing the ability of plants to absorb nutrients etc. The foremost benefit that nanotechnology can deliver is to enhance the efficiency of any material compared to its bulk material. But the outlook of nanoscience in agriculture is vague owing to a lot of grounds, for example, the unconstructive response from people towards genetically modified (GM) crops, need of a lot of required cleverness in government agricultural research and technology units for nano type of explorations and poorly-equipped new instruments and new-fangled technologies. There is a terrible call to slash down the jagged outline existing among the society, common man and budding scientific notions and if we achieve something in overcoming this line, then an unexpected bright and beneficial future will be at the door step of society.To address all the challenges ahead, we should think of an alternate technology such as "nanotechnology" to precisely detect and deliver the correct quantity of nutrients required by crops that promote productivity while ensuring environmental safety.

Reference

Amine, A., Mohammadi, H., Bourais, I. and Palleschi, G. 2006. Enzyme inhibition based biosensors for food safety and environmental monitoring. *Biosensors and Bioelectronics* **21**: 1405-1423.

Anonymous. (2010). A Manual on Integrated Farming Systems (IFS). Caribbean Agricultural Research and Development Institute (CARDI), Ministry of

Economic Development, Belize, 9th European Development Fund. (BRDP Call for Proposal No. 7; Agriculture Enterprise Development for Rural Belize (AED); 125993/M/ACT/BZ; Contract No. 002/09).

Argonide. (2005). NanoCeram filters. Argonide Corporation.

http://sbir.nasa.gov/SBIR/successes/ss/9-072text.html. Accessed December 10, 2013.

Chen, H. and Yada, R. 2011. Nanotechnologies in agriculture: New tools for sustainable development. *Trends in Food Science & Technology* **22**: 585-594.

Chinnamuthu, C.R. and Murugesa Boopathi, P. (2009). Nanotechnology and Agroecosystem. *Madras Agricultural Journal* **96**(1-6): 17-31.

Maity, A and Natarajan, N. 2014. Nanoparticles may enhance seed germination in cultivated fodders. Accepted for publication in IGFRI *Newsletter* published by Indian Grassland and Fodder research Institute, Jhansi.

Maity, A., Natarajan, N. and Vijay, D. 2014. Antimicrobial property of nanoparticles (NPs) during seed treatment. National Symposium on Climate Resilient Forage Production and its Utilization (November 13 - 14, 2014), Kalyani, West Bengal: 57 p.

NAAS. 2013. Nanotechnology in Agriculture: Scope and Current Relevance. Policy Paper No. 63, National Academy of Agricultural Sciences, New Delhi: 20 p.

OECD and Allianz. 2008. Sizes that matter: Opportunities and risks of nanotechnologies. Report in cooperation with the OECD International Futures Programme.

http://www.oecd.org/dataoecd/32/1/44108334.pdf.

Pramanik, P and Maity, A. 2013. Nanopollution- A Growing Issue. *National Environmental Science Academy Newsletter*. **16**(7): 2.

Subramanian, K. S. and Tarafdar, J. C. 2011. Prospects of nanotechnology in Indian farming. *Indian Journal of Agricultural Sciences* **81**: 887-893.

Tarafdar, J.C. 2012. Perspectives of nanotechnological applications for crop production. *NAAS News* **12**: 8-11.

Verma, A. K., Singh, V. P., and Vikas, P. 2012. Application of nanotechnology as a tool in animal products processing and marketing: an overview. *American Journal of Food and Technology* DOI: 10.3923/ajft.2012.

□□□

Chapter 38

Greenhouse Technology to Enhance Productivity of Fruits and Vegetables for Small and Marginal Farmers

SANJAY KUMAR SINGH, P.K. PATHAK, C.S. SAHAY AND AKRAM AHMED

ICAR - Indian Grassland and Fodder Research Institute, Jhansi-284003

Introduction

India is a land of small and marginal farmers sharing 83% of land holdings. It may be noted that 63% of land holdings belong to marginal farmers with less than 1 ha. This situation forces huge pressure on the land to multiply the productivity of fruits and vegetables for ever increasing population of the country. One of the important objectives of Integrated Farming Systems is to focus on increasing productivity of fruits and vegetables to help small and marginal farmers. During the era of green revolution, more emphasis was given on the productivity of cereal crops to feed the population. But after the liberalization policy of Indian Government and set up Plasticulture Development Centres at various institution/universities, focus was given on the productivity of fruits and vegetables along with the quality of the produce to meet the ever-growing food requirements. Both these demands can be met when the environment for the plant growth is suitably controlled. The need to protect the crops against unfavourable environmental conditions led to the development of greenhouses. A greenhouse is a framed or inflated structure covered with a transparent or translucent material in which crops could be grown under the conditions of atleast partially controlled environment and which is large enough to permit a person to work within it to carry out cultural operations. These are effective solar collectors which utilize solar radient energy to grow plants as well as for post harvest activities. Glasshouses, poly houses, rain shelters, shade houses and net houses are all different names given to greenhouses depending upon the type of material used and the utility of the facility.

Greenhouse Technology

Greenhouse technology is the most practical method of achieving the objectives of protected agriculture, where the natural environment is modified by using sound engineering principles to achieve optimum plant growth, yield and value added products. The environmental factors which affects plant growth includes air temperature, relative humidity, light, carbon dioxide concentration, soil temperature and moisture content of soil. By controlling these environmental factors, cultivation of vegetables during off season is possible under greenhouse, and, therefore, plants can be grown throughout the year instead of once during season only. The greenhouse may have heating, cooling and ventilating equipments for control over microclimates. Control over microclimate results in several fold increase in crop photosynthesis, and therefore, there is ample scope to double or triple and even in some vegetables like tomato yield can be enhanced more than 5 times.

Chandra *et al.* (2002) have proposed the design of a naturally ventilated greenhouse. Several other single and multispan greenhouses for use in various regions of India have been developed (N. C. P. A. H., 2001). Tiwari, G.N. (1984) analysed winter greenhouse. Effect of various parameters viz. Ventilation/infiltration, relative humidity, movable insulation etc. have been incorporated in the analysis and their effect on the performance of greenhouse were studied. He reported that temperature rise of the plant and room air is increased upto about 3-5 °C by covering the system with movable insulation. The performance of the greenhouse is improved by concrete north wall. It reduces the heat losses during the night. Tiwari, G.N. and Dhiman, N.K. (1986) presented a design and mathematical model for a winter greenhouse at Leh, Jammu and Kashmir in India. In this part of the country, the ambient air temperature, in winter, dips down to –30 °C and goes upto a maximum of about 25 °C in summer. Growing vegetation is difficult in this climate. Since the transportation of vegetables and other articles, to this place, is a problem. Greenhouses are essential for the growth of vegetables. Here, greenhouses are possible because solar isolation is available for almost 11 months of the year. Numerical calculations show that a glass wall on the south side and an insulated wall and roof on the north side give good results. Nimje *et al.* (1990) reported about greenhouse technology for vegetable crop production. The experiments on different vegetable crops under UV stablised plastic covered greenhouses were conducted at Central Institute of Agricultural Engineering, Bhopal. The results have shown that among the various crops evaluated, only okra and capsicum gave about three times higher yield inside the greenhouse as compared to open field condition. The fruit yields of pepper in greenhouse were two and half times higher under greenhouse. The biomass yields were also higher by 5 to 6 times. Selection of a proper variety of crops and time of planting was found to be a key to achieve maximum production in greenhouse. Sharma, P.K. *et al.* (1999) conducted a study on cultivation on strawberry in a greenhouse with the objective of extension of growing season and to obtain good quality produce. Strawberry, being a very susceptible crop to temperature, was protected from high temperatures through different modes of cooling viz. ventilation, evaporative cooling, misting and movable shade. They concluded that strawberry could be successfully grown inside a greenhouse during winter season. The average yield per sq. m. and per plant was found to 2.76 kg and 302 g respectively. Chandra and Garg (1993) developed a model of greenhouse economics with a view to study various options of greenhouse construction, environmental control and cropping activities in relation to maximized economic parameters. Girja Sharan (Indian Institute of Management) has developed earth tube heat exchanger as a cost effective method of heating and cooling. Coffin *et al.* (1988) measured the transmittance of solar radiation into scale models of multispan greenhouses for a complete year, under a wide variety of climatic conditions. The East-West greenhouse models had higher overall light levels than the North-South ones during the winter months. The insulated greenhouses had moderate reductions in light levels during the winter as compared to the conventional models.

Greenhouse technology as we know today has a history of over two centuries. During this period, many types of greenhouse for many different applications have been developed. Major greenhouse types based on materials of construction, shape, structure, cost, etc, are briefly described below.

a) Classification based on material

Before Second World War, most of the greenhouses were constructed using glass as the glazing material so that the term glasshouse and greenhouse became synonymous. With the development of plastic films and rigid sheets began the era of plastic greenhouses. Polyethylene, PVC, EVA, acrylic, polycarbonate, fiberglass, polyester, PVF are some of the plastic sheet materials. The classification may also be based on the material of the construction of the frame i.e. wood, bamboo, steel, aluminum, reinforced concrete etc.

b) Classification on the basis of shape

Shape of the greenhouse frame is another criterion of classification. Single span, ground and free standing greenhouses are synonymous. But, when a number of such single span greenhouses are joined together, then the houses is known as multispan or gutter connected facility. A greenhouse is said to be lean-to-type if its north side forms the wall of a residence. A greenhouse is uneven span if it is not symmetric about its long axis. A tower greenhouse is tall like a tower and is useful in areas where land is a constraint.

c) Classification on the basis of utility

The use, to which a greenhouse is put, can also be classification criteria. It should be retail or a whole sale greenhouse. Greenhouses for academic or research units, public park or gardens are termed as institutional greenhouses. An over wintering greenhouse is generally unheated facility as long as the air temperature in the greenhouse remains close to freezing. Some heating may be necessary if the temperature goes much below zero. A greenhouse is termed as cold frame if it is used to start or harden off seedlings in the spring or extend the growing season in fall. If such greenhouses are artificially heated, they are called hotbeds.

d) Classification based on cost

The per square meter cost of constructing a greenhouse may vary from low to high depending on the material used and facilities provided.

i) ***Low cost greenhouses:*** The low cost greenhouse is made of a supporting structure of GI pipe, angle iron, steel tubing, wooden or bamboo. For covering, ultraviolet stabilized plastic film of 200 micron is used. It does not have any control system.

ii) ***Medium cost greenhouses:*** The medium cost greenhouses have a cooling and heating arrangement and may have a double layer or ultraviolet stabilized plastic film.

iii) ***High cost greenhouses:*** It has many facilities like auto control mechanism of heating, cooling and humidification system, drip irrigation system, etc.

e) Classification based on climate control

i) ***High-tech or climate controlled greenhouses:*** This type of greenhouse is constructed to achieve higher degree of climate control to enhance the cultivation period of crops. Evaporative cooling and heaters are used to maintain required temperature inside the greenhouse space, whenever so required. The climate is usually accomplished through automation system. Such type of greenhouses are used for growing high value crops, like tomatoes, cherry tomato, sweet peppers and cucumbers for long period. These greenhouses are mainly being used in several European countries, Morocco, Israel, USA, Japan, etc.

ii) ***Semi-climate controlled greenhouses:*** In this kind of greenhouses, the structural frame is made of GI pipes, like the climate controlled greenhouse but only exhaust fans with evaporative cooling pads are provided to maintain favourable temperature and humidity during summer. Thses types of greenhouses are suitable for vegetable cultivation during mild winters and mild summers only. The basic cost of erection of these greenhouses is half of climate controlled greenhouses. These greenhouses are mainly being used in Turkey, Spain, USA, Morocco, Israel and Japan.

iii) ***Naturally ventilated or low cost greenhouses:*** These are simple construction greenhouse with low initial cost. The frame may be of GI pipes, wooden logs or steel pipes but no heating and cooling systems are provided for the structure. The top of the greenhouses is covered with plastics whereas the side walls have insect proof net from ground upto 5 to 6 feet height with or without manually rollable plastic over it. The initial cost of these greenhouses is less than half to that of semi-climate controlled greenhouse. These type of greenhouses are mainly used for crops like cucumber, offseason muskmelon, tomato, etc. China is using these on large scale where the area has rapidly increased during the last 10 years. Japan, Israel, Turkey, Morocco, etc are also using these greenhouses.

Prospects of Greenhouse Technology

The various countries began to practice the greenhouse technology, but the pace of progress could be increased several folds, if governments extend supports to protected cultivation efforts. Greenhouse cultivation of horticultural crops is reasoned to be next logical step to open field agriculture. One million hectares of area under greenhouse cultivation can very easily increase the production of vegetables by about 60-70 million tones. This can also create about one million jobs in the rural sector. Greenhouse vegetable production has tremendous scope in peri-urban areas of the country because of continuous increase in availability of upmarkets and continuous changing choice of consumers towards off season and high quality produce of different horticultural crops.

The vegetable growers who are having agriculture land 100-200 km around the metro cities like Delhi, Mumbay, Kolkata and Chennai have tremendous potential to start vegetable production under greenhouses for year round production of high quality vegetables to take the advantage of the available upmarkets in those cities. The farmers who are living 50-100 km around other big cities like Jaipur, Bangalore, Chandigarh, Hyderabad, Bophal, Raipur, Shimla, Lucknow etc can also start protected vegetable cultivation in big way to take advantage of off season availability of the produce.

Now-a-days nursery raising of vegetables has also become a specialized job with increasing susceptibility of vegetables to various biotic and abiotic stresses and very high cost of the seeds of hybrid varieties. The utility of greenhouse for raising virus free healthy seedlings of various vegetables during rainy and post rainy season and off-season nursery of cucurbits during winter months for out of season production of these crops may be a very successful and commercial venture for the farmers and unemployed agriculture graduates in different vegetable growing pockets and more specifically in peri-urban areas of the country.

Protected cultivation, undoubtly, offers several advantages to grow vegetables of high quality even under unfavourable climatic conditions. This production system, however, requires high input costs and good management. Whether it is economically viable depends on the cost incurred on the structures and the remuneration earned for the produce. Some discussions in the industries, government and with the farmers took note of the constraints in the large - scale adoption of this technology like limited indigenous know – how about greenhouse technology for off season vegetable production, and lack of technical manpower to manage the greenhouse.

Above all, to make it a successful venture grower need to make proper planning and vigil. Indigenous technological database need to be sounded to make adoption of protected agriculture sustainable. Some of the areas which need attention are given hereunder.

1. Cost effective design of greenhouses and their manufacturing based on location-specific need to be developed.
2. The package of practices including fertigation needs to be worked out for different agro-climatic situations.
3. Pest and diseases management is crucial in protected conditions which provide a favourable condition for their proliferation. Intensive cultivation in greenhouses give rise to the growth and development of numerous pathogens including nematodes. Effective and eco-friendly control measures are to be evolved.
4. Development of suitable varieties for protected cultivation requires an immediate attention and efforts to reduce dependence on imported seeds.
5. Protected cultivation is labour intensive. Therefore suitable design of tools, devices and equipments are required to reduce the cost of cultivation and enhance timeliness of operations.
6. A major concern is disposal of used plastics; therefore new generation biodegradable polymers need to be developed.

Greenhouse Structure

A greenhouse makes use of available sunlight to create suitable environmental conditions for plant growth. With this in view, a greenhouse is attempted to be made as transparent to sunlight as possible to admit the maximum sunlight during the periods of least sunshine. It is another matter that such greenhouses became least energy efficient in terms of thermal environment control. The following conditions could be set for a greenhouse frame under tropical conditions:

1. Adequate sunlight transmission for crop growth round the year.
2. Uniform distribution of transmitted light throughout the greenhouse growing areas

3. Adequate strength to withstand wind load, snow load, live load, crop load, etc.
4. The frame should not have excessive thermal mass.

There is a large amount of information now available for selecting greenhouse shapes, materials of construction, orientation, and environmental control features to suit a specific requirement. Computer models are available to provide further guidance in obtaining the desired product.

Site Selection and Orientation of Greenhouse

Ground slope for drainage is an important factor. Adequate provision should be made to divert surface water away from the greenhouse. It should be located at a place where unwanted shading from the surroundings is not present. Orientation of the greenhouse is another important factor for proper radiation environment and energy efficiency. An east-west oriented free standing greenhouse maintains better winter light as compared to a north-south oriented greenhouse. Gutter connected greenhouses should be oriented north south to avoid continuous shading of certain portions of the greenhouse due to structural members. A greenhouse attached to a building in northern hemisphere must be located on the south facing side of the building to get the maximum sun light.

Greenhouse Glazing

Glass has been the traditional glazing material. Other rigid materials include fibre glass reinforced plastics, plexy glass, polycarbonate, etc. Flexible materials include polyethylene, Poly vinyl chloride, EVA, PVF, etc. The criteria for the selection of a glazing material are quantity and quality of transmitted light, long wave radiation transmission, strength, weatherability, surface condensation properties and the cost. While total solar energy transmissivities are important from the standpoint of heating and cooling requirements, it is the transmission of photosynthetically active radiation (PAR, in the range of 400 nm – 700 nm) which is important for crop growth. The ratios of different segments of light, e.g., red to far-red ratio, are also important. The selected glazing material is adequately treated so that it does not get degraded fast by the ultra-violet rays in the terrestrial solar energy.

Greenhouse Frames

The greenhouse frame is the most important component of a greenhouse system. It provides support for glazing material and a place for fitting of environmental control equipment. Commonly used structural shapes are gable, quonset, lean- to and gothic arch with minor changes to suit local conditions. Greenhouse frames are designed to withstand wind, snow and crop loads with minimum obstruction to sunlight. Wind and snowfall data are available from meteorology departments for every part of India. For most of the locations a frame designed to withstand 100 km/h wind speed is sufficient with the option to use heavier sections in regions of exceptionally high wind speeds. Tubular steel sections are the most preferred structural members for greenhouse frames. Wood, bamboo and aluminium are other common materials. Steel pipes provide required strength at a competitive price and also an assured service life of over 20 years. It has been agreed that the frame should constitute about 5% to 10% of the total surface area of the greenhouse. Most of the transparent greenhouses have 60% to 75% overall solar transmission coefficient.

Greenhouse Environment Control

A greenhouse is essentially meant to permit at least partial control of microclimate within it. Depending upon the extent of environmental control, a greenhouse is classified as a cold frame, a heated greenhouse, an evaporatively cooled greenhouse or a sophisticated greenhouse with complete environmental control. The control of greenhouse environment means the control of temperature, light, air composition and nature of the root medium. Obviously, a control over all these parameters makes a greenhouse a completely controlled. However, in general, greenhouses with partial environmental control are more common and economical. In the following sections an effort has been made to present general details of controlling the environmental parameters under low and medium cost greenhouses.

1. Ventilation and Cooling

A greenhouse is ventilated for either reducing the very high greenhouse air temperatures or for replenishing carbon dioxide supply or for moderating the relative humidity in the greenhouse. Air temperatures above 35 ^{0}C are generally not suited to crops. It is quite possible to bring greenhouse air temperature below this upper limit during spring and autumn seasons by providing adequate ventilation for the greenhouse. The ventilation in a greenhouse could either be natural or forced. In natural ventilation, air becomes less dense as it gets heated and, thus, rises. This chimney effect could be used to move the warm air out and cool air in. Winds also permit creation of additional natural ventilation in the greenhouses. In case of small greenhouses (less than 6m wide) natural ventilation could be quite effective during spring and autumn seasons. However, fan ventilation is essential to have precise control over air temperature, humidity and carbon dioxide levels.

A fan ventilation system requires that the capacity for the fans be determined and proper fans be selected to fulfil the requirements. The locations for mounting the fans and the inlet window should also be determined. One of the recommendations for summer ventilation is to provide for two air changes per minute under tropical conditions. The fans selected for the ventilation should be able to deliver 7-9 cubic metres per second of air flow per horse power rating. The selection of fans for forced ventilation is made with the help of fan specification tables. Pressure drop of 3 mm of water gauge is allowed while selecting the fans for a given airflow rate. In order to obtain several stages of cooling, more number of smaller fans are selected as against a few large diameter fans. All fans should be equipped with louvers and fan guards for safety. Location and size of air inlets for the ventilation system are equally important. Fans are usually mounted on one end of the house and air inlets on the opposite end. Air inlets are designed to give an apparent mixing velocity of 3.5 m/sec.

2. Humidity control

A greenhouse is a closed space in which plants transpire and evaporation takes place from the floor. Some of this moisture, added to the greenhouse air, is taken away by the air leaving the greenhouse due to ventilation and/or leakages. Sensible heat inputs modify the relative humidities to some extent. In order to maintain desirable relative humidity levels in greenhouses, efforts are made to use humidification or dehumidification. For most crops the acceptable range of relative humidity is between 50% - 80%. However, for plant propagation work, relative humidities upto 90% may be desirable. Humidification in summers can be achieved in conjunction with greenhouse cooling by employing appropriate evaporative cooling methods such as fan-pad and fogging systems. Sometimes during winters when sensible heat is being added to raise the greenhouse air temperature during nights the relative humidity level might fall below the acceptable limit. In that situation, humidifiers might need to be operated to circumvent the problem.

Dehumidification is often a problem not amenable to simple solutions. During rainy seasons the ambient relative humidity is high along with that of the greenhouse. In this situation the ventilation can not lower the humidity of greenhouse air. But, when the ambient relative humidity is lower then ventilation could be practised to reduce the greenhouse relative humidity. Chemical dehumidification systems are technically feasible but expensive at present. Use of refrigeration systems (cooling coils) has also been made for dehumidification.

3. Greenhouse Heating

Traditionally, glass greenhouses have been heated by hot water systems. The procedure involved is heating water in an open kettle with a distribution pipe running along the length of a greenhouse. Use of a wood or a coal furnace with a horizontal brick chimney constructed through the greenhouse for heat transfer purposes has also been documented. Most larger commercial greenhouses are heated with some type of boiler system. Heat from steam or hot water is normally distributed through black steel pipes. Finned pipes and/or air heat exchangers may also be employed for heat distribution. Gas fired unit heaters for greenhouse heating, introduced in early 1950s, deliver heat at approximately half of the cost of the steam/hot water systems. The hot air is distributed through a perforated poly -tube running along the greenhouse length. Radiant heating has also been advocated in some cases. An overhead gas fired unit is placed as high in greenhouse as possible. This type of heater provides radiant energy directly to the plants. The problems of uneven temperatures have been noticed in the case of radiant heating. Electric heaters, both radiative and convective, are simple and convenient to control. But, the non availability of adequate electricity and its high cost may limit its use to small and/or experimental greenhouses. Use of heat pumps in refrigerated greenhouses could meet both heating/cooling requirements. The selected heating systems should not only raise the greenhouse temperatures but should also achieve uniformity of temperature distribution. Therefore, placement of heating units and the type of distribution system to be selected are important.

A number of active solar heating systems have been developed for greenhouses during the last two decades in order to reduce the dependance of greenhouse industry on conventional fuels. The normal components of a solar heating system are: solar collector, heat transfer medium, heat exchanger, and heat storage. Either water or air is heated in a solar collector and the heat transfer medium flows to the greenhouse where the heat is transferred to the greenhouse through suitable exchanger. Extra collected heat may be stored for night time or other cold periods. Expenditure of conventional fuels, such as electricity, oil or gas is required in such solar systems to move the air or water through the systems. A properly designed heat storage system is an essential component since solar energy is not always available for direct utilisation. Back up systems based on conventional fuels are generally recommended to take care of contingencies. Therefore, the solar systems at present only promise to reduce the consumption of conventional fuels.

Biogas, generated from agricultural wastes/residues, can be conveniently used for greenhouse heating. Since, the combustion of biogas (methane + carbon di oxide) will give off carbon dioxide, the flue gases could directly be admitted into the greenhouses with advantage.

Conclusion

Greenhouse production has tremendous scope in peri-urban areas of the country because of continuous increase in availability of upmarkets and continuous changing choice of consumers towards off season and high quality produce of different horticultural crops. The utility of greenhouse for horticultural production, raising virus free healthy seedlings and for several post harvest operations may be a successful and commercial venture for the farmers and unemployed youth. Above all, to make it a successful venture, growers need to make proper planning and vigil. Indigenous technological database need to be sounded to make adoption of greenhouse technology.

References

Chandra, Pitam and Garg, M. K. 1993. Economic analysis of greenhouse cultivation. Paper presented at the workshop organised by Indian Association for the Advancement of Science, New Delhi, March 1994.

Chandra, Pitam; Gupta, M. J.; Shrivastava, Ranjan; Dogra, A. K.; Singh, Arun Kumar; Singh, Awani Kumar; and Singh, Sanjay Kumar 2002. Establishment and use of a naturally ventilated greenhouse. Plasticulture Development

Centre, division of Agricultural Engineering, IARI, New Delhi-110012. p:48.

N. C. P. A. H. 2001. Greenhouse designs and environmental control. Bulletin published by National Committee on Plasticulture Application in Horticulture, Department of Agriculture and Co-operation, Ministry of Agriculture, Government of India, 2001.

Nimje, P. M., Wanjari, O. D. and Shyam M. 1990. Greenhouse technology for vegetable crop production. The Use of Plastics in Agriculture. Proceedings of XI International Congress, During 26th Feb. to 2nd March 1990. Nre Delhi, India, E-178-182.

Sharma, P.K. et al. 1999. Cultivation of Strawberry in a Greenhouse. Proceeding of National Conference on Commercialisation Aspects of Renewable Energy Sources, held at Udaipur during April 28-29, 2000.

Tiwari, G. N. 1984. Analysis of winter greenhouse. Int. J. Solar Energy. Vol. 3, pp. 19-24.

Tiwari, G. N. and Dhiman, N. K. 1986. Design and optimisation of winter greenhouse for Leh-type climate. *Energy Cover. Mgmt.,* **26**(1): 71-78.

□□□

About the Authors

Dr. Sunil Kumar is presently Head, Division of Crop Production, ICAR-IGFRI, Jhansi since November, 2010. His research areas includes cropping systems, soil fertility and fertiliser use, weed management, resource use in changing climate, GHG emission and C sequestration in fodder & grazing based livestock production system, conservation agriculture and forage based integrated farming systems. He has been bestowed with several awards viz. SAP Recognition Award, Best Team Award Krishi Vasant, Fellow, Range Management Society of India, Certificate of Honor, IGFRI, Certificate of Excellence, NRC Weed Science. He has been elected as Vice President, The Society of Agricultural Professionals & Indian Society of Agro Forestry; Secretary, Range Management Society of India. He has published more than 50 papers in reputed International & National journals including 2 books.

Dr. D. R. Palsaniya is working as Senior Scientist, Agronomy at ICAR-IGFRI, Jhansi. He has eleven years experience in the field of agroforestry, watershed management, Farming system research, wasteland development through silvipasture and weed management. He is a person with meritorious career throughout and has over one hundred and twenty publications to his credit. He was awarded with Bhamashah Award by Maharana Mewar Foundation, Udaipur in 2001, Excellent Services to On-farm Trials (Team Award) by NRCAF, Jhansi, twice in 2008 and 2010, Vasant Rao Naik Award of ICAR for Outstanding Research Application in Dryland Farming Systems – 2010, PS Deshmukh Young Agronomist Award 2010 of Indian Society of Agronomy, New Delhi, Best Paper Award of Indian Society of Agronomy for the year 2012 and IGFRI Best Team Award - 2014.

Dr. T. Kiran Kumar is agronomist and working as scientist at ICAR-Indian Grassland and Fodder Research Institute, Jhnasi since last five years. He is a recipient of IARI senior research fellowship and INSPIRE fellowship from DST, Government of India during his Ph.D. programme at IARI New Delhi. He also received best Ph.D thesis award from Indian Society of Agronomy. He is involved in research related to resource conserving technologies in rainfed forage based cropping systems and he has published more than 7 research articles in the reputed peer reviewed journals.

Dr Anoop Kumar Dixit is Senior Scientist (Agronomy) in ICAR-IGFRI Jhansi (UP). He has eleven years of experience in crop residue management, conservation agriculture and cropping system research. He has 22 research papers, 4 review papers, 12 popular articles 5 book chapters and 7 technical bulletins to his credit.

Dr M M Das is working as Principal Scientist in Plant Animal Relationship, Division at Indian Grassland and Fodder Research Institute, Jhansi (UP). He has 22 years of research experience in the field of Animal Nutrition. His main areas of research are improvement of low grade roughages and development of complete feed block for ruminants. He has successfully undertaken many research projects funded by ICAR, NATP, DST etc. in the field of animal nutrition. He has to his credit 70 research publication of National and International repute.

Dr. P.K. Ghosh is presently serving as Director, Indian Grassland and Fodder Research Institute. He is a renowned agronomist of country in the field of cropping system research, soil fertility and fertilizer use, crop nutrition and soil quality, conservation tillage and soil water conservation. He is fellow of National Academy of Agricultural Science and Indian Society of Agronomy, Indian Society of Plant Physiology, West Bengal Academy of Science and Technology, Member of International Grassland Congress (Representative of South East Asia), Chairman organising committee International Grassland Congress 2015, President of Range Management Society of India and is a recipient of many awards viz., ICAR Inter-disciplinary Team Award; ISA Gold Medal Awards , ISCA-Prof. Sushil Kumar Mukherjee Commemoration Lecture Award, Hari Kishan Shastri Memorial Award, Dr K.G. Tejwani Award, Groundwater Augmentation Award, Bioved Agri-innovation Award; IMPHOS-FAI Award, AAAS (Senior) Award, PPIC-FAI Award, Recipient of Best Scientific Oratory Award. Dr. Ghosh has 163 publications in the Journal of National and International repute including 6 books to his credit.